The Horned Dinosaurs

About the Author and Illustrators

PETER DODSON earned his B.Sc. in geology at the University of Ottawa in 1968, his M.Sc. in geology at the University of Alberta in 1970, and his Ph.D. in geology at Yale University in 1974. Since 1974 he has been a professor of anatomy in the School of Veterinary Medicine at the University of Pennsylvania, where he is also a professor of geology. He has studied Late Cretaceous dinosaur faunas of Canada and the United States for many years. Recently his dinosaur research has taken him to India, China, and Madagascar. He is a co-editor of the award-winning technical monograph *The Dinosauria,* published in 1990. He lives in Philadelphia with his wife and two children.

WAYNE D. BARLOWE was raised in a household of natural history illustrators. His passion for dinosaurs has been an undercurrent throughout his career. For twenty years his xenobiological illustrations have ranged over paperback covers, CD-ROMs, magazines, and toys. He is best known for his highly acclaimed illustrated books, *Barlowe's Guide to Extraterrestrials, Expedition,* and *The Alien Life of Wayne Barlowe.* His forthcoming titles include *Barlowe's Guide to Fantasy* and *A Pilgrimage to Hell. Publisher's Weekly* praised his last book with Peter Dodson, *An Alphabet of Dinosaurs:* "Barlowe's detailed, vibrantly colored paintings possess an almost photographic clarity." He lives in Rumson, New Jersey, with his wife and two daughters.

ROBERT WALTERS is an internationally recognized dinosaur restoration artist whose paintings and drawings have appeared in numerous books and magazines. His work is on display at the Smithsonian Institution, the Academy of Natural Sciences in Philadelphia, and the Creative Discovery Museum in Chattanooga, Tennessee. He has been seen on dinosaur television specials for PBS and The Learning Channel. He lives in Philadelphia.

The Horned Dinosaurs

A NATURAL HISTORY

Peter Dodson

Paintings by Wayne D. Barlowe

Additional illustrations and art editing by Robert Walters

PRINCETON UNIVERSITY PRESS

PRINCETON, NEW JERSEY

Library of Congress Cataloging-in-Publication Data

Dodson, Peter.
The horned dinosaurs: a natural history/Peter Dodson.
p.   cm.
Includes bibliographical references (p. –) and index.
ISBN 0-691-02882-6 (cl: alk. paper)
1. Ceratopsidae. I. Title.
QE862.065D64   1996
567.9'7—dc20   96-105

This book has been composed in Adobe Palatino

Princeton University Press books are
printed on acid-free paper and meet the guidelines
for permanence and durability of the Committee
on Production Guidelines for Book Longevity
of the Council on Library Resources

Printed in the United States of America by
Princeton Academic Press

10 9 8 7 6 5 4 3 2 1

Ad Majorem Dei Gloriam

Contents

List of Figures

CHAPTER THREE
THREE-HORNED FACE

CHAPTER FOUR
FIVE-HORNED FACE AND FRIENDS

CHAPTER FIVE
A BIG ONE ON THE NOSE

CHAPTER SIX
NEWER DEVELOPMENTS AND MODERN STUDIES

CHAPTER SEVEN
NO HORNS AND NO FRILLS

CHAPTER EIGHT
SISTERS, COUSINS, AND AUNTS

CHAPTER NINE
THE LIFE AND DEATH OF HORNED DINOSAURS

Preface

I AM A CONNOISSEUR of dinosaurs; I have been as long as I can remember. It has been a great privilege to devote my professional life to the study of these creatures, and I have long wanted to write my own dinosaur book. The dinosaurs I know best are the horned dinosaurs, with which I have worked off and on since I was a graduate student more than twenty years ago. They are among the most interesting of dinosaurs—surely grist for a good book.

Writing a book is almost a self-indulgence: a chance to savor the old literature, to survey the contemporary scene, to fluff a bit about my own contributions, and to add an opinion or two. This book is scholarly but it is also personal. I hope that everyone who loves dinosaurs will find something of interest in these pages. There is a lot of descriptive morphology here, for which I make no apology. The bones are the primary documents of my science. Indeed, paleontologists are scientists who love bones. We find them aesthetically appealing, complex, interesting, even sensuous objects. Certainly Georgia O'Keeffe found them so.

I have been privileged to work with two gifted and dedicated artists, my friends Wayne Barlowe and Robert Walters. Wayne and Bob help me to see with their eyes things that my own vision is too limited to see. I stand in awe of their artistic gifts, because these are so foreign to my own skills. Wayne and Bob both share my love of dinosaurs, and each strives to keep abreast of dinosaur science.

I am grateful to Shawna McCarthy, my agent, and to Jack Repcheck, my editor, for their respective contributions. I have learned over the past two years just how important their support is. In bringing this book to completion, Peter Strupp has lavished dazzling editorial skills, for which I am most grateful.

During my career, I have benefited from my relationships with my mentors—Dale Russell, Richard Fox, and John Ostrom—and my students, especially Cathy Forster and Paul Penkalksi, who chose to study horned dinosaurs, and Tony Fiorillo, who studied my Careless Creek site. Dale Russell put up with me for a year at the Canadian Museum of Nature and has supplied me with papers, measurements, and encouragement.

Extremely valuable to me has been a cadre of dedicated readers of chapter drafts. Their contributions range from the broadly conceptual to advice of the "you can't put that comma there!" variety. Susan Dawson is an English major as well as an anatomist, zoologist, and paleontologist. Joel Hammond is an editor as well as a zoologist. Other insightful readers were Peter Wilf, Anusuya Chinsamy, and Carl Sachs. Scott Sampson and Mark Goodwin read the manuscript in its entirety for the publisher and made helpful comments. I am grateful to colleagues who have provided me with manuscripts, theses, and illustrations: Paul Sereno, Dale Russell, Scott Sampson, Rolf Johnson, Cathy Forster, Phil Currie, Ken Carpenter, Gregory Paul, Dave Weishampel. Tess Kissinger drafted the maps. Pat Lieggi, Jack Horner, Phil Currie, and Chris McGowan were generous with their hospitality and friendship during my visits to their institutions. My secretaries, Judy Bennett and Marlene Mills, have cheerfully and efficiently handled a thousand ancillary tasks for me. Darren Tanke occupies a special place in my heart because of his unwavering dedication to horned dinosaurs and his generosity in sharing his knowledge.

When horned dinosaurs became paramount in my life in 1981, I was able to go forward because of the generosity of friends, especially Mrs. Edward Dillon, Mrs. Emilie de Hellebranth, Allen and Pamela Model, and Mrs. Margot Marsh. Charles Smart, Thomas Uzzell, Sam Gubbins, and Joel Hammond also took my dreams seriously. Eddie and Ava Cole and the Arthur J. Lammers family comprise an entire chapter in my life among the horned dinosaurs—a story that will be told on another occasion.

Finally, I thank my parents, who encouraged and nurtured my youthful interest in dinosaurs; my brother, Steve, who taught me what fossils were and how to collect them; my children, Jessica and Christopher, who have at various times participated with enthusiasm, watched with pride, or tolerated with bemusement as Dad was doing his dinosaur thing; and most of all my wife and best friend, Dawn, who, although sometimes doubting my sanity, has always been there.

The Horned Dinosaurs

With Horns on Their Faces

THE CERATOPSIA or horn-faced dinosaurs are exquisite creatures. They include large, exotic dinosaurs, the ceratopsids, with wondrous ornaments on their heads, including dazzling combinations of horns over the nose and eyes and lengthy frills behind the skull, often enhanced with rococo tracery and detailing. The ceratopsids are found only in western North America, although perhaps someday soon they will be found in Asia as well. Ceratopsians also include small, lithe protoceratopsids, shared equally between Asia and western North America. Protoceratopsids, which tend to be more tasteful and restrained in their adornments than ceratopsids, presumably include the ancestors of the ceratopsids. Recently admitted into the ceratopsian clan is the small *Psittacosaurus* from Asia, which is so generalized in its structure that it has neither horns nor a frill.

The central characters of this narrative are the horned dinosaurs themselves. We shall meet them all, learn everything about them, become friends on a first-name basis. Without question, the marquee player is *Triceratops* (Plate I). Everybody who can name a single dinosaur can name *Triceratops*. *Triceratops* was the first horned dinosaur known from a complete skull, described by O. C. Marsh in 1889. By now there are many others, some described long ago, others brand new, some famous, others obscure. The formal classification of horned dinosaurs is roughly as given in Table 1.1.[1]

Biologists name and classify living organisms in order to talk about them, to communicate to each other and to the public at large. Paleontologists are literally "paleo"-biologists: we strive to study extinct creatures as once-living beings, not as mere mineral formations exhumed from the ground. Thus we follow the same naming procedures that biologists do. Notice several generalizations that result from this hierarchical classification. Dinosaurs are a group of reptiles, even though this statement is sometimes disputed.[2] The term *ceratopsian* applies equally to *Psittacosaurus*, any protoceratopsid, or any ceratopsid in either sub-

3

TABLE 1.1
Classification of the Horned Dinosaurs

Class Reptilia	
subclass Archosauria	
superorder Dinosauria	
order Ornithischia	
suborder Ceratopsia	
infraorder Psittacosauria	
family Psittacosauridae	
Psittacosaurus	Osborn 1923
infraorder Neoceratopsia	
family Protoceratopsidae	
Leptoceratops	Brown 1914
Protoceratops	Granger and Gregory 1923
Montanoceratops	Sternberg 1951
Microceratops	Bohlin 1953
Bagaceratops	Maryańska and Osmólska 1975
Breviceratops	Kurzanov 1990
Udanoceratops	Kurzanov 1992
family Ceratopsidae	
subfamily Centrosaurinae	
Monoclonius	Cope 1876
Centrosaurus	Lambe 1904
Styracosaurus	Lambe 1913
Brachyceratops	Gilmore 1914
Pachyrhinosaurus	Sternberg 1950
Avaceratops	Dodson 1986
Einiosaurus	Sampson 1995
Achelousaurus	Sampson 1995
subfamily Chasmosaurinae	
Triceratops	Marsh 1889
Torosaurus	Marsh 1891
Diceratops	Lull 1905
Chasmosaurus	Lambe 1914
Anchiceratops	Brown 1914
Pentaceratops	Osborn 1923
Arrhinoceratops	Parks 1925

family. The term *neoceratopsian* applies to any ceratopsian except *Psittacosaurus*. The term *ceratopsid* applies only to the large, horn-bearing dinosaurs of western North America.

The classification also implies that there are several evolutionary levels within the Ceratopsia. We can describe a crude sketch of the evolutionary history of these dinosaurs. The psittacosaurs or parrot-beaked reptiles were small, two-legged plant-eaters that lived in Asia around 100 million years ago (henceforth abbreviated Ma). Psittacosaurs lacked both horns and frills. Protoceratopsids were small, four-legged predecessors of the ceratopsids that lived in both Asia and western North America from about 80 to 65 Ma; some had impressive frills over their necks and most lacked horns of any kind. The ceratopsids ranged in length from perhaps as little as 3.5 or 4 m to as much as 8 m. Currently they are known only from western North America, where they lived between 80 and 65 Ma.

Notice that I made these statements about psittacosaurs and proto-ceratopsids without resorting to the use of the word *primitive*. I also avoided using the word *advanced* for ceratopsids, particularly *Triceratops*. What is going on here? While valiantly resisting the assaults of postmodernism, scientific language, alas, is not immune to the dynamism of linguistic evolution. The terms *primitive* and *advanced* are no longer "politically correct" and so must be expunged from our vocabulary. They are to be replaced by the terms *basal* or *generalized* and *derived*, thereby avoiding unfortunate connotations of superiority or inferiority. Little *Leptoceratops* from latest Cretaceous beds of Alberta is a good example for this very rationale. There is no reason to think of it as an animal that was in any way inferior to *Triceratops*, with which it shared the ancient Alberta landscape. Otherwise it would have gone the way of all flesh (I can't say "dinosaurs" here, can I?) several tens of millions of years earlier than *Triceratops*.

THE HISTORY OF COLLECTING DINOSAURS

Another thing to notice about Table 1.1 is that our knowledge of horned dinosaurs, and of course of all dinosaurs, has a history—and a very colorful and interesting history it is, as we shall see. There was a time when no one knew anything about dinosaurs. Indeed, the recognition of fossils themselves as the remains of once-living plants and animals that inhabited an ancient world was a very difficult intellectual accom-

5

plishment that we too easily take for granted; we too readily heap scorn upon our ancestors who lacked this knowledge. Though fossils were observed and studied during the Renaissance, it was not until the late eighteenth and early nineteenth centuries that paleontology, the study of fossils, came into its own.[3]

One of the earliest historical records of a dinosaur is an illustration published by Robert Plot in a 1677 natural history of Cambridgeshire. It was claimed to belong to a giant human, but it appears to be part of the thighbone of a megalosaur.[4] Dinosaur remains were collected from the Normandy region of France as early as 1801, but the first dinosaur fossil formally described was collected in 1818 from near Stonesfield, Oxford-shire, England. It was described by Dean William Buckland in 1824 as *Megalosaurus*, a Middle Jurassic flesh-eater. Buckland (1784–1856) was a superb geologist and a notable eccentric who, among other oddities, kept a pet bear in his house at Oxford. In 1822, Gideon Mantell (1790–1852), a country physician, discovered several teeth in the Tilgate Forest in Sussex.[5] Mantell was sufficiently focused on his fossils that his medical practice suffered. Eventually, his house was so filled with fossils that his wife took the children and left him. In 1825, he named the dinosaur *Iguanodon*.

By 1842, several additional kinds of dinosaurs had been described, all based on fragments. All that was understood was that these were the remains of extinct reptiles, and that they were very large. It was the brilliant and irascible anatomist and paleontologist Richard Owen (1804–1892) who made the intuitive leap to conclude that these remains indicated the existence of a "tribe" of extinct reptiles of large size whose structure approached that of large modern mammals, and for which he coined the name *dinosaur,* meaning "terrible lizard."[6] Owen was one of the most important paleontologists of the nineteenth century, but he is often remembered today for his opposition to Darwin's theory of evolution. One interpretation is that he "invented" dinosaurs to prove the theory of progressive evolution wrong: reptiles had reached their peak (or "apotheosis," to use his term) in the Mesozoic and had gone down-hill since that time.[7]

The first dinosaur fossils from the United States came from the Nebraska Territory (now Montana) in 1855 and were described by Joseph Leidy at the Academy of Natural Sciences of Philadelphia in 1856. These too were fragmentary—a mere handful of teeth, as we shall see later. A very important skeleton came to light in 1858, not from the great fossil beds of the American West, which were yet to be discovered, but from

Haddonfield, New Jersey! Leidy described this skeleton as *Hadrosaurus foulkii*, the first duck-billed dinosaur. Leidy (1823–1891) is honored as the father of American paleontology, but he made great contributions in parasitology and anatomy as well. He is also remembered as the gentlest of men; he once walked 20 miles on the Sabbath to release a frog in its home pond.

Leidy's young protégé in Philadelphia was Edward Drinker Cope (1840–1897), a devout Quaker with a fiery temperament. Cope was rich, brilliant, and eccentric. He published more than 1,400 papers on living and fossil vertebrates in his short lifetime. Cope plays an important part in our narrative because it was he who discovered and named the first horned dinosaur, *Monoclonius crassus*, in 1876. Also of note for us is Cope's lifelong hatred of O. C. Marsh. Marsh (1831–1899) was the nephew of George Peabody, a wealthy Baltimore merchant who, like Marsh, was a lifelong bachelor. Peabody established a museum and endowed a chair of paleontology at Yale University for his nephew. Marsh reciprocated Cope's dislike of him. Both Marsh and Cope used their wealth to obtain large collections of fossils from the American West, and each strove to outdo the other. Marsh was a better politician than Cope, and for many years enjoyed the resources of the U.S. government to further his paleontological ambitions. Marsh was a great student of American dinosaurs, and for our purposes he is of paramount importance because he recognized horned dinosaurs for what they were, overcoming his initial blunder of mistaking *Triceratops* for a bison! He described *Triceratops* in 1889 and *Torosaurus* in 1891. It was Marsh who coined the name *Ceratopsidae* for the family of horned dinosaurs.

Neither Marsh nor Cope was himself a great collector, though each made several ventures into the western fossil beds—Marsh with a military escort, Cope without, true to his Quaker ideals. Many collectors provided Marsh with fossils, though John Bell Hatcher (1861–1904) was arguably the greatest of these. He was largely responsible for the marvelous collection of *Triceratops* and *Torosaurus* skulls from Wyoming that Marsh enjoyed. Hatcher once kept an expedition to Patagonia solvent with his poker winnings! It fell to Hatcher to write the greater part of the magnificent monograph on ceratopsians planned but never written by Marsh. Hatcher himself died tragically of typhus at age forty-two, stopping in midsentence. The monograph was completed by Marsh's successor at Yale, Richard Swann Lull (1867–1957), who named *Diceratops* in 1905. Published in 1907, it bears the names of Hatcher,

Marsh, and Lull in that order.[8] Cope employed the services of Charles Hazelius Sternberg (1850–1943), who accompanied him to the Judith River country of Montana on the 1876 expedition that netted *Monoclonius*. C. H. Sternberg had three sons—George F., Charles M., and Levi—each of whom learned his father's profession in the American West and became a great collector on his own.

After Marsh and Cope died, dinosaur paleontology languished in Philadelphia and New Haven as the twentieth century unfolded. As a student at what is today Princeton University, Henry Fairfield Osborn (1857–1935) fell under the spell of Cope's charm and was won over to paleontology. Osborn, born to wealth and privilege as the son of a nineteenth-century railroad baron, came to New York after graduation and awakened a sleepy institution, the American Museum of Natural History, with his ambitious plans. He hired Barnum Brown (1873–1963), who became one of the greatest dinosaur fossil collectors of all time. After collecting the *Tyrannosaurus* specimens that brought great fame to the American Museum, Brown proceeded to Alberta at the invitation of a local rancher from Drumheller and continued his record of collecting great fossil treasures. Beginning in 1910, a steady stream of magnificent skeletons flowed to New York. For our purposes we note *Leptoceratops* and *Anchiceratops*, both of which Brown described in 1914. The energetic Brown, despite a lack of formal training, became a great scholar of dinosaurs. Although he did not name further kinds of horned dinosaurs, Brown and his associate, E. M. Schlaikjer, wrote a very important monograph on *Protoceratops* in 1940.[9] Only one new horned dinosaur came from the United States at this time, the small, enigmatic ceratopsid from Montana, *Brachyceratops*, described by C. W. Gilmore in 1914.

Faced with the loss of paleontological treasures to the United States, the Canadian government in Ottawa responded in 1912 by hiring C. H. Sternberg and his three sons to come to Alberta and begin the hunt for dinosaurs. This they accomplished with great success. Initially Canadian fossils were described in Ottawa by National Museum of Canada paleontologist Lawrence M. Lambe (1863–1919). Lambe visited the Alberta fossil beds in 1899 and 1900. Lacking expertise in collecting methods, he had only modest success. He described *Centrosaurus* in 1904 on the basis of a fragmentary skull. Beginning in 1912, however, he had his own trove of splendid fossils, as the Sternberg fossils began to arrive in Ottawa. Descriptions quickly followed, notably of *Styracosaurus* in 1913 and *Chasmosaurus* in 1914. Lambe is remembered in the name *Lambeo-*

saurus lambei, a duck-bill named by W. A. Parks in his honor, as well as in the Lambeosaurinae, the subfamily of crested duck-bills. Lambe died suddenly in 1919, and his work was taken over by C. M. Sternberg (1885–1981), the greatest student of Canadian dinosaurs. Sternberg described new species of *Centrosaurus, Monoclonius,* and *Triceratops* in the 1930s and 1940s. In 1950, he described a bizarre new hornless horned dinosaur of large size, *Pachyrhinosaurus,* which has now been found at a number of sites in Alberta as well as Alaska. In 1951, Sternberg described three complete skeletons of *Leptoceratops* and named *Montanoceratops.*

Meanwhile, back in New York, the hunt for Canadian dinosaurs had ended during the war years, and as the 1920s dawned fresh challenges were needed. Osborn, who in 1923 described the horned dinosaur *Pentaceratops* from New Mexico, is best remembered for his studies of fossil mammals, especially elephants. As early as 1900, he predicted that fossil human ancestors would be found in Central Asia. The flamboyant adventurer Roy Chapman Andrews sold Osborn on the idea of mounting an expedition to Mongolia in 1922 to search for human ancestors. With Osborn's society connections and Andrews's worldly charm, a number of expeditions were financed with support from Manhattan high society. The expeditions failed to find human ancestors but instead brought back a marvelous cache of previously unknown dinosaurs, including dozens of *Protoceratops* skulls and skeletons and remains of the proto-horned dinosaur, *Psittacosaurus.* Both of these ceratopsians were described in 1923.

No further dinosaurs were brought back from Mongolia to New York after 1925. The Soviets mounted several expeditions to Mongolia after World War II, but for our purposes it is the Polish-Mongolian expeditions of 1965–1971 that are of great interest. For one thing, the expedition leader, Zofia Kielan-Jaworowska, as well as the principal scientists were all women. More to the point, new protoceratopsids were found. Most notable was *Bagaceratops,* described in 1975 by Teresa Maryańska and Halszka Osmólska, who the previous year had become the first (but not the last!) women to describe new kinds of dinosaurs. Recently, Soviet paleontologists have described additional new protoceratopsids: *Breviceratops* in 1990 and *Udanoceratops* in 1992.

No new kinds of long-frilled, chasmosaurine ceratopsids have been described since the description of *Arrhinoceratops* from Alberta by W. A. Parks in 1925.[10] After a long pause beginning in 1950, new short-frilled centrosaurines have now been discovered in Montana. I had the good fortune to discover and name *Avaceratops* from south-central Montana in 1986, and Scott Sampson described two new kinds, *Einiosaurus* and *Achelousaurus,* from north-central Montana in 1995.

This then is a thumbnail sketch of the history of the discovery of the horned dinosaurs that will be the subject of this book. A rich and interesting history it is, indeed. It is a history of fits and starts, of wrong turns and errors, of human vanity and wealth, of dogged determination and persistence. This is so because science is a human activity, not one done by machines. The fossil record is not simply an objective fact of nature. Fossils are objective documents, but the fossil record represents our knowledge of that document, and the way in which we have acquired that knowledge is a very human story. It takes wealth to acquire knowledge. Where would dinosaur paleontology be today had Marsh and Cope been required to earn their daily bread by the sweat of their brows? If Osborn had not been so well connected to affluent white Anglo-Saxon Protestant New York society? If Roy Chapman Andrews had not been so successful at charming money out of the privileged elite? I strongly suspect that this book, although it would have been written, would not have been written for another hundred years. Today we are the beneficiaries of a rich legacy of ceratopsian fossils. The task of making sense of these fossils falls to the paleontologist.

THE FOSSIL RECORD OF HORNED DINOSAURS

With ceratopsians, we are rather fortunate. Horned dinosaurs have one of the best fossil records of any group of dinosaurs. There are close to four hundred specimens in museum collections around the world. Twenty-three genera and perhaps thirty species have been described to date, with more to come.[11] There is an average of more than 30 specimens per genus. Even if the extremely abundant *Psittacosaurus* and *Protoceratops* were eliminated from this tally, we would still be left with a respectable 9.4 specimens per genus, which is well above the average of roughly 7 specimens per genus for all dinosaurs. Only six of the twenty-three genera (26 percent) are known from fewer than five specimens. By contrast, some 80 percent of all dinosaurs are known from fewer than five specimens, and 50 percent are known from only a single specimen.[12] Where the fossil record of horned dinosaurs really shines is in completeness of material. Nearly two-thirds of the twenty-three genera are based on essentially complete skulls and skeletons (whether or not these have been adequately described), and all of them have essentially complete skulls. By contrast, the giant sauropods have a fossil record that includes some forty-five genera, fewer than a dozen of which have skulls and only five of which are based on complete material.[13]

Why is ceratopsian material so abundant? Part of the answer is that these dinosaurs were restricted in time to the Cretaceous Period—in fact, except for *Psittacosaurus,* to the Late Cretaceous Period. A large volume of Cretaceous sediments has been preserved. This phenomenon has been termed the "pull of the recent," by which we mean that there has been less time for the physical processes of this dynamic earth we live on to wreak their havoc.[14] In effect, the closer we come to the present, the greater is the volume of sediment available, and the richer (potentially) is the fossil record. Many horned dinosaurs also had the good judgment to live in areas close to the sea, where active sedimentation was occurring. Another explanation is that horned dinosaurs simply seem to have been abundant, successful animals. As I shall discuss later, it is quite likely that many also were gregarious, social animals. So many of them were fossilized because there were so many of them, clustered in relatively limited areas, to fossilize.

A peculiarity of the fossil record of ceratopsians is that fossils are restricted to western North America and eastern Asia (Maps 1.1 and 1.2).[15] In the Late Cretaceous, some 65–80 Ma, the Rocky Mountains did not yet exist in their present form. A vast seaway extended from the Gulf of Mexico to the Arctic Ocean and at various times covered a large part of what is now the Great Plains. Giant marine reptiles frolicked in the waves, 4-m "sardines" ploughed the waters, and pterosaurs screamed through the skies over what is now western Kansas and southern Manitoba, and areas in between.[16]

The earth was stirring: volcanoes started to belch sulfurous soot into the air, and the proto-Rockies began to elevate. For several tens of millions of years, a titanic struggle ensued between land and sea, along a linear swatch of land that stretched from Alaska to northern Mexico. At times, the forces of land prevailed. The land surface rose to the west, and muddy streams and rivers flowed eastward, carrying life-giving, nutrient-laden sediment that sustained lush, swampy lowlands near the inland sea where dinosaurs and other living things flourished. At other times, the groans of the earth in labor quieted, and the land sank beneath the surface of the sea. Habitats that once nourished the dinosaurs were drowned as the sea pushed 500 km or more westward. Marine organisms—including unusual clams called inoceramids, and beautiful, straight or coiled mollusks with squid-like tentacles project-

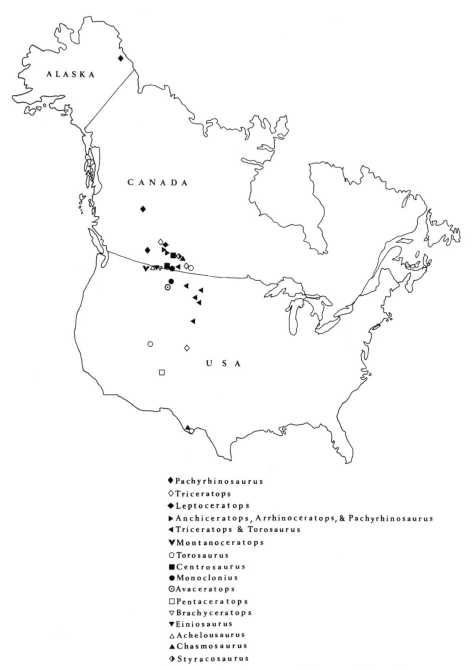

ALASKA

CANADA

USA

◆ Pachyrhinosaurus
◇ Triceratops
◆ Leptoceratops
▶ Anchiceratops, Arrhinoceratops, & Pachyrhinosaurus
◀ Triceratops & Torosaurus
▼ Montanoceratops
○ Torosaurus
■ Centrosaurus
● Monoclonius
⊙ Avaceratops
□ Pentaceratops
▽ Brachyceratops
▼ Einiosaurus
△ Achelousaurus
▲ Chasmosaurus
◈ Styracosaurus

MAP 1.1. Horned dinosaur localities, United States and Canada.

12

MAP 1.2. Horned dinosaur localities, eastern Asia.

ing from their floating dwellings—swept silently over the hardening tombs of the dinosaurs. Again Gaia stirred, and the seas retreated eastward. A fresh cast of dinosaurs, descendants of the previous ones, tenanted the verdant lowlands.

About 65 Ma, the cycle was broken, and the seas drained off the major part of the western land surface for good. Dinosaurs and many of the creatures with which they had shared the earth disappeared forever.

13

Marine and freshwater sediments of Cretaceous age hardened to rock (albeit often rather soft, readily eroded rock) deep beneath the sea. Eventually the land rose, and the sediments of western North America where we find dinosaurs today are frequently 1,000 m or more above sea level. When we collect dinosaur fossils today in the arid West, and the July sun beats pitilessly down on our backs, when the nearest cool beverage seems an eternity away, it is amusing to recollect that 75 Ma, the very spot we toil in was irrigated with ample fresh water and bathed in leafy, sun-dappled shade.

The Late Cretaceous dinosaur faunas of western North America represent a rather brief interval of geological time. The beds are arranged in three time-successive units. These formations are clearest in the Red Deer River Valley of Alberta. The Judith River Formation of Montana and Alberta is generally dated between about 77 and 74 Ma. The Horseshoe Canyon Formation of Alberta is a little younger, say 72–69 Ma, and the Scollard Formation is youngest, 68–65 Ma. The Judith River Formation of Alberta has the best-known vertebrate fauna of this time interval and has given its name to the interval: *Judithian*. Characteristic horned dinosaurs of the Judith River include *Monoclonius, Centrosaurus, Styracosaurus,* and *Chasmosaurus*. The intermediate time interval is not very well understood in the United States, although dinosaur-bearing formations in New Mexico and Texas are probably correlative. This time interval is named *Edmontonian,* in reference to the formation that was once called the Edmonton Formation in the Red Deer River Valley, but which is now called the Horseshoe Canyon Formation of the Edmonton Group. Its characteristic horned dinosaurs are *Anchiceratops* and *Pachyrhinosaurus*. The youngest time interval is well known in Wyoming from the Lance Formation and in Montana from the Hell Creek Formation. It is accordingly called the *Lancian* interval. Lancian beds are also known in Alberta, Saskatchewan, Colorado, and the Dakotas. Characteristic horned dinosaurs include the great *Triceratops* and *Torosaurus* and, ironically, the diminutive *Leptoceratops*. The terms *Judithian, Edmontonian,* and *Lancian* will frequently be used in this book.

In Mongolia, there are three time-successive formations that are roughly equivalent in time to the Judithian, Edmontonian, and Lancian intervals. Only the oldest of these, the Djadochta Formation, has horned dinosaurs in abundance, notably *Protoceratops,* a possible ancestor of North American ceratopsids. The oldest horned dinosaur of all, *Psittacosaurus,* comes from Mongolian deposits dated at about 100 Ma, from the end of the Early Cretaceous and the beginning of the Late

Cretaceous. *Protoceratops* and *Psittacosaurus* are both found in China as well as in Mongolia.

In North America, we can document a variety of habitats for horned dinosaurs. In the Judith River Formation of Montana and Alberta, dinosaurs lived in wet lowland environments within a few tens of kilometers of the sea. In the Horseshoe Canyon Formation of Alberta, marine influence was very strong, and the sea was close. In the Hell Creek Formation of Montana, water energy was lower than in the Judith River Formation, judging by the finer grains of sediment, and low-energy, flat, coastal floodplains are envisioned.[17] In the Two Medicine Formation of Montana, environments were higher and drier. Drought may have been an agent of mortality.[18]

In Mongolia, proto-horned dinosaurs lived 1,000 km or more from the sea, and aridity was a significant factor. There was water, but lakes may have been alkaline. Some dinosaurs were buried in dune sands. In fact, protoceratopsid fossils are not found in the Nemegt Formation, whose pale sediments were laid down in well-watered conditions that most resemble the Judith River Formation, but in the dry beds of the Djadochta Formation. At Bayan Mandahu in nearby Chinese Inner Mongolia, small dinosaurs appear to have been overcome by blowing sand.[19]

LIFE-STYLES OF THE LARGE AND FAMOUS

We have surveyed the animals themselves, their distribution, and the diverting history of their discovery. Few amateurs worry at night about whether *Stegoceras* is a suitable outgroup for the Ceratopsia or whether the lack of parietal fenestrae in *Triceratops* is a retained basal character or a character reversal. Children and adults alike really want to know what manner of beasts were the horned dinosaurs. We really yearn to know them as once-living, breathing, behaving, socializing, reproducing animals. To understand them as living animals, the modern analogue that springs to mind immediately is the rhinoceros. Reminding ourselves that no grasses carpeted the Cretaceous savannahs, we still feel that this is a reasonable first approximation. African white rhinos reach 4.0 m in length and nearly 1.9 m in height and weigh up to 2.3 metric tons.[20] Published estimates for the weights of horned dinosaurs range from 177 kg for adult *Protoceratops*, to 190 kg for *Leptoceratops*, to 3.7 to 3.9 metric tons for *Styracosaurus*, to 8.5 metric tons for *Triceratops*.[21] Thus it seems that the Judithian horned dinosaurs *Styracosaurus, Mono-*

15

clonius, Centrosaurus, and *Chasmosaurus* were roughly similar in size, though larger than, the African white rhinoceros. Small, basal ceratopsians were more the size of certain African antelope. The very largest horned dinosaurs, the giants *Triceratops, Pentaceratops, Torosaurus,* and maybe *Pachyrhinosaurus* as well, were easily three times the bulk of the largest living rhinoceros.

Rhinos are large-headed herbivores, but they were nothing compared to ceratopsians. Large head size is characteristic of all ceratopsians beyond the level of *Psittacosaurus. Pentaceratops* reaches the ultimate condition, with a skull greater in total length than the vertebral column from the first vertebra to the pelvis! In other ceratopsians, the total length of the skull ranges from 60 to 85 percent of the length of the backbone.[22] Even disregarding the frill, which is an "add-on," not really part of the true skull, the head is large. The basal length of ceratopsian skulls ranges from 30 to 45 percent of the length of the backbone.

And the brain inside the skull? That is another matter. As much as I would like it to be otherwise, it was not large. Casts of the brain have been studied in *Triceratops* and *Anchiceratops.* The volume of the brain cast in *Triceratops* measures about 300 cm^3. James Hopson of the University of Chicago studied comparative brain sizes among dinosaurs and uncovered two major findings about ceratopsids. One is that the brain size of ceratopsids compared to estimated body weight is slightly less than that expected in an alligator of ceratopsid size. The second is that ceratopsids had the largest relative brain size of all four-legged herbivorous dinosaurs. Only two-legged dinosaurs had larger relative brain sizes.[23]

R. S. Lull analyzed ceratopsid brains. He reported that a cast of the brain of *Triceratops* showed a well-developed olfactory region (for smell), small cerebrum ("which gives evidence of an extremely low grade of intelligence, compared with mammalian standards") and relatively large cerebellum, which coordinates movement (Fig. 1.1).[24] All vertebrates have a sense of balance (through the vestibular system) that is related to the sense of hearing, both being innervated by the eighth cranial nerve, the vestibulocochlear nerve. The vestibular system uses a set of three semicircular canals, arranged in three perpendicular planes, that are sensitive to motion in each of the three planes (pitch, roll, yaw). In the skull of *Anchiceratops ornatus,* described by Brown in 1914, the semicircular canals were beautifully preserved. In 1928, John Tait and Barnum Brown further considered the significance of semicircular canals. They determined that the horizontal canal indicates that the skull

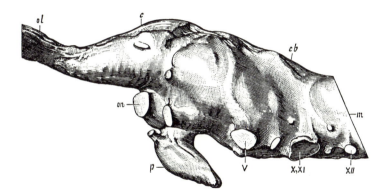

FIG. 1.1. Brain cast of *Triceratops*. Many regions of the brain can be identified, for example the olfactory region (ol) for the sense of smell, the cerebrum (c), the cerebellum (cb), the medulla oblongata (m), and the pituitary (p). Some of the twelve pairs of cranial nerves are designated by Roman numerals (V, X, XI, XII), except for cranial nerve II (on), the optic nerve (From Hatcher et al. 1907.)

of *Anchiceratops* was habitually dipped forward, such that the tips of the postorbital horn cores were about 8 cm higher than the bases. They inferred the same head orientation for *Triceratops*. The two vertical canals on each side were more or less oriented toward the four legs, making the animal sensitive to tripping or stumbling and thus protecting it from falling. The horizontal canals are larger and thus more sensitive to movement. Tait and Brown concluded that the head was tipped about the long axis and elevated or depressed in a vertical plane more than it was swung sideways in a horizontal plane. These dinosaurs did not groom their flanks. Their visual fields were sideways, not forward. "By turning their head on one side, like a hen viewing a hawk, they looked upwards with the upper eye."[25]

Rhinos have rather poor eyesight. This perhaps might be viewed as a luxury permitted by the fact that African rhinos weigh nearly ten times more than their largest potential predator, the lion. Adult rhinoceros, like elephants and giraffes, essentially have no natural predators. It is mainly young, sick, or injured animals that fall to predators. The world of the ceratopsians was not so benign. Ceratopsians had to confront predators that essentially were their equals in body size. North American ceratopsids that lived between 76 and 70 Ma had to deal with *Albertosaurus* and *Daspletosaurus*, and their young with a guild of smaller predators headed by *Troodon*, followed closely by *Saurornitholestes* and several other little nasties (Fig. 1.2). The giant Lancian cera-

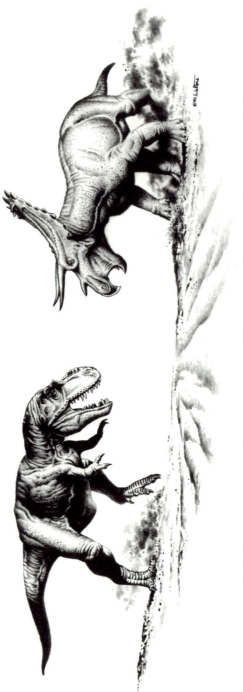

FIG. 1.2. *Anchiceratops* confronts *Albertosaurus*, Late Cretaceous of Alberta, circa 70 Ma. (Robert Walters.)

topsids *Triceratops* and *Torosaurus* (68–65 Ma) were faced with the greatest predator ever to walk the earth, the 12-m-long *Tyrannosaurus rex*. That big bully did not have an easy time of it, however. No wonder there were so few of them. A healthy bull *Triceratops* was undoubtedly an extremely dangerous animal, and a *Tyrannosaurus* so unwise or inexperienced as to engage one in frontal combat was likely to have paid with his life (Plate I). I would wager that meat-eaters generally did not enjoy a great taste for fresh ceratopsid drumstick. I suspect that more often the treat was savored as carrion.

The Lilliputian version of the struggle was played out in Asia between *Protoceratops* and *Velociraptor*. An adult *Protoceratops* may have outweighed *Velociraptor* by a factor of three. *Protoceratops* had no particular weapons other than a sharp beak and an arched nose, and one might have predicted an easy meal for *Velociraptor*. However, a famous specimen preserved in Mongolia shows both animals, apparently locked in mortal combat, dead upon the playing field. Evidently *Velociraptor*, nowhere near as intelligent as portrayed in recent print and cinematic fiction, had seriously underestimated the odds for an easy snack. Some artists have portrayed ceratopsids as taking up a defensive circle, musk-oxen-like, to protect themselves and their young against a marauder. Such a posture is within the license permitted an artist, but we must not fall into the trap of believing that this is a *scientific* representation of known ceratopsian behavior; it is not.

It is easy to gain the impression that ceratopsians had good eyesight. The width of the bony eye socket routinely ranges from 80 to 100 mm in *Centrosaurus* and *Chasmosaurus* to 120 mm in *Triceratops* to 166 mm in a specimen of *Torosaurus* as reported by Hatcher! Even little *Leptoceratops* had an orbit that measured 85 mm in width (but only half that amount in height). It is probable that the actual eyeball inside the bony orbit was significantly smaller, but nevertheless the eyeballs were probably large and the sight was probably good. We can also infer that this is so from the elaborate visual signals found all over the skull. Not only was there a showy frill, but there was also sculpture on the frill to enhance its already striking appearance: scallops, hooks, knobs, and processes. Moreover, there were the horns themselves in variable pattern (Figs. 1.3 and 1.4). Yes, some horns were dangerous weapons. The bony horn cores were extended varying distances by keratin sheaths like those found on cows. But other horns, such as the floppy nose horn of the bizarre *Einiosaurus* of north-central Montana, were hardly dangerous; instead, they were probably species-specific display structures.

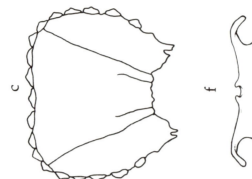

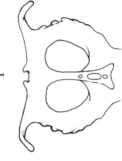

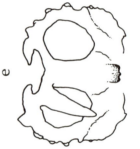

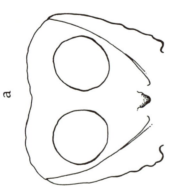

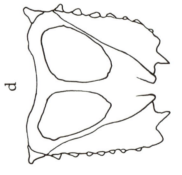

Fig. 1.3. Representative frill types in ceratopsids (dorsal views). (a) *Torosaurus*; (b) *Styracosaurus*; (c) *Triceratops*; (d) *Chasmosaurus*; (e) *Centrosaurus*; (f) *Pachyrhinosaurus*. (Robert Walters, after various sources.)

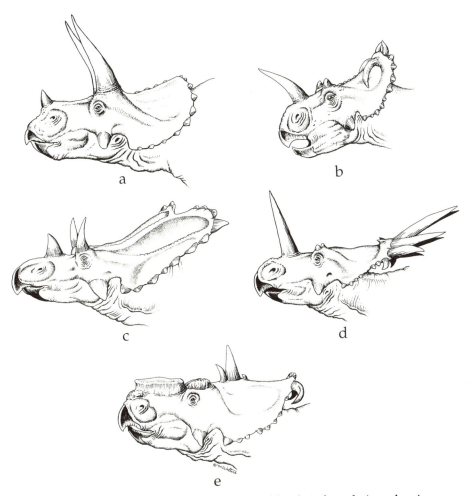

FIG. 1.4. Reconstructions of various ceratopsid heads in lateral view, showing horns and frills. (a) *Triceratops*; (b) *Centrosaurus*; (c) *Chasmosaurus*; (d) *Styracosaurus*; (e) *Pachyrhinosaurus*. (Robert Walters.)

Ceratopsids were walking billboards, with their names written all over them. The extravagances on the head sent the message; the acute vision of other members of the species received the message. Taking this line of reasoning one step further, it may also be correct to infer that these same structures indicate that ceratopsians were gregarious animals. Rhinos with their poor vision are not gregarious animals, and their horns, though impressive, are not particularly elaborate as display structures go. As a general rule, the greater the population density and

21

the greater the diversity of a community, the more elaborate the display structures.

Horned dinosaurs probably had an advantage that rhinos do not. Like most mammals, rhinos are color blind. How about dinosaurs? There is a simple way to approach this question. If you want to see colors at the zoo, where do you go? You go to the reptile house or the bird house (yes, you could go to the aquarium instead). Dinosaurs are phylogenetically bracketed by reptiles and birds, and it is therefore most economical to assume that they too had color vision. Inasmuch as they had structures with large surface areas that were meant to be seen, I believe that it is reasonable to infer that the conspicuousness of the structure meant to be seen was enhanced by the use of color. If not necessarily the whole animal, at least the particular display structure— be it the plates of stegosaurs, the cranial crests of duck-bills, or the frills of ceratopsians—was thrown into relief by its coloration. When a *Torosaurus* dipped its head, it revealed a huge, shimmering display that could have made a male peacock look drab—think about it!

We all know that the crests of horned dinosaurs are defensive structures, don't we? Maybe so, but it is instructive to consider crests at their origins among the protoceratopsids. *Protoceratops* had a bony crest that was eggshell thin—literally translucent. What kind of defense could an eggshell offer? Moreover, the showy crest of *Protoceratops* came in two different models: one taller, wider, and more showy, the other narrower, flatter, and less showy. Furthermore, the two models correlate with distinct patterns of arching of the face over the nose (the "proto-horn"), thickening of the jugal (cheek) horns,[26] and other features. These two patterns plausibly represent males and females (or vice versa). Are we to believe that one sex needed better defenses than the other, or is it wiser to consider whether the crest actually originated as a *display structure*?

Surely horns are prima facie evidence for active, aggressive behavior against mortal enemies? There again, the report is mixed. If the object of the horns is purely defensive—to rebuff, repel, maim, or kill a would-be predator—one might expect more uniformity of weaponry. Not all horn patterns are equally effective, and in general longer, sharper horns may be more effective than shorter, blunter ones. The long-horned *Chasmosaurus kaiseni*, for example, would seem to be much better protected than the short-horned *Chasmosaurus belli*; yet the former seems to have been much less common than the latter. Colbert found the situation puzzling: "Why should some ceratopsians have a single nasal horn, while others have large brow horns in addition to a nasal horn? Why should the horns be straight in some and curved in others? Why should the frill be solid in some, perforated in others, and why should the frill be so variously decorated with spikes and scallops and knobs?"[27]

If we consider the horn patterns of African antelope today, we notice great variation. Some are short, some are long; some are straight, some are spirally twisted; some bend outward, some bend backward. But how does an antelope defend itself from a lion or a cheetah or a leopard? Not by fighting but by fleeing! It seems that for modern horn-bearing mammals the precise pattern of the horns serves to permit recognition *within* the species. Members of the same species are able to recognize each other by their horns. Furthermore, the horn pattern may serve as an indicator of the type of struggle between males to achieve breeding rights, as discussed in the next section.

SEX IN THE CRETACEOUS?

Given the emphasis on display and male interactions, did both males and females have the same patterns of horns and frills? In my view, it is almost an a priori expectation, based on our knowledge of living horned and antlered animals, that there should be differences in pattern between males and females among horned dinosaurs. Yet in the past, paleontologists were loathe to adopt this interpretation of ceratopsian morphology. Lull was characteristic of his era in 1933 as he dismissed the possibility of sexual differences: "Little or no sexual distinction is provable among ceratopsians and . . . apparently both males and females of a species possessed both the horns and crests in equal degree."[28]

This statement is made despite the fact that there is so much variability in ceratopsian skulls that cries out for interpretation in sexual terms. The variability is seen not in structures that relate to the biomechanical demands of such vital physiological systems as the respiratory, circulatory, nervous, or masticatory systems, but rather in the frills or ornaments of the skull—the finery, the haberdashery (Fig. 1.5). Did these guys spend so much time in their scholarly work that they never attended a society ball? Did they not notice that visually oriented, color-receptive but black-clad males respond to females in bright colors and plumes with pectoral displays?

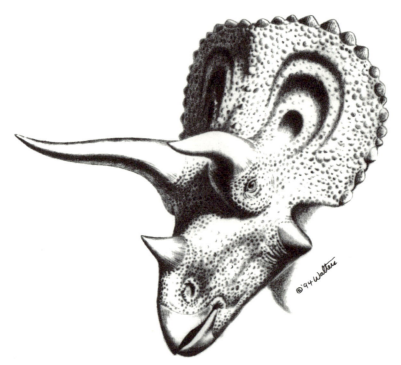

FIG. 1.5. *Triceratops* in full display. (Robert Walters.)

One reason that sexual dimorphism went unrecognized is that it was hidden by taxonomy. If a peahen and a peacock are classified as separate species, then there is no sexual dimorphism in peafowl! In 1975, I pointed out that in duck-billed dinosaurs not only sexual dimorphism but also growth was disguised by the historical artifact of taxonomy.[29] In *Triceratops*, it would seem that the evidence had already been uncovered in terms of two evolutionary lineages. Were these *really* animals that lived together at the same time and in the same place, but were of separate evolutionary lineages? This seems an awfully elaborate construct to avoid letting sex raise its ugly head in the Cretaceous. Is this an example of Victorian prudery—cultural baggage, as Stephen Jay Gould would put it—affecting science? In 1976, I published my interpretation of sexual dimorphism in *Protoceratops*. This was a relatively easy case in which only a single species had been recognized. I was able to demonstrate dimorphism in showy characters that plausibly represent secondary sexual display characters. The interpretation of males and females from this pattern was no great intellectual leap.[30] In 1990, Tom

Lehman of Texas Tech University reviewed the species of chasmo-saurines, and he interpreted the historical taxonomy in thoroughly modern terms of sex differences between male and female.[31] His inter-pretations are not beyond dispute. To my mind, however, the important thing is not whether Lehman is correct in every detail but rather the openness of his interpretation to this very fundamental aspect of cerat-opsian biology.

In summary, is it possible to recognize differences between males and females among the Ceratopsia? Yes! Not only is it possible, I maintain that it is *required*. In general, females resembled males but differed in the degree to which certain characteristics were expressed. The differences are sometimes subtle, and we may not always be correct in our interpre-tation of them, but the differences are definitely there.

The Canadian zoologist Valerius Geist in 1966 outlined stages in the evolution of horn-like organs. One stage is represented by animals with small, simple horns that engage in potentially dangerous combat. His example was the Rocky Mountain goat (*Oreamnos*, a goat-antelope, not a true goat). The small horns of this animal have little display value, and the unbranched horns can have a lethal effect on an opponent. A second stage of horn development is represented by cattle (*Bos taurus*) or bison (*Bison bison*), in which cranial display is not elaborate, but the horns can be locked together safely so that combat among males is a test of strength, not a struggle to the death. The third stage of horn develop-ment involves the development of elaborate, species-specific display structures on the head, usually among males (except for caribou, all female deer lack antlers; but among horn-bearers, females may or may not imitate males). Accompanying the elaborate display structures are behaviors that allow rivals to assess each other's rack of antlers or curl of horns or other display structure. These ritual displays serve to estab-lish dominance hierarchies and avoid damaging combat. Males that are deficient in display structure may elect not to engage a more dominant male. Combat occurs between males equally convinced of their own prowess.[32]

Jim Farlow and I applied Geist's model of horn-like organs and behavior to horned dinosaurs in 1975. We noted the correlation be-tween short frills and dominant nasal horn cores in centrosaurines, and between long frills and dominant postorbital horn cores in chasmo-saurines. The shorter frill has less impact as a display structure, and a single nasal horn is a dangerous organ that cannot be readily engaged in a safe manner by a rival male. We therefore postulated that animals such as *Centrosaurus, Monoclonius,* and *Styracosaurus* were more solitary animals with less elaborately ritualized behavior. We even went so far as to suggest that they were more solitary animals than their twin-horned contemporaries. (The fossil record of bonebeds that has come to light during the past decade suggests that the hypothesis of solitary habits, especially for *Centrosaurus,* is almost certainly incorrect.) Chasmo-saurines, by contrast, have long, interesting frills that had marvelous display value when the head was dipped and the vast expanse of the frill was exposed as a peacock-like vertical fan, the head rocking gently back and forth to make sure the message was received. At the same time, the dipped head presented the points of the postorbital horns in an unmistakable attitude for combat.[33]

If a furious rival bull *Pentaceratops* would not back off but presented in a similar fashion, the opponent's paired horns could be joined. A grunting, twisting, writhing wrestling match could ensue. The strain-ing bodies would heave first one way and then another, back and forth across a clearing, oafishly trampling delicate ferns and weedy flower-ing plants underfoot until, minutes or even hours later, the weaker male would at last capitulate. He would disengage, flatten his head, meekly extend his neck, and lower his chin toward the ground in his most conciliatory posture, then slink off into the gloom of the dawn redwood forest as his rival strutted, stomped his feet, and snorted loudly in triumph.

Did ceratopsians injure each other? There is some debate on this score. Darren Tanke of the Royal Tyrrell Museum of Palaeontology in Alberta is quite definite in his belief that bone injuries documented in ceratopsid bonebeds from Alberta are quite rare, and that ceratopsids therefore were not pugnacious animals.[34] However, the contrary opin-ion is very persistent.[35] There are a number of chasmosaurines that show unusual openings in the squamosal bone on the side of the frill, often in just one squamosal. These include *Chasmosaurus, Pentaceratops, Arrhinoceratops,* and *Triceratops.* The openings are often large, measur-ing 10 cm or more in length. The bone surrounding the openings rarely looks diseased, and it is not beyond question that the openings repre-sent horn punctures. But if they are not, then what are they, and why do they occur with such frequency in animals that we can otherwise judge to be armed and dangerous? Put another way, we have the weapon, we have the motive, so why doubt the result? Whatever the cause of the lesions, they were not fatal in any of the cases recorded. The victims parried the thrusts and lived to go about their business. Perhaps that is an advantage of being an animal of large size and small brain. One is not very vulnerable. Injuries to the squamosal are uncommon in centrosaurines, and I do not know of a hole through a squamosal in this subfamily. However, a lovely skull of *Centrosaurus* at the University of Alberta shows a dent in the squamosal that may be the result of a horn thrust.

PROSPECTUS

In this chapter we have seen what this book is about. The plan is to understand the horned dinosaurs, to learn where they lived, how they were discovered, and what we know about them. I approach this narrative as a scientist who loves horned dinosaurs and has studied them for longer than he cares to admit. I want to share my joy and excitement, but I also want to convey what it is that we scientists actually study. The plan of the narrative could proceed in several ways. A strictly historical approach could be taken, beginning with the first discoveries in the American West in 1855 and doggedly moving forward. Paleontologists are very fond of the history of their field, and such an approach is not without appeal. This book will contain much history. By itself, though, the historical approach would provide too sparse a framework for the dinosaurs themselves.

A second approach would be to take a strictly phylogenetic approach and discuss the earliest, most generalized ceratopsians first, then successively more derived forms in their turn. However, ceratopsians do not form a neat phylogenetic ladder; it is doubtful that any group of organisms does. Branching is a more appropriate expression of evolutionary geometry, without the implication that one branch is more highly evolved than the others. I will, however, discuss dinosaurs within evolutionarily coherent groups. In this case, there are three major groups: the centrosaurine or short-frilled ceratopsids, the chasmosaurine or long-frilled ceratopsids, and the protoceratopsids. For reasons of geography, history, and convenience, *Psittacosaurus* will be discussed with the latter.

I will begin with *Triceratops*. This approach has a number of advantages. It is a dinosaur that everyone knows and loves, and historically it is the first horned dinosaur known from a complete skull. I will discuss *Triceratops* and its relatives in the Chasmosaurinae in historical order (Chapters 3 and 4; Plates I and II), then *Monoclonius* and its relatives in

27

the Centrosaurinae (Chapters 5 and 6; Plates III and IV). *Protoceratops* and its relatives in Mongolia and North America form the topic of Chapter 7 (Plates V and VI). In the final chapters, 8 and 9, we put this information together to decipher relationships and study the biology of these magnificent creatures. However, before any of this can happen, it is necessary to learn some skeletal anatomy. Because the subject of this book is bones and what we can learn from them, Chapter 2 will be devoted to the skeleton of horned dinosaurs. Are you ready?

Skin and Bones

THE ANATOMY OF A HORNED DINOSAUR

ANATOMY is the study of the parts that make up the whole organism, plant or animal. Anatomy is a very old subject, probably about as old as human curiosity. Anatomists today often use light microscopes or electron microscopes and tend to call themselves cell biologists, structural biologists, electron microscopists, or anything else but anatomists in order to sound modern. Anatomy is vitally important for doctors and veterinarians, and it is the first subject that students training for the healing professions study.

When paleontologists think about anatomy, we are usually thinking of skeletal anatomy. We sometimes have to remind ourselves that there are other body systems, most of which are what we call "soft parts." Soft parts themselves rarely fossilize, although under favorable circumstances impressions of them may. We have no direct knowledge of the stomach, intestines, reproductive or urinary organs, heart, or lungs of horned dinosaurs. However, we are confident that these organs were once present because all land-living vertebrates have them.[1] It is an interesting exercise to try to infer what these organs may have been like.

Muscles are another matter of concern to the paleontologist. It is a naive thought that paleontologists routinely reconstruct muscle patterns on fossils, for this is actually a difficult enterprise. It is true that some muscle scars are prominent. For instance, in reptiles a major muscle sweeps off the tail (cauda) and inserts on the inner surface of the thighbone (femur), where it leaves a major scar. The caudifemoralis muscle may be seen in dissection of an alligator or lizard; it is an important muscle that pulls the thigh backward during walking or running. The scar for this muscle can almost always be recognized on a dinosaur femur. Yet for every muscle that leaves a definite scar, there are four or five muscles that leave no scar, either because they are too small or because their area of attachment to the bone is too diffuse.

29

Some muscles are found in virtually all tetrapods, for instance the well-known biceps and triceps muscles of the upper arm, or the gastrocnemius muscles of the calf. We feel pretty confident that horned dinosaurs were no exception. Other important muscles familiar to us mammals are completely absent in both alligators and birds, as well as in other reptiles, and so were almost certainly absent in dinosaurs. The gluteal (rump) muscles are an example. Therefore we cannot rely on muscle scars alone to do a major muscle reconstruction, and we cannot rely on our knowledge of mammalian anatomy, either human or veterinary.

Is there no basis at all for reconstructing muscles in dinosaurs? Of course there is. The best sources for learning muscles are crocodiles and birds. Together they constitute the *extant phylogenetic bracket* for dinosaurs; that is, they are the living animals most closely related to dinosaurs.[2] Crocodiles (including alligators and caimans) are the closest living reptilian relatives of dinosaurs, and birds appear to be direct descendants of small meat-eating dinosaurs (to some scientists, they *are* dinosaurs). Thus alligators and ostriches (or even chickens and turkeys) are extremely useful organisms to dissect. It is surprising how few published accounts there are of detailed muscular anatomy. The reconstruction of dinosaur muscles is a very sophisticated and intellectually demanding undertaking. Anyone who wants to reconstruct muscles on a dinosaur must first spend many hours dissecting to understand how muscles work. It is always appropriate to inquire after what animals a reconstruction is modeled. Any comprehensive muscle reconstruction does not merely flow from understanding the dinosaur's skeleton but encompasses significant conceptual inferences, which may or may not be justified. For instance, if mammalian muscles are applied to dinosaur skeletons, as was sometimes done prior to 1950, significant errors may be introduced. Any muscle reconstruction necessarily imposes something from the present onto the past.

COVERING THE BODY

Skin is the first body system that we notice in a living animal, but it is also the first part of a buried animal to disappear. Even though the skin in a large animal may form a tough shield several centimeters thick in places, it usually rots away eventually after death. For some Ice Age mammals, such as giant ground sloths, wooly mammoths, or bison, skin with its hair is, under special conditions, actually preserved, but these fossils are at most only a few tens of thousands of years old.[3] Regrettably, this is not the case for dinosaurs; no actual skin survives. Was ceratopsian skin furred, feathered, bare, or scaled? Fortunately, in some cases, the carcass of an animal may make an impression in fine-grained mud before the ravages of decay proceed. That is the case with the Canadian horned dinosaur *Chasmosaurus,* for which an impression of a patch of skin was preserved with a skeletal specimen.[4] The skin, which comes from the pelvic region, might be best described as reptilian in nature. Large circular plates up to 55 mm in diameter are set in irregular rows at a spacing of 50 mm from each other. Between the large round plates are irregular polygonal plates a centimeter or less across. Unfortunately neither *Chasmosaurus* nor any other ceratopsian shows the pattern of skin over the skull or frill.

There is one other matter concerning the body covering or integument, and that has to do with the covering of the horns. Living bare bone is not exposed to the environment.[5] It is in point of fact an inference that horned dinosaurs had horn—not a risky inference, but a real one nonetheless. That is because horn refers to a specific anatomical tissue that is never preserved. Horn is composed of a protein substance called keratin (Greek: made of horn). What we call horns on the skulls of horned dinosaurs are in fact horn cores. The true horn is a keratinous sheath that covers the bony horn core. In modern horn-bearing mammals, the bovids (such as sheep, cattle, and antelope), the horn sheath extends a variable, and often considerable, distance beyond the tip of the horn core. Horn is nonliving material; it is insensitive to pain and lacks blood supply. The horn core underneath is certainly living bone, and it is also covered by a deep layer of skin called dermis, the site of the blood and nerves that indirectly provide sensitivity to the horn. New horn is added around the base, and old horn may be worn off the tip by rubbing, a behavior that is common among horn-bearing animals. I am certain that the horn cores of horned dinosaurs were covered with true horn to protect the living bone underneath. Ceratopsian horn cores show grooves and channels that suggest that a blood supply ran underneath the horn, as expected.

Not only farmers are familiar with horn: all of us experience horn on our fingers and toes. Nails are essentially horn, that is, keratin. We are familiar with the flexibility, insensitivity, and growth of our nails. Nails are specialized flattened keratin structures covering the ungual bones (the terminal bones of the hands and feet) of primates. Claws are more generalized structures covering the pointed ungual phalanges among a

31

FIG. 2.1. Fleshed-out reconstruction of *Chasmosaurus belli*. (Robert Walters.)

wide range of vertebrates, including reptiles, both herbivorous and carnivorous. It is clear enough that many dinosaurs had claws that were similar to those of living claw-bearers, whether reptiles or mammals. Horned dinosaurs did not have claws; the unguals instead were broad and rounded. To me they look like nothing so much as a horse ungual. I do not think, however, that there was an exposed hoof as in horses. The feet of ceratopsians, like those of virtually all dinosaurs, were digitigrade. That is to say, the toes were flat on the ground, and were for the most part invested in dense skin. I believe the unguals of horned dinosaurs were covered in a somewhat hoof-like covering on both the upper and lower surfaces. As I sit here at my desk writing, I am examining the ungual of what may be a *Chasmosaurus* (although one cannot tell for sure from an isolated bone, a "spare part"). It is coffee-brown and heavy, unlike the light, ivory-colored horse ungual beside it. But the two bones share some striking similarities. In both, I see a pair of holes near the proximal articular surface where the arteries that run down the leg and continue onto the toe entered the bone to form a terminal arterial arch. And along the crescentic tapered edge I see a series of holes where blood from the terminal arch left the spongy bone of the toe bone to nourish the dermis underneath the hoof. In fact, I see more similarities than differences between the ungual of the horse and the ungual of the horned dinosaur.

The final stage of anatomy is to answer the question: what did the animal look like? In a real sense this is not the role of the scientist but of the artist (Fig. 2.1). Twice blessed is the scientist who is also an artist, for example, David B. Weishampel of the Johns Hopkins University; Dave was my student. There are also artists who contribute to the scientific literature, notably Gregory Paul.[6] For most of us paleontologists, restoring the appearance of the animal is beyond our skills. Our work typically ceases with description and interpretation of skeletal remains. Often the bones then repose in the tranquillity of a museum drawer. Sometimes there is sufficient interest that they are then mounted by highly skilled museum technicians, who also possess skills that many of us scientists lack. It may also come to pass that we have the opportunity to work with an artist to restore the appearance of our favorite animal. Such collaboration between scientist and artist has long been fruitful. When Richard Owen coined the name *dinosaur* in 1842, it had no impact on the public—none whatsoever. But twelve years later, when he teamed up with the eccentric sculptor and artist Benjamin Waterhouse Hawkins, the dinosaur at last took tangible form and became accessible to the Victorian public—and dinomania was born!

THE SKELETON: FANTASIA ON A THEME

The major body system the paleontologist works with day in and day out is the skeletal system, the bones. It is to the skeleton that we must now turn our attention. Knowledge of the skeletal system forms the great divide between those who know a few dinosaur names and can recognize *Triceratops* in a museum and those who are really eager to learn paleontology. (I still remember the thrill I felt on Christmas Day many years ago when the *Tyrannosaurus* model my parents gave me came with the bones named—I really felt like a paleontologist then!)

I have always wanted a ceratopsian skeleton in my living room. I finally found one, a high-quality epoxy resin cast of *Chasmosaurus belli,*[7] in a catalog. The price seemed almost too good to be true, so I quickly ordered one from a mail-order house that I had never heard of: Big Bob's Bargain Bone Barn. In due course a large and rickety crate appears on my doorstep. I barely manage to get it inside when the box collapses and an unbelievable cascade of bones tumbles across my living room floor—more than three hundred of them in all sizes and shapes! I look in vain amid the chaos for assembly instructions or a key for identifying the bones. I indignantly call Big Bob's number, only to find it disconnected—Big Bob is no more. It was too good to be true, or at least too good to last. What to do?

I decide not to panic. Although the number of bones in a *Chasmosaurus* skeleton is large, it is finite, and many shapes are very distinctive,

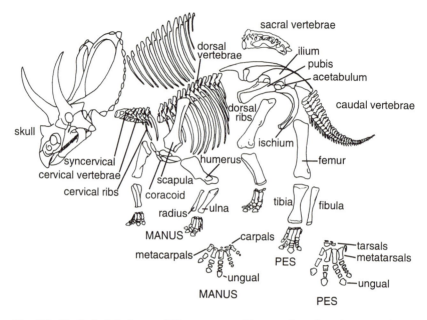

FIG. 2.2. Exploded skeleton of *Chasmosaurus*. The text describes the magnificent *Chasmosaurus belli* skeletons at the Canadian Museum of Nature in Ottawa, but the skull shown is that of *Chasmosaurus mariscalensis* from Texas, which better serves to illustrate the separate skull elements. (Bruce J. Mohn/Robert Walters.)

so I decide to try to sort them into groups according to their shapes. There are still several hours left before my wife is to return home. I also have a couple of good books in my home library. The best one for this task is a thick and daunting technical treatise called *The Dinosauria*, which has detailed chapters on each group of dinosaurs.[8] I turn to the

FIG. 2.3. Representative vertebrae of *Triceratops*. (a) Syncervical or fused vertebrae of the neck. The first vertebra is only a small vestige, but the second, third, and fourth vertebrae are completely and inseparably joined. In this specimen, the fifth vertebra is joined by sediment but is not part of the syncervical. (b and c) Second dorsal vertebra in side view and front view. (d and e) First caudal vertebra in side and front views. (f, g, and h) Distal caudal vertebra in side, front, and bottom views. The parts of the vertebrae are labeled as follows: a, cranial articular surface of the vertebral centrum or body; p, cranial articular surface of the vertebral centrum or body; hypophysis (h) and diapophysis (t) are facets for attachment of the rib head and tubercle, respectively; t also denotes the transverse process; n, nerve or spinal canal; r, rib; s, spinous process; z and z', cranial and caudal articular processes, respectively. (From Marsh 1891a.)

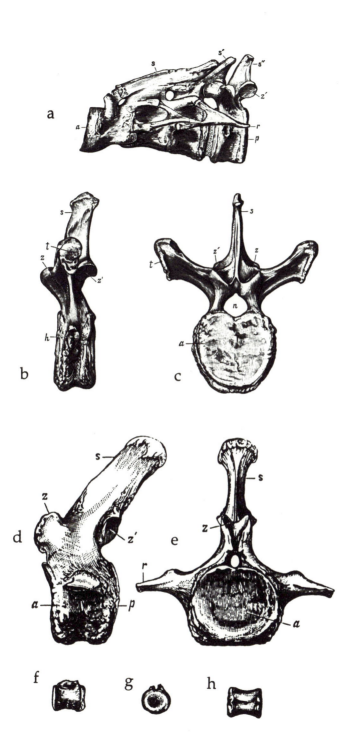

35

last chapter of the book, entitled "Neoceratopsia."[9] It looks like it will be a useful guide. So I sit on the floor with an ethereal smile on my face and sort the bones into piles: big bones and little ones, limb bones and vertebrae, ribs and toe bones, and just plain odd bones (Fig. 2.2).

Vertebrae are easy to recognize. They have large spool-shaped bodies, with blade-like neural spines on top, enclosing a hollow canal where the spinal cord once sat. The vertebrae differ in size, but I assemble a pile of sixty-three separate vertebrae. There are also two units where the spools are joined together and cannot be separated. One consists of what seem to be three vertebrae, with a deep cup at the end. The other is a long unit with ten spools joined together, with flat faces at both ends. I learn that the first, called the *syncervical*, which is 28 cm long,[10] actually represents a joining or fusion of the first four neck or *cervical* vertebrae that helped support the heavy head (Fig. 2.3a). The deep cup on the face of the first vertebra forms a socket in which the ball at the back of the skull sat. The second set of fused vertebrae is called the *sacrum*. It is 72 cm long and represents the connection of the vertebral column to the pelvis.

I am able to arrange the vertebrae between the syncervical and the sacrum into a series in which the bodies, bony processes, and canal change gradually from one to the next. The separate vertebrae of the neck region, six in number, have round faces about 12 cm wide; the canal for the spinal cord is fairly wide; and the neural spines are rather low. The next region of the body is called the *dorsal* region, which includes the thoracic and lumbar regions of mammals. Most dinosaurs did not have a rib-free region in front of the hips, which is called the lumbar region in mammals. Twelve vertebrae belong to the dorsal region. Here the neural spines grow tall, and the transverse processes become prominent, swept-up wings to which the ribs attach (Figs. 2.3b,c). At their tallest in the mid-dorsal region, the vertebrae are about 40 cm tall. At the front, the vertebrae are rather erect, but farther back they have a backward lean to them, and the neural canal for the spinal cord is decidedly narrower than it was farther forward. The vertebrae have narrowed to about 10 cm in width. The sacrum is long and heavy, and the neural spines and transverse processes form almost continuous surfaces at right angles to each other (see Fig. 2.5b). In addition, heavy specialized ribs form a bar for attachment to the bony pelvis. There are forty-five tail or *caudal* vertebrae. They start at the base of the tail as tall, erect vertebrae with long spines and strong horizontal transverse processes, but with short bodies (Fig. 2.3d). Proceeding backward, the spines quickly decrease in height, and the transverse processes dwindle and disappear halfway down the tail. Toward the end of the tail the vertebrae are simple spools, longer than they are high (Fig. 2.3e).

I feel pleased with the results of this sort. As I join the vertebrae together with their interlocking articular processes (zygapophyses) and lay them out on my floor, they stretch 4.1 m in length, including 1.5 m from the sacrum to the front of the spinal column and 1.5 m from the back of the sacrum to the tip of the tail.

Ribs also seem reasonable to deal with. I have a pile of forty-two bones or twenty-one pairs, many of which I am certain are ribs, since they are up to 75 cm long, thin, and curved like the staves of a barrel. Others, however, are short, only 15 cm long, straight, and forked at one end like a slingshot, or like an asymmetrical lowercase y. In between are ribs of intermediate shapes and lengths. In fact, they all fit together into a continuous series, with no sudden jumps in size between them. This makes me certain that they are all ribs. The short slingshot-shaped ones attach perfectly to the vertebrae at the front end of the neck. In fact, there are clear points of attachment; one fork of the slingshot attaches to the side of the spool, the other to the transverse process. When attached to the vertebrae, the neck ribs point backward. The neck ribs closest to the chest increase in length and, though still straight, no longer look like slingshots at their upper ends. The first several chest or dorsal ribs are T-shaped, and both knobs at the upper end of the rib attach to the transverse process of the vertebrae; the ribs point down, not backward. Most chest ribs are long and become curved like barrel hoops. The last ribs next to the pelvis are shorter than the typical dorsal ribs (Fig. 2.4).

Another group of bones seems to be related to the vertebral column in some way. They too are slingshot-shaped, but quite symmetrical, which suggests to me they are situated on the midline of the animal, and therefore single rather than paired. There are thirty-two of them, the longest about 75 mm long and the smallest a quarter of that size. I determine that they are the *chevron* bones or *hemal arches*. They hang beneath the tail, the counterpart of the neural spines above the tail vertebrae. The tail vertebrae show facets on their undersurfaces where the chevrons attach. The canal framed by the slingshot encloses blood vessels to the tail, accounting for the descriptor *hemal*, which refers to blood.

Now I turn my attention to the long bones. I make a pile of twenty bones. There are also four others that are only 30 cm or so long. They too might belong with the long bones. Some of them I am certain are leg

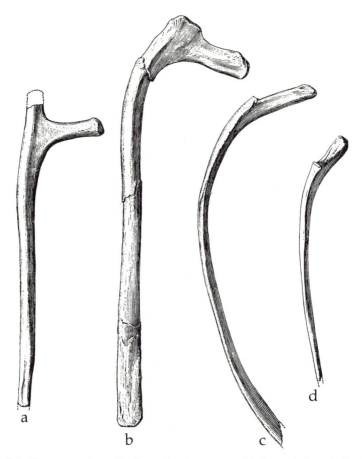

FIG. 2.4. Representative ribs from *Brachyceratops*. (a) Second dorsal rib; (b) a middle dorsal rib; (c) another middle dorsal rib; (d) final dorsal rib. (From Gilmore 1917.)

bones, but others are either flat and broad or curved, so I am not quite so certain, except they are as long as the bones about which I am certain.

I can make a pretty good guess at the pelvic bones. I know that there are three on each side, the *ilium*, the *ischium*, and the *pubis*. The ilium is dorsal; the ischium and pubis are the two ventral bones.[11] Mammals, birds, crocodiles—in fact all land vertebrates—have the same three hip bones. The ilium is a long, somewhat flat bone. At 96 cm in length, no other bone in the skeleton equals it in length. It has a blade that projects forward (cranially), another that projects backward (caudally), and a shallow notch between them that faces ventrally and defines part of the hip joint. The cranial blade lies horizontally and forms an overhanging shelf above the hip. On the internal (medial) surface of the ilium are a set of ten scars that show the points of attachment of the ten sacral vertebrae to the ilium. The ischium is very distinctive in horned dinosaurs, because its long, simple shaft points backward underneath the tail and curves downward. It measures 70 cm in length. The forked upper end touches the ilium above and the pubis below. The arc in between defines more of the hip joint. The hip joint is completed by the third bone, the pubis, on the cranioventral aspect of the pelvis. The pubis consists primarily of a vertical blade about 45 cm long, which spreads laterally from the hip joint toward the ribs. It has a small rod caudally that parallels the ischium for a short distance (Fig. 2.5).

The longest pair of columnar straight bones measures 75 cm in length. These have to be the thighbones or femora. Each *femur* has a heavy, ball-shaped head leaning inward, and at the lower end a pair of curved surfaces or condyles that form a roller surface at the knee joint. There is a prominent depression with a bony tab next to it on the inside surface of the middle of the femur, which is the site of attachment of the major muscle, the caudifemoralis, which we have already met. Also distinctive is a pair of somewhat shorter but robust bones that measure 53 cm in length. These are the tibiae or shinbones. Each *tibia* reminds me of a bowtie, being broad at the upper and lower ends and forming a slender shaft in the middle. The expanded upper end points forward and forms a prominence at the knee, whereas the lower end is expanded side to side and forms a hinge for the ankle joint. Each tibia also has a long slender rod of bone beside it, called the *fibula*. I now have a good idea of the size of the hindlimb of *Chasmosaurus*, because I know the foot is not going to add that much to its length. I am struck by the thought that because the shin is so much shorter than the thigh, these dinosaurs must not have been terribly swift runners (Fig. 2.6).

Before I turn to the feet, I decide to work on the front legs. There is a pair of long flat bones with a slight curve to them. These are not leg bones, but they must have something to do with the legs. At 68 cm in length, each bone is longer than any of the true leg bones of the front limb. I recognize it as the *scapula* or shoulder blade, quite similar in form to that of a chicken. It is gently curved inward to conform to the shape of the rib cage. The lower end has an attachment surface for another bone, the dinner-plate–sized *coracoid*. The lower end of the coracoid lies near the *sternum* or breastbone, which consists of a pair of flat, kidney bean–shaped plates that sit on either side of the midline of

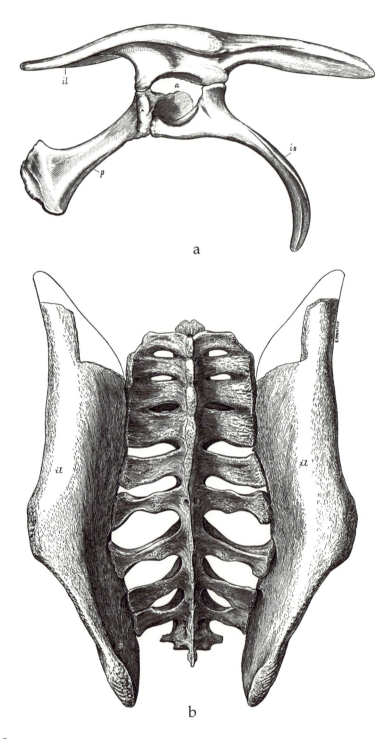

a

b

40

the body on the lower or ventral surface of the chest. This arrangement is totally unlike that of mammals. A prominent notch shared by the scapula and the coracoid on the caudal side represents the shoulder joint. The biggest of the remaining bones is the *humerus*, the bone of the upper arm, which is about 50 cm long. It has a bulbous head and a wide blade that extends halfway down the shaft before it narrows, then expands modestly at the elbow. As in the hind leg, there are two bones for the middle segment of the front leg. The larger of the two is the *ulna*, which is 43 cm long and has a prominent process on its upper end. This process corresponds to the point of our elbow, the funny bone. The other bone is the *radius*, about 32 cm long, which is a rather nondescript shaft (Fig. 2.7).

The job has been fairly straightforward up to now. I have been avoiding dealing with the foot bones because they are the smallest bones, and there are so many of them. I have a collection of ninety-two bones, which range in length from shafts of 21 cm to little nubbins 13 mm long.[12] After lengthy contemplation I manage to sort them into two groups: a larger, more robust set of forty-eight bones representing the hind feet and a slightly shorter, more lightly built set of forty-four bones representing the forefeet. I work on the front feet first. The longest bones of the front foot are the five *metacarpals* (which correspond to the knuckle bones of our hand), two of which are about 13 cm long and the other three of which are 8–10 cm long. They are labeled metacarpals I–V, with Roman numerals usually designating the digits (fingers) from medial to lateral (that is, from thumb to pinkie). Metacarpal (MC) III is the largest, corresponding to the central axis of the forefoot. It is barely longer than MC II but is much wider at its proximal end.[13] MC II is barely shorter but more slender. Metacarpals I and IV are both robust, but MC I is 2 cm shorter than MC IV. MC V is much less substantial than the other four.

I am quite confident about assembling the metacarpals and feel proud of what I have done. The *phalanges* are quite another matter. I find it very useful to have prior experience with the pattern of the

FIG. 2.5. Pelvis of *Triceratops* in (a) left lateral view and (b) dorsal view. In (a), the ilium (il) is dorsal, the pubis (p) points down and forward, and the ischium (is) curves down and back. All three hip bones meet at the hip joint or acetabulum (a). In (b), the broad, shelf-like ilia (il) join the elaborate fused sacrum. (From Marsh 1891b.)

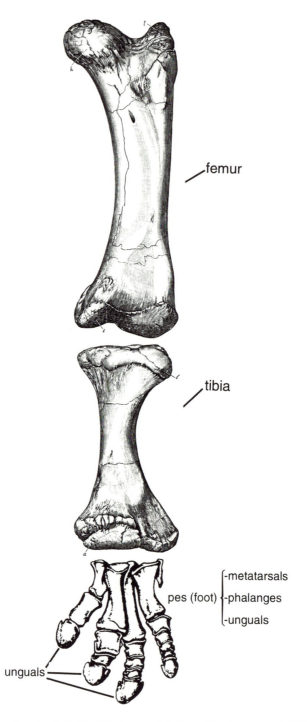

FIG. 2.6. Ceratopsian left hind limb, cranial view. The femur and tibia are of *Triceratops;* the foot is based on *Centrosaurus.* a, astragalus; h, head of femur; t, greater trochanter. (From Marsh 1891b and Brown 1917.)

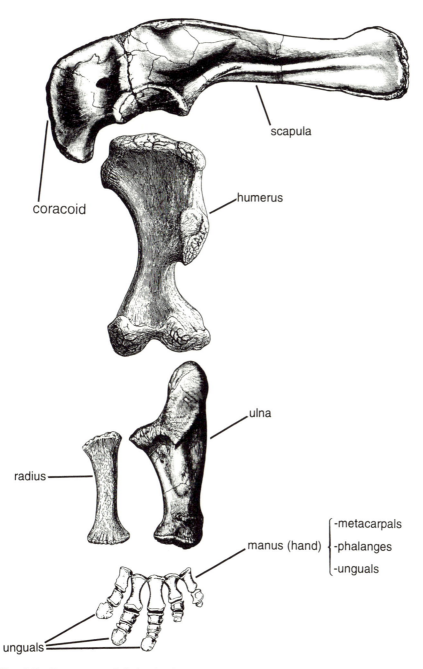

FIG. 2.7. Ceratopsian left forelimb, cranial view. Scapula, humerus, and ulna based on *Triceratops*; radius and manus based on *Centrosaurus*. (From Marsh 1891b and Lull 1933.)

digits. I read that the ceratopsian hand, or forefoot if you prefer (*manus*, the anatomical term, covers both designations), departs a little from the norm, having fourteen phalanges arranged in a formula of 2-3-4-3-2, which indicates that the outer digits are somewhat reduced in size.[14] The most obvious feature of the phalanges are the *unguals*, the terminal phalanges or hoof bones. These are the flattened, crescent-shaped last segments of the first three toes. The unguals of horned dinosaurs are quite similar to those of hadrosaurs or duck-billed dinosaurs, and even similar to those of modern horses (except that horses have only one on each foot). This suggests to me that horned dinosaurs had some sort of specialized hoof-like structure enclosing both the upper and lower surfaces of each toe. The outer two toes end in little nubbins of bone. Otherwise the remaining eighteen phalanges are not very distinctive. I have no chance of separating left from right. The first or proximal row of phalanges are a little longer than the more distal ones; those of digits II and III are wider than the others. I try my best to make each foot look like the picture in the book. I also place three small flattened disks, the *carpal* or wrist bones, at the upper ends of the metacarpal bones. They are smaller than the surfaces of the metacarpals, so no precise fit is possible. It seems that there were cartilage, dense connective tissue, and tendons, all of which contribute to a definitive fit. As these materials do not fossilize, we are left with a bit of a gap.

My work still isn't perfect, but it is the best I can accomplish with a jumble of bones. It is much nicer when an entire skeleton is found completely articulated, that is, in life position, with no guessing required. But I think you would be impressed by my results. Now the hind feet do not seem so daunting. The *metatarsals* are longer and heavier than the metacarpals that they otherwise resemble. There are four major ones that range between 12 and 21 cm in length. A fifth metatarsal (MT V) we describe as *vestigial*: at less than 8 cm in length, it is a very much reduced and somewhat useless version of a functional metatarsal. It has no toe bones associated with it, and, like the splint bones of a horse leg, would not have been visible in a living animal. As in the forefoot, the central bone, in this case MT III, is the longest, and the second, MT II, is next. MT IV is robust but 5.5 cm shorter than MT III. MT I, at 12 cm, is slightly shorter but still more robust than the longer metacarpals. The phalangeal formula of the foot is 2-3-4-5-0. This means we only have to worry about phalanges for four toes instead of five. They are all wider than the manual phalanges (i.e., those of the manus or hand), and, as in the hand, the proximal row of phalanges is longer than any of the others. Each of the four digits bears a full ungual phalanx; there are no nubbins in the foot. One characteristic of the fourth digit is that because it has the most phalanges (five), it follows that each of these is relatively short; in fact, the lengths of all the nonungual phalanges decrease from the first digit to the fourth digit. This is true not just of horned dinosaurs or even of dinosaurs, but also of birds and all vertebrates that have a phalangeal formula of 2-3-4-5. These characteristics are sufficient to allow me to put the feet together pretty convincingly.

I have a few other bones left over. These are the *tarsal* bones. The largest one, the *astragalus,* is 18 cm wide and forms a roller surface that caps the lower end of the tibia and constitutes a major component of the ankle joint.[15] The smaller, block-like *calcaneum* extends the ankle joint surface laterally over to the lower end of the fibula. There are three irregular, flattened disks constituting the distal tarsals that in life were probably attached by bands of connective tissue to the proximal ends of the metatarsals. I can only guess at their exact positions.

I am now the proud possessor of a semiarticulated *Chasmosaurus* skeleton. It measures 4.1 m (13 ft 6 in.) in length without the skull, laid out along my living room floor and snaking into the dining room. When I prop the skeleton[16] up in a quasi-reasonable posture, it measures 1.5 m (5 ft) high at the hips. How is it that I feel that it is not going to be a permanent fixture in my living room? If I had a drawing room, it might go there. Sadly, however, I do not live such a life-style. I know my daughter would helpfully suggest I put it in the freezer, beside the frozen alligator legs—teenagers can be so cheeky! Maybe it will have to go into the garage. I refuse to consider the backyard. Anyway, my wife has long since come and gone. At least she said nothing instead of what she really thought. It is past midnight. I have identified and positioned 255 bones. I am exhausted. The skull bones can wait.

The next day I am bright-eyed and bushy-tailed, eager to get back at it. Again I am confronted with a jumble of bones, albeit a smaller pile than before. These skull bones are much more irregular than the bones that form the skeleton—not much by way of cylinders, tubes, or rods here. Some have conspicuous holes, grooves, or channels that probably permitted nerves or blood vessels to pass through. Some have interesting textures or patterns on the surface. Most bones of the skull come in pairs, with the bone on the right side of the animal being a mirror image of the one on the left.[17] Some bones on the midline of the body are single and symmetrical. The longest bone in the pile is a whopping 87 cm

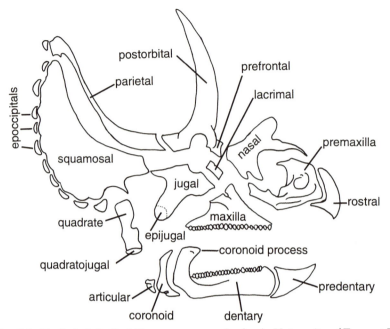

FIG. 2.8. Exploded skull of *Chasmosaurus mariscalensis,* University of Texas at El Paso, viewed from the right side. (After Lehman 1989. Bruce J. Mohn/Robert Walters.)

long, longer than the longest limb bone. Can it really be a skull bone? Its symmetrical shape suggests it belongs on the midline, and the fairly slender shape of the processes shows that it could not have borne the weight of the body as the limbs must do. The smallest bones are just a couple of centimeters long. It might be tough solving the whole puzzle, but certain bones stand out. The four jaws with teeth are obvious enough, two uppers and two lowers. To lay them out is an obvious starting point. Also obvious are the two bones that form the beak at the front of the skull, and three bones that bear horns on the top of the skull. A good guess is that the long T-shaped bone and a pair of other, somewhat fancier bones, only a little shorter, belong to the crest that sticks out behind the skull. Laying out these few bones gives a very useful but rough shape to the skull. Now I can open my book and get down to work (Fig. 2.8).

The upper jaws are called the maxillae (singular *maxilla*). Each is rather triangular in form, with a long, straight base corresponding to

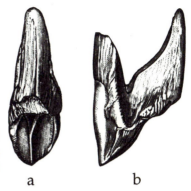

FIG. 2.9. Upper or maxillary tooth of *Triceratops*. (a) Lateral view, showing enamel ridge; (b) side view, showing split root. (From Marsh 1891a.)

a   b

the position of the tooth row, and with the apex pointing dorsally. The maxilla is about 36 cm long and half that high. Twenty-eight *teeth* can be counted (Fig. 2.9). The bone in front of the maxilla is called the *premaxilla*—the logic for this name being refreshingly clear. It is 36 cm long and 22 cm high. The paired premaxillae are half round and frame a large part of the opening for the nostrils (external nares), which are usually large in horned dinosaurs. In front, the two bones are in broad contact with each other, but behind the two bones diverge somewhat. A rod of premaxilla ascends along the front surface of the maxilla. There is a distinctive dimple or pit in the bone in front of the nostril, and a tab of bone projects into the floor of the nostril. Horned dinosaurs had a turtle-like toothless beak composed of two bones, the *rostral* bone dorsally and the *predentary* ventrally. Both bones are robust, have sharp keels along the midline, end in points, and have sloping cutting edges; the edge on the rostral bone continues backward along the ventral edge of the premaxilla. The rostral and predentary bones each have a pronounced texture that suggests that they were covered in life with a horny material such as that which covers the beak of a turtle or a bird. The rostral bone is found only in horned dinosaurs, but the predentary is found in all the ornithischian or bird-hipped dinosaurs.

The *nasal* bone is easy to recognize. It is a good-sized, saddle-shaped bone 29 cm long that covers the external nostril and bears the horn core over the nose, the *nasal horn core*. The horn core is rather low and blunt. At only 10 cm in height above the bridge of the nose, it really could not have been too menacing. Behind it, the nasal bone sits on the premaxilla and maxilla, forming both the roof of the nasal cavity inside and the bridge of bone between the eyes and the nose horn. The orbital horns over the eyes are well preserved but, interestingly enough, are of rather different heights, the left one being 14 cm high over the eye socket, the right one only 8 cm high. The left horn core ends in a somewhat blunt point and curves backward (it is said to be recurved). The right horn core happens to be much blunter. Again, neither horn core seems particularly threatening. Each horn is an outgrowth of the *postorbital* bone, an extensive element that forms much of the upper and back borders of the eye socket or orbit. A variety of names are used to designate the horns over the eyes. Among them are "brow horn," "supraorbital horn," and "postorbital horn." They refer, respectively, to the brows, the position of the horn above the eyes, and the bone that composes the major portion.

I am pleased that each postorbital has some other, smaller bones fused to it so that I do not have to struggle with them in order to fit them into the whole skull. These smaller bones, including the *lacrimal* bone (the tear bone, because it contains the tear duct), the *prefrontal,* and the *supraorbital,* are hard to fit. Collectively these bones can be termed the *circumorbital series,* because they surround the orbit. Another circumorbital bone forms a major component of the cheek as well. This bone is called the *jugal.* In *Chasmosaurus* it measures 27 cm by 27 cm, and it is anything but rectangular. It is notched above, where it forms the lower rim of the orbit, and ends below in a tapering point. A second notch along the rear border defines a portion of the opening called the infratemporal fenestra.[18] In some horned dinosaurs, especially *Pentaceratops,* the point is so thick and prominent, accentuated by a small bone termed the *epijugal* ("on top of the jugal") that may fuse to the tip of the jugal, that it is described as a jugal "horn" (hence the name *Pentaceratops,* which means "five-horned face," instead of the usual three-horned-faced design of most ceratopsids).

The *quadrate* bones are complex and distinctive bones that are braced on the back of the skull to support the lower jaws.[19] In lateral (side) view of the skull each quadrate is somewhat covered by the jugal, and it is seen as a sloping rod that is expanded transversely at the lower end, where it attaches to the lower jaw. The upper end is a surprisingly thin blade, and it also has a wing-like process that attaches to delicate bones of the palate (roof of the mouth). The bone measures about 20 cm in length, and the expanded lower end is about 8 cm wide. When the skull is viewed from behind, the quadrate bones are prominent vertical props just inside and underneath the frill. The quadrate does not actually touch the jugal, but a spacer element is wedged between the quadrate and the inner surface of the jugal. This element is called the *quadrato-*

jugal, named for the two bones it separates. It is more than 3 cm thick ventrally but thins dorsally as it wraps around the shaft of the quadrate.

The bones of the frill are unmistakable.[20] There are only three bones, one parietal and two squamosals, making up the bony frill, which projects behind the skull and overhangs the neck. It is a terribly distinctive, extravagant structure. The frill of *Chasmosaurus* has huge, paired open spaces in it, called *parietal fenestrae,* measuring about 60 cm in length by 30 cm in width. The central bone of the frill is called the *parietal* bone. It runs from the postorbitals to the back of the skull. It measures 87 cm in length and has something of a T shape. Its striking features are a central bar 4 or 5 cm thick and a symmetrical crosspiece measuring about 1 m from end to end. Beside the front of the bar, a thin shell of flat bone slopes down toward the *supratemporal fenestrae.* Adjacent to the supratemporal fenestrae the surface is very smooth. A well-developed channel leads from the fenestra toward the frill. I infer that jaw-closing muscles exited the inside of the skull through the fenestrae and occupied part of the frill, but not necessarily the whole thing.

Besides the parietal, a pair of *squamosal* bones comprise the frill.[21] In ceratopsid ancestors, the squamosal bones were small bones only a centimeter or two long at the top corners of the skull that stabilized the quadrate bone and thus indirectly contributed to the support of the lower jaws. In ceratopsids generally there is an exaggeration of the squamosal bones, and in *Chasmosaurus* and its relatives (the chasmosaurines) the squamosal is greatly elongated behind the real skull. If the skull is considered a bony box enclosing the brain, the nose and mouth cavities, and the jaw-closing muscles, the frill is literally an add-on behind the skull, an extravagant come-on, as we saw in Chapter 1. In *Chasmosaurus* the squamosal measures 76 cm in length. It is 27 cm wide, thick, and flat and tapers toward the free end of the frill. At the front the squamosal is keyed firmly to the rest of the skull, especially to the postorbital and to the jugal. The infratemporal fenestra is enclosed by the squamosal and the jugal. The squamosal is constricted to a narrow point just behind the infratemporal fenestra and then flares to its widest point and sweeps backward and upward as it tapers to a point. The free edge of the squamosal is festooned with a series of scallops, seven or more in number, which tend to increase in prominence caudally. These ornamentations probably were separate bones, ineptly named *epoccipitals,* that fused to the squamosal in adult animals. The squamosal contacts the parietal along its length and extends clear to the back corner of

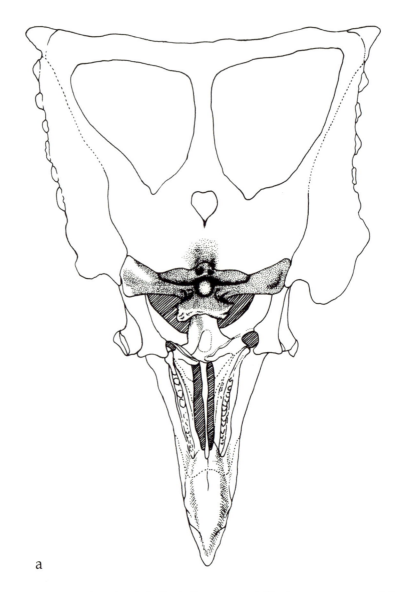

a

FIG. 2.10. (a) Underside of skull of *Chasmosaurus belli*, showing position of the occipital condyle, at the back end of the "true" skull. The frill is an "add-on" that doubles the length of the skull. (b and c) Occipital condyle and braincase of *Centrosaurus*: (b) caudal view; (c) lateral view. The foramen magnum (fm) is the opening through which the spinal cord leaves the braincase. The roman numerals indicate the holes by which the cranial nerves leave the braincase. bo, Basioccipital; fo, fossa ovale; lca, opening for internal carotid artery. (From Dodson and Currie 1990. Donna Sloan. Courtesy of the University of California Press.)

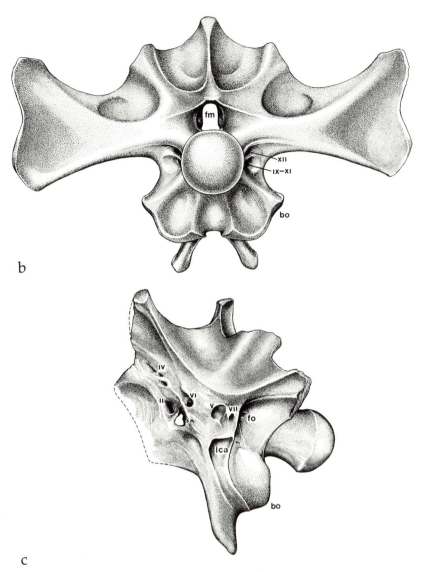

b

c

FIG. 2.10. (*continued*)

the frill. On the medial or internal surface of the squamosal there are several distinct ridges that provide slots to stabilize the head of the quadrate and the lateral processes of the braincase, but these joints do not seem all that firmly constructed.

Fortunately for my puzzle-building skills, the *braincase* presents as a single compound structure rather than as its separate constituents. The braincase is a box that encloses the brain, fits within the greater box of the skull, and also attaches the skull to the vertebral column and the muscles of the neck—a tall order for a single structure. What is most evident about the braincase viewed from behind are the large opening at the back, the spherical knob, and the wings or lateral processes. The opening is the *foramen magnum* (literally the large hole) by which the spinal cord leaves the brain and travels through the body in the neural canal of the vertebrae. The knob is known as the *occipital condyle,* and it is another highly distinctive feature of ceratopsids (Fig. 2.10).[22] No other kind of dinosaur has such a large, perfectly spherical structure that projects so prominently away from the skull. The structure of the condyle speaks of a high degree of mobility or maneuverability of the head. The width of the condyle in our specimen is 67 mm, only slightly smaller than a tennis ball. The braincase consists of both unpaired and paired elements, difficult to distinguish in adult specimens. The condyle consists of three bones, including the paired *exoccipitals* and the unpaired *basioccipital,* each contributing about one-third to the whole structure. The exoccipitals constitute much of the occipital surface and also form a pair of wing-like or even fan-shaped lateral processes that meet the squamosal and support the upper end of the quadrate. The basioccipital is an unpaired midline bone that lies underneath the brain. Besides contributing to the condyle, it has robust processes for attachment of muscles from the neck. Two pairs of modest openings for nerves are located on each side of the condyle.[23] The basioccipital continues forward underneath the brain as the *basisphenoid,* also unpaired. The major paired bones forming the side of the braincase are the *prootic* ("in front of the ear") caudally and the *laterosphenoid* rostrally. These bones enclose a series of openings for cranial nerves and vessels. They contact the underside of the skull roof dorsally and ensure that the brain is well protected.[24]

The paired bones of the palate are thin, complex, hard to understand, and rarely seen because they are inside the skull. We shall content ourselves with knowing they exist and leave it at that. They are the

pterygoids and the *palatines*. The long, unpaired *vomer* is rarely identified, but our *Chasmosaurus* has a nice one.

After all this brain work it is nice to switch to the lower jaws, which
are pretty straightforward. Come to think of it, straight forward is the
direction they run, from the jaw joint at the back to the beak in front.
Each jaw is quite separate from the other, and each is articulated
(joined) with the rather heavy predentary bone in a joint that may have
allowed some degree of independent movement. The principal bone of
the lower jaw is the *dentary*, the tooth-bearing bone.[25] In *Chasmosaurus*
the dentary is 42 cm long, and at the back it has a heavy, erect process,
the coronoid process, that stands 16 cm above the lower border of the
jaw. The coronoid process is a lever on which the jaw-closing muscles
insert. The body of the dentary is rounded laterally, whereas the inner
surface has a series of vertical grooves that represent the positions for
each of the vertical columns of teeth. There are twenty-eight tooth
positions in both the upper and lower jaws. We will have more to say
about teeth in the last chapter. The other bones of the jaw are much
smaller and are joined rather loosely; if an isolated lower jaw is found
as a fossil, the smaller bones are usually missing because they have
fallen off. The small bones are a bit of a jigsaw puzzle. The *splenial* is a
long, thin bone that covers an open groove on the lower inner side of
the dentary. The *articular* is a wide, cup-shaped bone that forms the joint
with the skull.[26] The cup is a socket to receive the condyle of the
quadrate. The articular is very important, because it forms the hinge by
which the jaw rotates open. The articular is supported underneath by
the *angular*, and the prominent bone that links the articular and angular
to the dentary is the *surangular* ("on top of the angular"), which is a tall
bone that forms the rear portion of the coronoid process.

There—we have done it. The skull we have before us measures an
impressive 147 cm in length. This size is routine, not exceptional, for
horned dinosaurs, which had very large skulls. Of course the "true"
skull, that is, the part common to all dinosaurs, was only half that
length. The great length of these skulls is accounted for by the frill, the
advertising structure. When the skull is joined to the skeleton, the total
length of the animal is 4.9 m, or 16 ft 2 in. We have in fact assembled an
entire skeleton from a pile of 312 bones (Fig. 2.11). It was, I admit, hard
work. Even though I have studied dinosaurs for many years, I have
never done a job like this before. I admit I could not have done it
without the help of some excellent books and technical papers. Newton

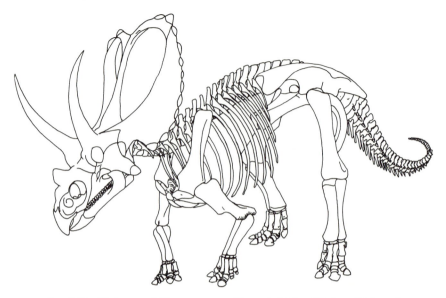

FIG. 2.11. Skeleton of *Chasmosaurus*. (Bruce J. Mohn/ Robert Walters.)

was not just kidding when he stated, with modesty that was completely out of character for him, that the reason he saw farther was that he stood on the shoulders of giants. Science really is an additive process; we benefit from the discoveries and errors of those who have gone before us. It is not fair to criticize the gaffes of scientists from an earlier era who did not have the benefit of all that we know today.

There tends to be a sense of wonder that paleontologists are able to reconstruct skeletons at all. Admittedly it does require a base of knowledge, but the thing to remember is what vertebrates have in common with each other rather than what differs. The overwhelming pattern of a common body plan among all vertebrates is part of the evidence that convinced Charles Darwin that evolution had occurred; that is to say, the wonderful diversity of living things has arisen by the modification over time of a single common ancestor. Most of the bones of the dinosaur skeleton are found in our own bodies as well! If we can recognize a bone as a humerus, then we know, whatever animal it comes from, that it articulates with the shoulder joint at one end and at the elbow at the other. A skilled paleontologist can usually determine what part of the body a bone comes from, even if he or she has never seen that kind of animal before. We are not geniuses (at least I am not)—we just use the clues before us.

So we have completed a rather detailed survey of anatomy. Are you still with me? Don't worry—you won't be tested! But we have learned a very important vocabulary that will make communication much easier. The effort is not wasted. We do not have to learn a separate vocabulary when we speak of *Triceratops* or *Protoceratops,* or of *Tyrannosaurus,* or of alligators, or even of mammals. The features held in common among all vertebrates are far more important than the differences. But as the French say, *vive la différence*!

CHAPTER THREE

Three-Horned Face

MONOCLONIUS! *Styracosaurus*! *Triceratops*! These names evoke such images of large, powerful, dare I even say *sophisticated* plant-eating dinosaurs. Their huge heads, bristling with sharp spikes and luxuriant bony frills, their solid, four-legged bodies, all suggest rhinoceros-like body plans run amok in the Cretaceous. Clearly these were animals to be reckoned with, not meek victims ready to bare their necks to the glistening teeth of a predator. Horned dinosaurs of the family Ceratopsidae are uniquely North American, as far as we know. They ranged from Mexico to Alaska but have not yet been found outside the bounds of our North American continent. Ceratopsids conveniently, and for the most part unambiguously, divide themselves into two groups or subfamilies, which we may designate as the Centrosaurinae, or short-frilled ceratopsids, and the Chasmosaurinae, or long-frilled ceratopsids. As *Triceratops* is a chasmosaurine, we might say that the chasmosaurines are slightly better known. The centrosaurines enjoyed their greatest diversity slightly earlier than the chasmosaurines and are primarily found in Alberta and Montana, whereas the chasmosaurines showed their greatest diversity slightly later in time and extend from Mexico to Alaska. If we treat the chasmosaurines before we consider the centrosaurines, there is no implication that the latter are more derived—the course of the evolutionary history of the ceratopsids presents a fork in the road, not a ladder of progress.

The first known and greatest of the "horn faces" was of course *Triceratops* itself. This mighty plant-eater is one of the all-time favorites among connoisseurs of Mesozoic saurians.[1] It was also the beneficiary of some powerful public relations. "Three-horned face" is known as one of the last of all dinosaurs, and it was a fitting contemporary of that redoubtable predator and stalker of children's dreams, *Tyrannosaurus rex* (Plate I). *Triceratops* was also exceptionally abundant, or at least its remains were preserved with a high frequency. Barnum Brown estimated that he had seen "no less than five hundred fragmentary skulls

56

and innumerable bones referable to this genus" while working in Montana between 1902 and 1909.[2] His statement may have been slightly hyperbolic, but it remains true that, even with only about one-tenth that number of specimens in museums, this is still one of the most abundant dinosaurs that we know. *Triceratops* is of surpassing importance because it was the first horned dinosaur to be described on the basis of a complete skull. Prior to the description of *Triceratops* in 1889, it was not clear, despite nearly two decades of finding ceratopsid fossils, that there were great reptilian herbivores with horns on their faces. The concept of horned dinosaurs nucleated around *Triceratops*. Yet scales did not drop suddenly from the eyes of pioneering paleontologists. *Triceratops*, like many dinosaurs before and since, was born in controversy and error.

The Chasmosaurinae are generally recognized by the prominence of the paired horns over the eyes combined with a modest horn over the nose, and by having a long frill with a long squamosal bone. The position of *Triceratops* within the Chasmosaurinae has been controversial until recently, as we will see, because its frill is much shorter relative to basal skull length than that of any other chasmosaurine. The solid frill of *Triceratops* is also a very unusual feature.

WHERE THE BUFFALO ROAM

Othniel Charles Marsh (1831–1899) was Yale University's great nineteenth-century vertebrate paleontologist (Fig. 3.1). He enjoyed the benefits of a great family fortune, earned by his maternal uncle, George Peabody, a Baltimore textile merchant, who founded the Yale Peabody Museum and endowed a professorship for his nephew. No child prodigy, Marsh did not settle down at Yale to his serious scientific career until 1866, in his thirty-fifth year, although he began publishing five years earlier. He was unquestionably intelligent and productive as a scientist, though in neither of these qualities was he the equal of E. D. Cope of Philadelphia. Also unlike Cope, he possessed superb political skills, as manifested by his presidency of the National Academy of Sciences (1883–1895), his appointment as vertebrate paleontologist of the U.S. Geological Survey (1882–1892), and his honorary curatorship of vertebrate paleontology at the Smithsonian Institution beginning in 1886. He was an excellent organizer, using others' monies—both his Uncle George's and our Uncle Sam's—to provision two museums with magnificent fossils. Marsh was a lifelong bachelor and a somewhat cold

FIG. 3.1. Othniel Charles Marsh (1831–1899). (Robert Walters.)

character. Whereas Cope's portrait (Fig. 5.2) shows eyes that burn like coals, Marsh's portrait shows cool, glinting eyes.

Henry David Thoreau mourned that he found so few people who still made their own houses. I sympathize with Thoreau—a similar problem prevails in paleontology. It is difficult to find those who both spend lengthy field seasons out in the fossil beds and sit at their desks and write the papers they should. Most of us are forced to choose between spending summers collecting as we so dearly love to do, and staying at home and writing. One solution is to have a staff of collectors. Marsh employed a veritable stable of paid (albeit poorly paid) collectors. Among the greatest of these was John Bell Hatcher (1861–1904), principal author of the marvelous 1907 monograph *The Ceratopsia,* the greatest opus ever written on horned dinosaurs.

Hatcher entered Marsh's employ after graduating from Yale in 1884 and was immediately sent to the fossil beds of western Kansas. Hatcher was bright, energetic, imaginative, impatient, ambitious, and highly effective at everything he did. As we shall see, Hatcher was a consummate collector of horned dinosaurs and other fossils for Yale. He worked for Marsh until 1893, when he accepted an appointment as curator of vertebrate paleontology at Princeton University. While at Princeton, he organized highly successful expeditions to collect Cenozoic mammalian fossils in desolate regions of Patagonia where Charles Darwin had collected sixty years earlier. In 1900 he moved to the Carnegie Museum in Pittsburgh. Marsh had died the year before, after a long and productive career. He had prepared a number of elegant plates to illustrate his proposed monograph but had written no text. Henry Fairfield Osborn (1857–1935) was asked by the director of the U.S. Geological Survey to oversee the publication of Marsh's intended monograph, for the plates had been prepared at government expense. He wisely asked Hatcher to shoulder the task. Hatcher set to work in 1902 and worked six hours a day for two years. He was on page 147 of the revised manuscript, that is to say about three-quarters of the way through, writing on *Triceratops flabellatus*, when he literally stopped in midsentence, left his typewriter, and, several days later, on July 3, 1904, died of typhus. Hatcher's death at age forty-two was a tragedy. He wrote his first paper in 1893 and completed a total of fifty-two between that date and 1904.

For our purposes, we remember Hatcher as a collector of horned dinosaurs. Marsh sent him to the Judith River country in the summer of 1888 to follow up on the earlier discoveries of Leidy and Cope, published in 1856 and 1876, respectively (see Chapter 1). The trip was not a smashing success, but Hatcher obtained, among other things, an occipital condyle and a pair of prominent horn cores. By December, Marsh had written a paper in which he described the find as *Ceratops montanus*.[3] This unpromising specimen, housed in the Smithsonian under the catalog number USNM 2411, is the specimen on which the entire family of horned dinosaurs, the Ceratopsidae, is based. Did that mean that Marsh had a clear concept of what was to follow? Not quite. His description contains the following interpretation: "The present genus appears to be nearly allied to *Stegosaurus* of the Jurassic, but differs especially in having had a pair of large horns on the upper part of the head. These were supported by massive horn cores firmly coossified with the occipital crest."[4]

Mercifully we are spared any pictorial representation of what such a chimaera might look like, but neither should we be harsh in our judgment. At least Marsh realized that there were dinosaurs with horns, an understanding that Cope—who we now know had had ceratopsid remains in his possession since 1872—had not achieved on his own. The *Ceratops* specimen drew renewed attention to the fact that interesting animals were preserved in the Judith River Formation. The occipital condyle of *Ceratops* measured 71 mm in diameter, an average sort of size for Judithian ceratopsids, and the orbital horn cores measured 220 mm in height. The respectable size of the horn cores suggests that their owner was probably a chasmosaurine. Marsh referred what is evidently a centrosaurine squamosal to *Ceratops*, but Hatcher was embarrassed by this referral, for the squamosal came "many miles from the locality which furnished the type."[5] By 1904, it was clear to Hatcher that *Ceratops* bore some resemblance to the new material from the Judith River of Alberta that Lambe was describing, notably *"Monoclonius" canadensis*. Currently, no further material having been described, *Ceratops montanus* is regarded as a doubtful name, there being too little material to characterize the species.

A year earlier, in 1887, Marsh had been sent a pair of huge horn cores 600 mm high and 160 mm in diameter collected by a Mr. George Cannon from Green Mountain Creek, just west of Denver, Colorado, and south of Golden. Geologist Whitman Cross sent the specimens to Marsh with the report that they came from a Cretaceous sandstone. Marsh nonetheless thought they were bison horns, and accordingly, despite their massive size, he named them *Bison alticornis*, the "high-horned bison." He described it as "one of the largest of American bovines, and one differing widely from those already described."[6] Hatcher delicately described the situation:

> It is now well known that Professor Marsh erred in referring these remains to the bisons, and that they are in reality the supraorbital horn cores of one of the larger Ceratopsia. Nor is Professor Marsh's error to be wondered at, but on the other hand it is quite excusable, since at that time nothing was known regarding the structure of the skull in these strange dinosaurs, and in size, surface markings, and form these horn cores more nearly resembled those of certain extinct bisons than of any other known animals, while the very imperfectly petrified nature of the remains might very readily be taken as indicative of the Pliocene or Pleistocene age of the deposits.

My, what a lengthy sentence! Hatcher continued in a philosophical tone:

> Indeed, this mistaken identification is a striking example to show how occasionally one may the more readily be led into error through a complete familiarity with his subject, for all that was then known of compar-

ative osteology, as well as the superficial structure and general character of these remains indicated that these horn cores pertained to a very large extinct bison. Had the remains fallen into the hands of any other vertebrate paleontologist probably the same error would have been made, for prior to the discovery of complete skulls, which occurred a few years later, nothing short of a microscopical examination of the minute structure would have revealed their reptilian nature.[7]

Once a Yale man, always a Yale man—but also not unfair to his mentor. Perhaps more to the point, it was Marsh himself who realized his error. Curiously, when he recognized the dinosaurian nature of his animal, he referred the species not to *Triceratops*, which is what we believe today to be the correct assignment, but to his Judithian genus *Ceratops*. Thus Marsh in 1889 recognized *Ceratops alticornis*, which we regard as a doubtful species. It does, however, serve the purpose of drawing attention to the existence of *Triceratops* as far south as Denver, even today the southernmost extension of its range.

TRICERATOPS—THE MEASURE OF ALL HORNS

Hatcher was only warming to his task. On his return from the Judith River country of Montana in the autumn of 1888, he passed through southeastern Wyoming. He stopped in the town of Douglas and was introduced to a rancher, Mr. Charles Guernsey, owner of the Three-Nine cattle ranch. Guernsey was a skilled amateur collector, and he showed Hatcher an impressive collection of fossils from the ranch, including a piece of a very large horn core measuring nearly 50 cm in length and 20 cm in diameter. It turned out that the specimen, embedded in a hard sandstone concretion, had lain at the bottom of a canyon 50 km north of Lusk. Guernsey himself wrote a hair-raising account of the discovery of the specimen by ranch foreman Edmund Wilson, who "had reported to him excitedly that during a beef roundup the boys had seen a large head 'sticking out midway of a bank on one side of a deep dry gulch,' with 'horns as long as a hoe handle and eye holes as big as your hat.'"[8]

An attempt at recovery of the skull (allegedly with a lariat looped about a horn!) failed, and it rolled to the bottom of the gulch.[9] Nonetheless, the horn was clue enough. When Hatcher returned to New Haven and saw for the first time the horns of *"Bison" alticornis* and compared them with those of *Ceratops* and those he had recently seen in Wyoming, he realized the significance of the latter. Guernsey sent the horns to Yale for Marsh to examine, and Marsh immediately sent Hatcher back to Wyoming to collect the skull. The skull, weighing nearly 500 kg, arrived in New Haven in May 1889.

Hatcher continued to collect in what he termed "the *Ceratops* beds of Converse County" (now known as the Lance Formation) until 1892, when the loss of government funds forced Marsh to curtail field operations. His success was legendary, for he had sent in the remains of some fifty ceratopsians, thirty-three of them more or less complete skulls. The largest weighed more than three tons and was dragged out of a deep ravine and hauled 65 km by wagon over rough trackless prairie to the nearest railroad. On May 20, 1889, Hatcher sent Marsh a packet with a few specimens of Cretaceous mammals that he hoped Marsh "would not despise." Marsh was so excited by these rarities that he ordered Hatcher by telegram to stop collecting ceratopsians and to go after mammals. Hatcher did as he was bid, protesting in response to entreaties for more that they were very rare and recoverable only at a rate of two per day. However, he proved himself remarkably resourceful. He discovered that ants on the windswept prairies had the interesting habit of carrying small hard objects to form mounds about their nests. The anthills were veritable mines of fossils, including the jaws and teeth of fossil mammals! Moreover, Hatcher discovered that sieves formed excellent tools for screening nest material, allowing for sand grains to run through, leaving behind a concentrate enriched in fossils. (To this day, paleontologists use screens for large-scale recovery of tiny fossils, especially those of mammals.) In one banner day, he collected eighty-seven fossil mammals! Previously surpassingly rare, Cretaceous mammals now numbered more than eight hundred specimens in the Yale collection. Marsh became an expert in this area as well.

Hatcher's first ceratopsian skull was no sooner in the front door of the Peabody Museum than Marsh wrote a brief, unillustrated, preliminary account of *Ceratops horridus*, in an issue of the *American Journal of Science* dated April 1889, in which he stated:

> The strange reptile described by the writer as *Ceratops montanus* proves to have been only a subordinate member of the family. Other remains received more recently indicate forms much larger and more grotesque in appearance. They also afford considerable information in regard to the structure of these animals, showing them to be true Stegosauria, but with

the skull and dermal armor strangely modified and specialized just before the group became extinct.

Marsh was very impressed with the horn cores, one of which measured 60 cm in length and 40 cm in girth at its base. He estimated that the head of his new animal must have weighed fifty times more than the head of the largest sauropod.[10]

With some cleaning up of the imperfect skull, further details were evident, and by August 1889, Marsh had transferred his species to a brand-new genus as *Triceratops horridus* ("three-horned face, projecting or standing up").[11] He glowed with enthusiasm: "The remarkable reptiles which the writer recently described and placed in a new family, the Ceratopsidae, prove to be more and more wonderful as additional specimens are brought to light."

What Marsh now noticed was a third horn, so that the name "three-horned face" was obvious. He also noted the rostral bone in front of the premaxillaries, never before observed, "forming a projecting beak, like that of a tortoise. Over all there was, evidently, a huge horny covering, like the beak of a bird." There was a "huge occipital crest, extending backward and outward. In the present specimen this is bent downward at the sides, like the back part of a helmet, thus affording in life strong protection to the neck." The skull was about 2 m long. The diameter of the occipital condyle, as later reported by Hatcher in 1907, was a whopping 116 mm!

In this second paper of August 1889, Marsh still had not illustrated *Triceratops*, but he added two new species. Marsh succumbed to the bad habit of taxonomic splitting, of naming excessive numbers of species. This malady has plagued the history of *Triceratops*. *Triceratops flabellatus* ("like a small fan") was described in a scant fourteen lines, though it was nearly 2 m in length, with a frill "like an open fan" 1.2 m in width. The margin of the crest was "armed with a row of horny spikes, supported by separate ossifications," which Marsh soon came to refer to as epoccipitals. The horns were 90 cm long. Marsh rhapsodized: "These dimensions far surpass those of any of the *Dinosauria* hitherto known, and indicate to some extent the wonderful development these reptiles attained before their extinction at the close of the Cretaceous."[12] A partial skeleton was also recovered, but nothing was said of it.

Two years later, apparently rather gratuitously, Marsh abruptly created a new genus for his species. He wrote: "A third genus, which may be called *Sterrholophus*, can be readily distinguished from the other two by the parietal crest, which had its entire posterior surface covered with ligaments and muscles supporting the head."[13]

Hatcher was not impressed. He politely but firmly expressed his opinion:

> In writing the above lines Professor Marsh appears to have forgotten that the parietal crest of *Ceratops* was quite unknown, and that therefore it was uncertain as to whether in that genus the parietal crest was free and protected by a horny covering or covered over with ligaments and muscles. Another point which does not seem to have been sufficiently considered by Marsh in establishing this genus is the immature nature of the skull upon which it was based. Considering the youth of the individual it does not appear at all improbable that if the parietal crest had been free, as in *Triceratops*, it would have shown those rugosities and other features so prominent on the surface of these bones in the skulls of older and more mature individuals. Then, again, if these characters are present in some and absent in other skulls of adult animals should they be considered as of generic or even specific importance or as sexual characters?[14]

No one has since referred to the species in question as anything but *Triceratops flabellatus*.

The second species discussed in the August 1889 paper, *T. galeus* ("helmeted"), came from Colorado, not Wyoming, and was described on the basis of a nasal horn core 71 mm in height. Despite the meagerness of these remains, Marsh considered it to be an animal of much smaller size, about 25 ft long. This estimate is curious, as he had not yet taken the measure of *Triceratops*. This "species" has not attracted much attention since its description but has enjoyed well-deserved obscurity.

By December 1889, Marsh had illustrated a skull for the first time, a reconstruction of *T. flabellatus* in lateral and dorsal views (Fig. 3.2). He also began to synthesize his finds. Marsh described the *Ceratops* beds as stretching some 1,300 km along the eastern flank of the Rockies, "freshwater or brackish deposits, which form part of the so-called Laramie."[15] It was clear in his mind that the Ceratopsidae were "the most important of this assemblage." The skulls of *Triceratops* were enormous. Based on the size of a comparatively young individual nearly 2 m in length, he estimated the skull in an old individual to reach 2.4 m in length. In retrospect, although some of the contemporaries of *Triceratops* did reach this size, *Triceratops* itself probably did not. Nonetheless, his claim that only whales had larger skulls remains true. He was greatly impressed

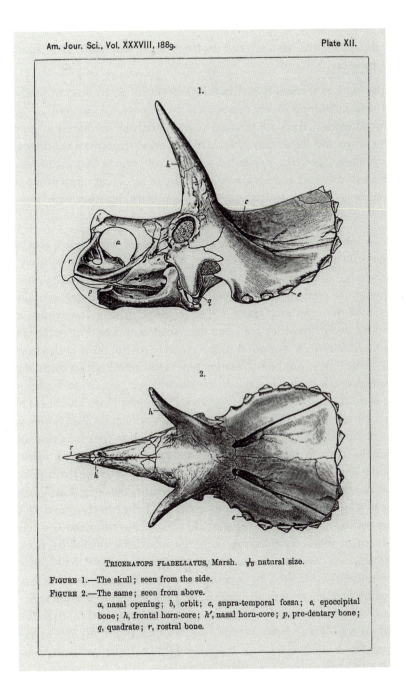

TRICERATOPS FLABELLATUS, Marsh. $\frac{1}{20}$ natural size.

FIGURE 1.—The skull; seen from the side.
FIGURE 2.—The same; seen from above.
        a, nasal opening; *b*, orbit; *c*, supra-temporal fossa; *e*, epoccipital
        bone; *h*, frontal horn-core; *h′*, nasal horn-core; *p*, pre-dentary bone;
        q, quadrate; *r*, rostral bone.

FIG. 3.2. *Triceratops flabellatus* skull. Reproduction of a page from the *American Journal of Science,* December 1889. (Marsh 1889c.)

65

by the defensive and offensive armature of the skull. It is hard to improve on his words:

> The skull is wedge-shaped in form, especially when seen from above. The facial portion is very narrow, and much prolonged in front. . . . In the frontal region, the skull is massive, and greatly strengthened to support the large and lofty horn-cores, which formed the central feature of the armature. The huge, expanded parietal crest, which overshadowed the back of the skull and neck, was evidently of secondary growth, a practical necessity for the attachment of the powerful ligaments and muscles that supported the head.

It is interesting to recall that, although Marsh was highly articulate about the strength and power of *Triceratops,* he had absolutely no knowledge of the mighty contemporary with which *Triceratops* had to contend, the redoubtable *Tyrannosaurus rex.* The first skeletal remains of this monster predator were not discovered until three years after Marsh's death, and they were not described until a year after Hatcher's death. Marsh drew attention to the massive rostral bone at the front of the upper jaws, "not before seen in any vertebrate," which formed part of a turtle-like beak. He gave an excellent description of each region of the skull, only highlights of which will be noted here. The nasal horn core was described as a separate center of ossification that unites with the nasal in adults without trace of suture, but which is also quite variable in form in different species. The postorbital horn cores are hollow at the base "and in form, position, and external texture, agree closely with the corresponding parts of the Bovidae" (e.g., cattle, sheep, goats, and antelope). The bony orbit underneath the horn is surrounded by a very thick margin. He correctly recognized the central element of the occipital crest as the parietal, surrounded by peculiar ossicles, the epoccipital bones, that fuse with the parietal in old animals. The parietal is flanked by squamosals. The occipital condyle is very large and nearly spherical, "indicating great freedom of motion." "The brain-cavity is especially diminutive, smaller in proportion to the skull, than in any other known reptile." Marsh thus found an abundance of characters to define the Ceratopsidae. He concluded with some interesting thoughts about the evolutionary fate of his Ceratopsidae:

> Such a high specialization of the skull, resulting in its enormous development, profoundly affected the rest of the skeleton. Precisely as the heavy armature dominated the skull, so the huge head gradually overbalanced

the body, and must have led to its destruction. As the head increased in size to bear its armor, the neck first of all, then the fore limbs, and later the whole skeleton, was specially modified to support it.

This kind of naive evolutionary thought, evolutionary inertia leading to extinction, does not find favor today. Nonetheless, this was an admirable paper.[16]

By January 1890, it was another paper, two new species of *Triceratops* in two brief pages of text, with no illustrations. The first of these was *T. serratus* ("serrated"). In Marsh's words, "The present skull is more perfect than any hitherto found, and exhibits admirably the strongly marked characters of the genus. It is likewise of gigantic size, being nearly six feet in length (1.8 m), although the animal was not fully adult." Marsh was impressed by a series of bumps along the midline of the parietal frill, as well as a second, less prominent, series along the squamosal, which gave a serrated appearance. The nasal horn core was poorly developed or not preserved in the specimen. Marsh believed that it had not fused with the underlying nasal bones (Fig. 3.3).

The second species was *Triceratops prorsus* ("straight on"). It was about the same size as the skull of *T. serratus*. Though somewhat distorted, it showed an excellent, strong nasal horn core, directed nearly horizontally forward. Hatcher later reported that the nasal horn core was 21 cm long and the orbital horn cores 55 cm long.[17] Marsh found the surface of the nasal horn core to be "rugose from vascular impressions, indicating that it was covered by horn, thus forming a most powerful weapon." The orbital horn cores are more massive than those of *T. serratus*. He also found a contrast with the skull of *T. horridus*, in which the nasal horn core pointed upward instead of forward. Hatcher pointed out that the skull, at 1.5 m in length, is on the small side for *Triceratops*; its occipital condyle measured only 84 mm in diameter (Fig. 3.4). Also found with the skull were some cervical vertebrae. For the first time, the fusion of the first several vertebrae of the neck was recognized. "This union, unknown hitherto among the *Dinosauria*, was evidently rendered necessary to afford firm support for the enormous skull."[18]

In May 1890, Marsh illustrated another paper on *Triceratops*. On this occasion he named *Triceratops sulcatus* ("furrowed"). This was a large but poorly preserved skull. He noticed that the upper halves of the postorbital horn cores had deep grooves on their caudal surfaces— hence the specific name. He also described a deep groove on the ventral surface of the caudal vertebrae, a character that was later observed in

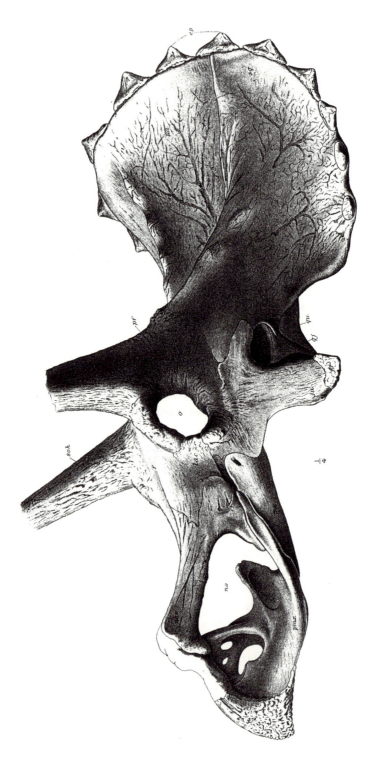

FIG. 3.3. *Triceratops serratus*. ej, Epijugal; ep, epoccipital; f, frontal; ju, jugal: lac, lacrimal; mx, maxilla; nas, nasal; no, nostril or external naris; o, orbit; pa, parietal; pf, postorbital; pmx, premaxilla; prf, prefrontal; qj, quadratojugal; qu, quadrate; soh, supraorbital horn core; sq, squamosal. (From Hatcher et al. 1907.)

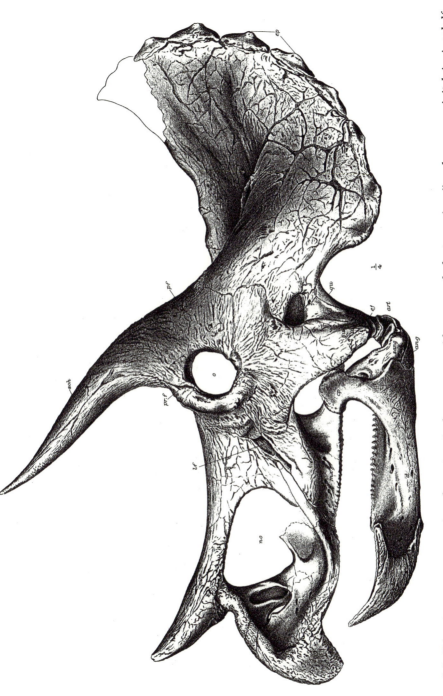

Fig. 3.4. *Triceratops prorsus.* ang, Angular; art, articular; cp, coronoid process; d, dentary; ej, epijugal; ep, epoccipital; ju, jugal; lf, antorbital foramen; mx, maxilla; no, nostril or external naris; o, orbit; pa, parietal; pf, postorbital; pmx, premaxilla; prf, prefrontal; qj, quadratojugal; qu, quadrate; sang, surangular; soh, supraorbital horn core; sq, squamosal. (From Hatcher et al. 1907.)

other horned dinosaurs of the long-frilled variety. When Hatcher reviewed this species in 1907, he was not impressed:

> In view of the fact that as shown above, grooves similar to those described by Marsh as characteristic of the present species may occur at various places on the supraorbital horn cores of the Ceratopsidae, it does not seem advisable to consider either the presence or position of such grooves as of specific importance. It is probable such grooves have, in most instances at least, had their origin in an infolding or thickening of the horny sheath with which in life the horn core was incased, and that their position, form, and depth were determined by the place, nature, and amount of thickening, or infolding of the horny substance. Such being their origin, as appears not improbable, they are likely to appear in any of the various genera and species, and should not be considered as of specific importance.[19]

This paper, however, represents a continuation of the synthesis of data on the newly emerging Ceratopsia, a name that Marsh coined as the name for what is now a suborder, the taxonomic category at which horned dinosaurs, considered most inclusively, have been generally recognized ever since.[20] Adding to the epoccipitals, already recognized, Marsh reported "another ossification has been found attached to the lower extremity of the jugal bone. This is a separate element, like the epoccipital bones, but in very old animals it is coössified with the jugal, on which it rests." This he called, somewhat more logically than the epoccipitals, the epijugal. He illustrated an internal cast of the brain and showed its position in the skull. He also figured the snout of *Triceratops prorsus*, drawing attention to the rostral bone, the nasal horn core, the toothless premaxilla, and the structure of the external nostril (Fig. 3.5). For the first time, he recognized the structure of the teeth as significant, by virtue of their having two distinct roots. Previously he had compared the teeth of *Triceratops* with those of hadrosaurs, which, like those of all other dinosaurs, are single-rooted (Fig. 3.6). Marsh described the arrangement of the teeth: "The teeth form a single series only in each jaw. The upper and lower teeth are similar, but the grinding face is reversed, being on the inner side of the upper series, and on the outer side of the lower series. The sculptured surface in each series is on the opposite side from that in use."

In addition to illustrations of the skull and teeth, Marsh figured the fused neck vertebrae, so distinctive of horned dinosaurs, an ungual bone of the toe, and the pelvis. Finally, he paid tribute to the collector:

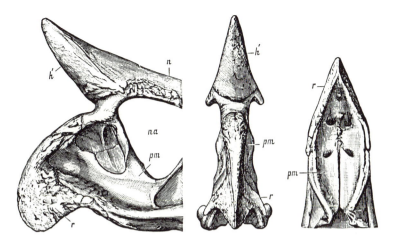

FIG. 3.5. Snout of *Triceratops prorsus,* showing nasal horn, external nostril, rostral bone, and beak-like premaxilla (side view, front view, and ventral view). h′, Nasal horn core; n, nasal; na, naris or external nostril; pm, premaxilla; r, rostral bone. (From Marsh 1890b.)

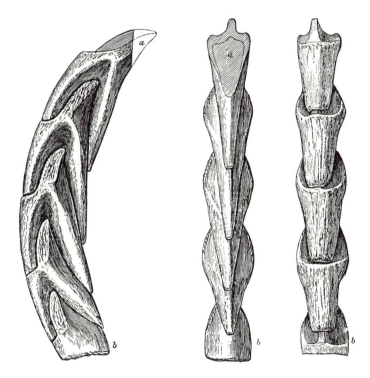

FIG. 3.6. *Triceratops* lower teeth in front view, external view, and internal view. Only the top tooth in each column is in use. (From Hatcher et al. 1907.)

71

"For the discovery of the specimens here described, belonging to this order, science is mainly indebted to the writer's able assistant, J. B. Hatcher, whose genius has done so much to bring to light the rare fossil vertebrates of the West." This was high praise, as Marsh was very guarded with accolades for those in his employ.[21]

In February 1891, Marsh published a further synthesis of "The gigantic Ceratopsidae, or horned Dinosaurs, of North America," the text of an address he had delivered the previous September to the British Association for the Advancement of Science in Leeds. Once again, he reviewed the anatomy of *Triceratops*, repetitively, sometimes verbatim, but with further details emerging. The illustrations now ran to ten handsome plates. The figures of *T. flabellatus* and *T. prorsus* from the earlier publications are repeated, but there are also lovely new figures of the vertebrae and limbs. One skeletal highlight is the large, robust humerus, whose length (nearly as long as the femur) indicated to Marsh that the animal walked on all fours, a conclusion that has been well received. (The thought of a bipedal *Triceratops* strains the imagination of even the most fancifully inclined!) Marsh believed that "various spines, bosses and plates" found in the *Ceratops* beds represented a body armor. We know, however, that these are actually remains of ankylosaurs and pachycephalosaurs, animals with which he was unfamiliar, although he had already named *Nodosaurus* on the basis of scanty remains. He elaborated on the efforts of Hatcher,

> who has done so much to bring to light the ancient life of the Rocky mountain region. I can only claim to have shared a few of the dangers and hardships with him, but without his skill and energy little would have been accomplished. If you will bear in mind that two of the skulls, represented in the diagrams before you, weighed nearly two tons each, when partially freed from their matrix, and ready for shipment, in a deep, desert cañon, fifty miles from a railway, you will appreciate one of the mechanical difficulties overcome. When I add that some of the most interesting discoveries were made in the hunting grounds of the hostile Sioux Indians, who regard such explorations with superstitious dread, you will understand another phase of the problem.[22]

Triceratops *Restored*

In April 1891, Marsh took a very important step. He published a careful skeletal restoration of *Triceratops*, his first attempt to make a com-

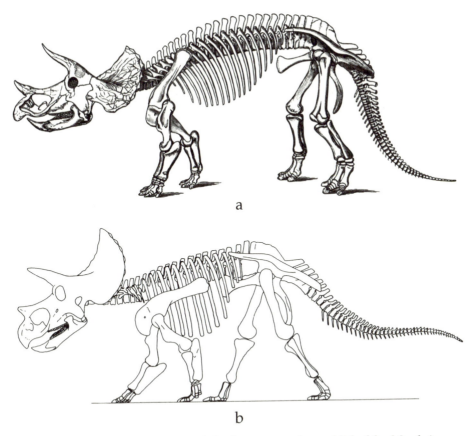

a

b

FIG. 3.7. (a) *Triceratops prorsus* skeletal reconstruction published by Marsh in 1891. The body is too long and the limbs are too stiff by modern standards, but for its day this was a pioneering reconstruction. (From Marsh 1891b.) (b) Modern skeletal reconstruction of *Triceratops horridus*. (Kenneth M. Carpenter.)

prehensive, accurate restoration of a dinosaur (Fig. 3.7). (In the same paper he restored "*Brontosaurus*.") The restoration was a composite, based on the skull of *T. prorsus*, plus skeletal parts from a second specimen that he referred to the same species. Other missing parts were found in the same region. He restored *Triceratops* as an animal "twenty-five feet in length and ten feet in height" (i.e., 7.6 m long and 3 m in height).[23] For a first attempt, it was a good one, although the body has too many vertebrae and so is too long. The limbs are also too straight, particularly the front legs. Marsh was a pioneer in making visual reconstructions of dinosaurs, and by 1895 he had published reconstructions of a dozen dinosaurs, including four European ones.[24]

Does this mean that Marsh was a leader in the exhibition of dinosaur skeletons? Far from it—he was adamantly opposed to the practice. The first time a dinosaur skeleton was reconstructed and placed on public exhibit anywhere in the world was in 1868 at the Academy of Natural Sciences in Philadelphia. Joseph Leidy, working with artist Benjamin Waterhouse Hawkins, restored the missing parts and exhibited the duck-billed dinosaur *Hadrosaurus foulkii*, correctly representing it as an upright, two-legged animal. As popular as this exhibit and copies shown at other museums around the country were, Marsh was scornful, and no dinosaur was exhibited at the Peabody Museum during his lifetime. He wrote in a letter to Spencer Baird, Secretary of the Smithsonian, in 1875:

> I do not believe it possible at present to make restorations of any of the more important extinct animals of this country that will be of real value to science, or the public. In the few cases where materials exist for a restoration of the skeleton alone, these materials have not yet been worked out with sufficient care to make such a restoration perfectly satisfactory, and to go beyond this would in my judgment almost certainly end in serious mistakes. Where the skeleton, etc., is only partly known, the danger of error is of course much greater, and I would think it is very unwise to attempt restoration, as error in a case of this kind is very difficult to eradicate from the public mind. . . .
>
> A few years hence we shall certainly have the material for some good restorations of our wonderful extinct animals, but the time is not yet.[25]

His conservatism in some respects is to be commended. Whatever the shortcomings of his work, Marsh is not remembered in history for some of the anatomical gaffes that have plagued other paleontologists. His prediction of events "a few years hence" was certainly realized. But his prejudice against public exhibit was not softened by the passage of time. Twenty years later he saw fit to heap delicious scorn upon the concrete monsters of the Crystal Palace. The concrete sculptures, Victorian delights still to be viewed today in London, were the first attempt to bring dinosaurs to life and make them accessible to the public. They also represent a historically important interaction between a scientist, Richard Owen, and an artist, the aforementioned B. Waterhouse Hawkins. Of these creations Marsh wrote:

The dinosaurs seem . . . to have suffered much from both their enemies and their friends. Many of them were destroyed and dismembered long ago by their natural enemies, but, more recently, their friends have done them further injustice in putting together their scattered remains, and restoring them to supposed lifelike forms. . . . So far as I can judge, there is nothing like unto them in the heavens, or on the earth, or in the waters under the earth.[26]

Perhaps Marsh turned in his grave in 1902, when the hind legs of "*Bronto-saurus*" were erected in the old Peabody Museum. Certainly he must have when the complete skeleton was exhibited in the new hall at Yale in 1931.

More Species of Triceratops

In September 1891, Marsh named *Triceratops elatus* ("elevated") on the basis of an enormous skull that measured 1.9 m in length. The skull, which was not figured, had a blunt nasal horn core that pointed upward. The postorbital horn cores were directed forward and measured 74 cm from the dorsal rim of the orbit to the tip. Marsh found that the crest was quite long and much elevated, which is the reason for the name he gave. The crest was fully a meter in breadth.[27] Hatcher gave the diameter of the occipital condyle as a respectable 106 mm.

There followed a hiatus in the taxonomic history of *Triceratops*. Marsh returned to *Triceratops* in the twilight of his life, when he was called upon to transfer specimens of fossils collected at government expense from New Haven to Washington, D.C. He sent a dozen ceratopsian skulls to the Smithsonian and named two of them as new species, neither of them illustrated. The first is *Triceratops calicornis* ("cup horned"), a very large skull, measuring nearly 2.2 m in length according to Hatcher. The nasal horn core shows a peculiar cupping on its caudal surface, which "somewhat resembles the bottom of a horse's hoof." The second species is *Triceratops obtusus* ("blunt"), another species based on the nature of the nasal horn core. This horn is described as being 14 cm in width but only 2.5 cm high.[28]

TRICERATOPS AFTER MARSH

Marsh died at the age of sixty-seven on March 18, 1899, after a long, genuinely distinguished, and productive scientific career. Not surprisingly, he left many projects unfinished, among them the projected monograph of the Ceratopsia, for which he had overseen the preparation of a large number of superb lithographic plates and illustrations, again at government expense. As mentioned earlier in this chapter, Henry Fairfield Osborn enjoyed an honorary appointment as vertebrate paleontologist to the U.S. Geological Survey, and it was his task to ensure publication of the ceratopsian monograph. He in turn asked J. B. Hatcher to be the author. Hatcher reviewed all of the *Triceratops* material, which was very dear to his heart. In the course of doing so, he recognized two new species. As we have already discussed, Hatcher himself died in 1904, leaving the manuscript incomplete. Hatcher's material accounted for 157 pages of text, leaving 10 pages unrevised. Richard S. Lull, Marsh's successor at Yale, saw the monograph through to completion as third author and was responsible for the final 34 pages of text.[29] As was acceptable practice in those days, the text on Hatcher's new species was extracted from the unpublished manuscript and published separately, in this case in the *American Journal of Science*, in 1905.

The name of the new species was *Triceratops brevicornus* ("short horn"). The skull was a fine one, and there was a complete series of vertebrae in front of the pelvis and a partial pelvis as well. The skull was an ample 1.65 m in length, with an expansive frill 1.1 m in breadth. The occipital condyle measured 88 mm in width, and the horns were comparatively short. The nasal horn was erect. The supraorbital horns were only about 50 cm high but quite robust, with circular rather than laterally compressed bases. Hatcher documented six epoccipitals on the squamosal and at least nineteen on the parietal.[30]

The status of the second species is a little unusual, in that the nominal author of the paper is Hatcher, but, as Lull explained, Hatcher had failed to designate a name. Therefore Lull provided (and receives credit for) the name. It was construed as a new genus, which Lull named *Diceratops hatcheri* ("Hatcher's two-horned face"). This somewhat unusual, imperfectly preserved skull has had a history of controversy. It is a large skull, 2 m long. It incontrovertibly has a "peculiar" look about it, oddly short-faced, with a very weak, blunt nasal horn core (said by Lull to be missing), resembling that of *T. obtusus*, with very erect, nearly vertical horn cores. The squamosals on the sides of the frill were pierced with large openings. The parietal is thin, and there appear to be small parietal fenestrae. The better preserved one of the right side measures 15 by 5 cm. If this is correct, this would be an extremely unusual character for *Triceratops*. Lull was rather guarded in his assessment. He concluded the paper:

> While I believe *Diceratops* to be a valid genus, I am not inclined to lay stress upon the parietal and the squamosal fenestrae which Hatcher does, as they may possibly be pathologic. Those of the squamosal bones, which are found in no other form among the Ceratopsia, are not of the same size, while only one is known in the parietals for the sufficient reason that the bone is broken away upon the left side where the fenestra would come if present, and it is quite possible that it may never have existed.[31]

Given Lull's stated reservations, and the unique status of the specimen, no others having come to light subsequently, the tendency has been either to ignore this species altogether or to regard it as a species of *Triceratops*. By 1933, Lull had reaffirmed his belief that the species was valid but referred it to the genus *Triceratops*, subgenus *Diceratops*. The skull he regarded as that of an aged individual. In 1933 he gave the length as "about 6 feet 1 inch in overall-length, and therefore slightly below average for *Triceratops* skulls." It had unaccountably shrunk in size in the intervening years. He elaborated on the holes in the frill: two in the left squamosal, one in the right, and one in the parietal. They were irregular in size and symmetrical. The bone in front of the large hole in the left squamosal is "tumid, thickened, and irregular, which heightens the impression that all of these perforations are pathologic and may be cystic openings. . . . as such the result of disease, or injury, or both. Surely they can have no taxonomic significance."[32]

Two further species of *Triceratops* were described independent of the Hatcher-Marsh-Lull triumvirate. All of the *Triceratops* skulls collected by the Hatcher parties came from a small area of Converse County in eastern Wyoming. In fact, most of the skulls came from within a few kilometers of each other, a span of 20 km accounting for all of the legitimate types. E. M. Schlaikjer collected for Harvard University in 1930 a poorly preserved skull of small size. The site was in Goshen County, Wyoming, 150 km south of the classic *Triceratops* localities. The horizon is the Torrington member of the Lance Formation, which is regarded as slightly younger than the *Triceratops* beds of Converse County. Schlaikjer studied this specimen and in 1935 described it as a new species, *Triceratops eurycephalus* ("wide head"). The skull had an estimated length of 1.39 m, with a reconstructed breadth of the frill almost as great (129 cm). The condyle is a puny 66 mm in breadth, but the separate parts seem to have fused to form the definitive adult state. As reconstructed, the skull has a peculiar appearance, and it has been difficult to assess its status. *T. eurycephalus* has rarely been illustrated and often it is simply ignored. The nasal horn core is low and blunt. The orbital horn cores seem especially elongate and measure some 63 cm in length on a small skull. As illustrated and described, the squamosal is long and smooth bordered. It shows no indication of the scalloped shape so characteristic of typical *Triceratops*. The parietal is very broad, and grooves for blood vessels were not very prominent, again a character that does not accord well with mature *Triceratops*. Schlaikjer noted that a femur was found with the specimen that is "practically as long as the femur of the large *T. elatus* (?) skeleton in the American Museum, but it is far less massive." This suggests that *T. eurycephalus*, whatever it is, "was almost as large as any of its Lance relatives but was indeed a more slender-bodied and perhaps a more agile form."[33]

Triceratops is also found in Canada. C. M. Sternberg found an incomplete skull of *Triceratops* in the Red Deer Valley of Alberta, northwest of Drumheller, in 1946. This find constituted the first record of *Triceratops* in Canada. Subsequently, *Triceratops* has been found in Saskatchewan as well. In 1949, Sternberg described the specimen as *Triceratops albertensis* ("from Alberta"). The specimen is incomplete, missing its beak and jaws, but it appears to be a very large skull. The preserved portion measured over 1.9 m; Sternberg estimated the complete length to have been 2.4 m, by far the largest skull of *Triceratops* ever described. The nasal horn core was not preserved, but the orbital horn cores are erect, even slightly inclined toward the back of the skull. The horn stands 72 cm high, and its massive base measures 34 cm from front to back. The squamosal and parietal appear typical of *Triceratops*.[34] Although *Triceratops* clearly existed in Canada, it seemingly was an uncommon faunal component.

There is further evidence of *Triceratops* in Saskatchewan. The Frenchman Formation of southwestern Saskatchewan is an equivalent of the Scollard Formation of Alberta and the Hell Creek and Lance Formations of Montana and Wyoming, respectively. Sure enough, *Triceratops* is found there as well. A large skull measuring 2.1 m in length was collected near Shaunavon, Saskatchewan, in 1967. It resides in the Royal Saskatchewan Museum in Regina. A second, incomplete specimen was collected nearby in 1936. This specimen is housed in the Eastend Museum, Eastend, Saskatchewan. Both specimens have been referred to *Triceratops* cf. *T. prorsus*.[35] The designation *cf.* indicates a degree of cautiousness, to the effect that the specimens are "more like Marsh's *T. prorsus* than they are like any other species."

HOW MANY SPECIES OF *TRICERATOPS*?

Thus it was that the last new species of *Triceratops* was described in 1949 (Table 3.1). It is somewhat ironic that Barnum Brown never described or named skulls of *Triceratops*, because he claimed to have seen five hundred in the field, and he collected several fine ones from the Hell Creek beds of Montana, where he began his career as a collector on his own. It is by no means true that no further specimens were collected after 1949. Far from it. Although many specimens come from Montana, none are name bearing. Many specimens of *Triceratops* have been collected since that time; these reside in museums from Buffalo to Milwaukee to St. Paul to Bozeman to Los Angeles. Unfortunately, few of these have been introduced into the scientific literature. It is usually purely by chance that one finds out about these specimens.

A revolution occurred in evolutionary biology during the 1930s and 1940s. This has come to be known as the modern synthesis, a reconciliation of the genetic research of the laboratory with the field studies of population biologists, who saw ample evidence for variation within species of wild plants and animals. Several paleontologists were leading proponents of the modern synthesis, the most influential by far being the great George Gaylord Simpson, Yale trained and a curator at the American Museum of Natural History. Other contributors were E. C. Olson of the University of Chicago and Glenn Jepsen of Princeton University. None of these paleontologists was a student of dinosaurs, but the research agenda for all of paleontology—indeed all of systematic biology—shifted as a result of their work, and Darwinism at last had an impact on the daily lives of those who studied organisms.

Paleontology had changed. The taxonomy of *Triceratops* came to be seen as a problem, a mess, a quagmire, an anachronism, even an embarrassment. The inclination to describe and illustrate, to document the primary data of paleontology, had passed. Lush illustrations came to be viewed as a luxury, not a necessity.

The desire naturally arose to make some sense of the proliferated species of *Triceratops* (Fig. 3.8). The first attempt to discern a pattern was in the 1907 monograph, but it was written by Lull and not by Hatcher, who (no criticism intended) was too close to his mentor and too rooted

TABLE 3.1
Summary of *Triceratops* "Species"

"Species"	Date	Author	Locality	Feature(s)
Triceratops horridus ("projecting or standing up")	1889	Marsh	Wyoming	Very large skull
Triceratops flabellatus ("like a small fan")	1889	Marsh	Wyoming	Large skull, jaws, partial skeleton
Triceratops galeus ("helmeted")	1889	Marsh	Colorado	Nasal horn core
Triceratops serratus ("serrated")	1890	Marsh	Wyoming	Large skull
Triceratops prorsus ("straight on")	1890	Marsh	Wyoming	Medium skull, jaws vertebrae, partial skeleton
Triceratops sulcatus ("furrowed")	1890	Marsh	Wyoming	Large skull, poorly preserved, with some vertebrae and skeletal parts
Triceratops elatus ("elevated")	1891	Marsh	Wyoming	Large skull
Triceratops calicornis ("cup horned")	1898	Marsh	Wyoming	Large skull
Triceratops obtusus ("blunt")	1898	Marsh	Wyoming	Large skull
Triceratops brevicornus ("short horn")	1905	Hatcher	Wyoming	Large skull, vertebral column
Diceratops hatcheri ("Hatcher's two-horned face")	1905	Lull	Wyoming	Large skull
Triceratops eurycephalus ("wide head")	1935	Schlaikjer	Wyoming	Poorly preserved smaller skull
Triceratops albertensis ("from Alberta")	1949	Sternberg	Alberta	Very large partial skull

Note: This table is comprehensive and in historical order. Figure 3.8 shows only those "species" that have been figured.

in history to strive toward a modern understanding. Hatcher's detailed descriptions of all of the species of *Triceratops* are extremely valuable and generally far more informative than Marsh's original descriptions. Hatcher's writings form an excellent starting place for any consideration of our three-horned friend. The monograph has beautiful plates,

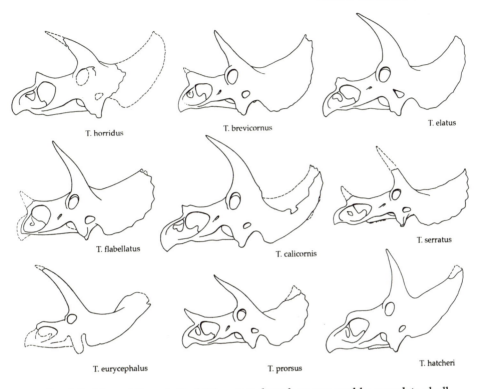

FIG. 3.8. Named "species" of *Triceratops,* based on reasonably complete skull material. (Robert Walters after Ostrom and Wellnhofer 1986.)

which Marsh had prepared in his lifetime but did not publish before his death. I am also most appreciative of the extensive measurements of each skull that Hatcher provides.

In considering phylogeny, Lull placed *Triceratops* in a lineage with *Monoclonius,* a view we do not hold today. Lull recognized ten species of *Triceratops,* including "*T. (Bison, Ceratops) alticornis.*" The inclusion of the latter is a little surprising; generally the species is either forgotten or remembered only as a classic paleontological gaffe. The only species he eliminated was *T. galeus,* on the grounds that it was simply too fragmentary. He arranged the named species into two groups:

Group 1	Group 2
T. horridus	*T. elatus*
T. prorsus	*T. calicornis*
T. brevicornus	

81

He did not state his basis for the two groups and noted that *T. serratus* and *T. flabellatus* would each stand alone, even though they resembled each other in some features. The remaining species were too fragmentary for Lull to consider.[36]

In 1933 Lull published a monograph, "A Revision of the Ceratopsia." His review includes few fresh data but is a useful summary of the subject material, though it lacks the gorgeous artwork of the grand old 1907 monograph. He emphasized the size of the various skulls and offered the helpful datum that the average skull of *Triceratops* measures 1.93 m in length, the range being from 1.55 m in *T. prorsus* to 2.1 m in *T. calicornis*. He eliminated four species as based on inadequate material: *T. alticornis, T. galeus, T. sulcatus,* and *T. maximus*—a step forward! He also presented a chart illustrating for the first time a tentative phylogeny of the horned dinosaurs. As before, he recognized the same two lineages of *Triceratops*, plus a third one, consisting of *T. obtusus* and *T. (Diceratops) hatcheri*. He still had no placement for *T. serratus* and *T. flabellatus*. The lineage of *T. prorsus–T. brevicornus–T. horridus* he regarded as "the most conservative," and this sequence showed an increase in skull size and a decrease in the size of the nasal horn. The second group, that of *T. elatus–T. calicornis,* is characterized by "great brow horns and curiously reduced nasal horn." He admitted that the latter species may be a variant of *T. elatus*. The third group, consisting of *T. obtusus* and *T. hatcheri,* was characterized by a nasal horn that was almost vestigial. Lull added many cautionary notes, pointing out that the sequences he presented did not correspond to the actual order of appearance of the species in the stratigraphic section, but his was nevertheless a noteworthy attempt.[37]

Following Lull's attempted phylogeny, the matter of the species of *Triceratops* and their phylogenetic arrangement mainly just sat there. Sternberg in 1949 added Schlaikjer's species, *T. eurycephalus,* to the diagram. He suggested that it was common to both the *T. obtusus–T. hatcheri* lineage and the *T. elatus–T. calicornis* lineage, and that these two were more closely related to each other than either was to the third lineage that culminated in *T. horridus*. Major synthetic papers by Colbert in 1948 and Ostrom in 1966 moved on to other matters, ignoring the question of species-level relationships for the legitimate reason that this was not an interesting question at that time.

More recently, a biological species concept that emphasizes morphological variation as an expected attribute of nature has come to the fore, with many examples drawn from the living world being applied to the fossil record. Furthermore, powerful statistical tools have become part of the paleontologist's bag of tricks, as computers have become essential fixtures on everyone's desk. Inelegant? Possibly. Effective? Definitely.

The most important taxonomic study of *Triceratops* since 1907 was published in 1986 by John Ostrom, with the co-authorship of Peter Wellnhofer of the Bavarian State Museum of Paleontology and Historical Geology in Munich. John Ostrom is the distinguished successor to O. C. Marsh and R. S. Lull at the Yale Peabody Museum, as well as the scientific heir to the Yale half of Marsh's great collection of horned dinosaurs (the other portion residing at the Smithsonian). Horned dinosaurs are not common fossils in Europe; in fact, they are nonexistent. However, a fine specimen of *Triceratops*, the type and only specimen of *Triceratops brevicornus*, was traded to the Bavarian Museum in 1964 for permanent exhibition there. Ostrom was then a young professor at Yale. He had completed doctoral research on duck-billed dinosaurs at the American Museum of Natural History, and he quickly turned his attention to horned dinosaurs at Yale. His focus in 1964 and 1966 was on understanding the biology and functional morphology of horned dinosaurs, particularly the way the jaws and teeth worked. He was then distracted from horned dinosaurs by his discovery in Montana of a marvelous small meat-eater, which he named *Deinonychus* in 1969. *Deinonychus* was the dinosaur that gave birth to the modern concept of warm-blooded dinosaurs. As if this were not enough, Ostrom then visited Europe and recognized the first new specimen of the earliest bird, *Archaeopteryx*, that had been seen since 1877. It was natural that his attention did not return to horned dinosaurs for more than a decade. Wellnhofer is not a dinosaur expert but a student of pterosaurs, flying cousins of dinosaurs that are particularly abundant in Germany. It seemed natural to him to invite Ostrom to come to Munich to join him in a description of the *Triceratops brevicornus* specimen.

Ostrom and Wellnhofer's joint paper is a beautiful monograph with lush illustrations and very useful summaries of the history of *Triceratops*. They succeeded admirably in their goal of describing the specimen. They chose also to review and interpret the species of *Triceratops*, and they reached a novel and important conclusion. In their judgment, there is only *one* valid species, *Triceratops horridus!* All other nominal species are simply biological or even preservational variants of a single species. They illustrated morphological variation in two species of modern African horn-bearing mammals, the hartebeest (*Alcelaphus buselaphus*) and the buffalo (*Syncerus cafer*), which are convincing mod-

83

els for variation in size and shape of the horns. Although sexual dimorphism could potentially contribute to the historical taxonomy, the authors declined to consider this factor.[38]

Thomas Lehman of Texas Tech University accepted the Ostrom and Wellnhofer position that there is only a single species of *Triceratops*, *Triceratops horridus*, but he added a detail. Lehman posited that Lull's lineage of *T. brevicornus–T. prorsus–T. horridus* constitutes females of *Triceratops*, whereas the *T. calicornis–T. elatus* lineage constitutes the males of the same species. By Lehman's reckoning, the *T. obtusus–T. hatcheri* lineage consists of aged and pathological males. Males have taller, more erect horns and attained larger size than females, which have shorter, more forward-directed horns. The small but mature *T. prorsus* thus is a female, whereas large, immature skulls (e.g., *T. flabellatus*) have male morphology.[39]

The Ostrom-Wellnhofer-Lehman solution to the old problem of *Triceratops* species is intuitively appealing. That there was only one species is consistent with our understanding of the biology and ecology of living large-bodied animals. In Africa today we see only one species of giraffe, one species of elephant and hippopotamus, and two species of rhinoceros. There are sound ecological reasons why we don't see sixteen species of these animals. Their food requirements would be so similar that they would cause each other's extinction. It seems almost an expectation that large, horn-bearing herbivores should show sexual dimorphism. However satisfying this single-species interpretation is, it too may not be correct, although it is certainly far closer to the truth than the sixteen-species model.

An independent approach was chosen by Catherine Forster in 1990. Cathy is a cheerful, high-spirited young woman from Minnesota. She plays a mean game of ice hockey, but I have never known her to high-stick an opponent deliberately, either in hockey or in paleontology. Cathy loves fieldwork, which she has pursued with success from Montana to Argentina to Africa, but for her doctoral dissertation at the University of Pennsylvania she chose to study *Triceratops*. It was my privilege to look over her shoulder during this work, as I was her doctoral supervisor. I would like to say that I taught her everything she knows, but that claim would be at variance with the facts. With my blessing, Cathy (who now teaches anatomy at the State University of New York, Stony Brook) took an absolutely independent approach to the study of *Triceratops*.

With a minimum of preconceived notions, Cathy undertook both a specimen-by-specimen analysis of character distribution and a quantitative analysis of skull shape. She measured as many skulls of old "three-horned face" as she could and included a number of specimens from Montana, which had not figured in the analysis of Ostrom and Wellnhofer. She used several different techniques of multivariate analysis to look for groupings independent of the historical taxonomy. The analysis of character distribution suggested to her that *T. brevicornus–T. prorsus* stands separate from the rest of *Triceratops* specimens. The multivariate computer analyses encouraged the separation of *Triceratops* specimens into two groups. One group, in addition to *Triceratops horridus*, includes *T. flabellatus*, *T. serratus*, *T. elatus*, *T. calicornis*, and *T. obtusus*. The second group consists only of *T. brevicornus* and *T. prorsus*. Forster's interpretation is that, because specimens of the *T. horridus* group greatly outnumber those of the *T. prorsus* group, these are two separate species, *Triceratops horridus* and *T. prorsus*. I prefer the interpretation that these are male and female (or is it female and male?), but I recognize that the two-species approach is eminently defensible. Lehman's male and female types are both subsumed within Forster's *Triceratops horridus*, rendering the story of sexual dimorphism somewhat ambiguous.[40]

One specimen, however, did trouble Forster: the one that Lull described as *Diceratops hatcheri*. It is a large skull, about average for *Triceratops* at about 2.0 m.[41] It has several peculiarities that are usually explained away as pathologies or variants within the expected range of *Triceratops*. The poorly developed nasal horn core that gives *Diceratops* its name is one example, the large openings in the squamosal another. However, several other characters seem less easily waved away. Where *Triceratops* should have a thick parietal lacking in fenestrae, *Diceratops* has a thin parietal that shows evidence of parietal fenestrae. Although this is a typical character for horned dinosaurs with few exceptions, it is completely unexpected for *Triceratops*; however, it may possibly be found in a species ancestral to *Triceratops*. In addition, the squamosals are completely unlike those of *Triceratops*. The typical squamosals are short compared to the length, broad and thick. In *Diceratops*, the squamosals apparently extend farther back on the frill and are comparatively thin and narrow, somewhat reminiscent of a separate genus of horned dinosaur, *Torosaurus*, to be discussed in the next chapter. Forster believes that *Diceratops* should be considered as separate from *Triceratops*, in a more basal position in phylogeny, possibly as the nearest outgroup or relative. This conclusion is aesthetically unappealing because the specimen is solitary (no others having been found), appears

FIG. 3.9. Reconstruction of *Triceratops*. (Robert Walters.)

to have some frank pathologies, and is poorly preserved to boot. None-theless, I cannot refute it, and I accept it provisionally, hoping that further finds may resolve the matter.

In summary, the conclusion that there was a single dominant species of *Triceratops*, which we may designate as *Triceratops horridus*, seems well founded (Fig. 3.9). This species is recognizable from Montana, South Dakota, Wyoming, and Colorado, at least. I accept Lehman's position that large male types with erect horns are separable from smaller female types with more forward-pointing horns. This solution did not appeal to Forster because it suggested that females were more common in the fossil record than males by a ratio of two to one. I believe that this is biologically reasonable. For many large mammals today, for example zebras and elephants, females are gregarious and males more dispersed and solitary. It is possible that there were other species of *Triceratops* in Montana; these have been less thoroughly studied, and the museum collections are geographically more dis-persed than the original Wyoming specimens. I cannot even speculate on the taxonomic status of *Triceratops* in Canada. The specimens docu-mented are very large. Two are of the female *Triceratops prorsus* type, and one is of the male type, possibly larger still. Barring any unreported gems in Canadian collections, that is as much as can be said at present.

HOW BIG WAS *TRICERATOPS*?

The size of *Triceratops* skulls is well documented. But how big was the whole animal? The answer to that question is less clear, for the simple reason that no complete skeleton has ever been reported! *Triceratops* is among the most common of dinosaurs, but its fossil record is strongly biased toward skulls. Skeletons of *Triceratops* are exhibited at a number of museums—among them the American Museum of Natural History, the Smithsonian, the Field Museum in Chicago, and the Science Museum of Minnesota—but *all* of these skeletons are composite, composed of the remains of two or more individuals. Descriptions of skeletal reconstructions have been published on the *Triceratops* mounts by Osborn in 1933 and Erickson in 1966. Pride of priority goes to the Smithsonian for the first mount of *Triceratops,* that of *T. prorsus,* completed in 1904.[42]

A common theme of the displays is the composite nature of the skeletal reconstructions. The American Museum mount of *Triceratops elatus,* for example, as described by Osborn, consists of parts of four individuals from two states. The skull is from Wyoming and the skeleton from Montana. The skeleton includes the complete series of dorsal vertebrae and ribs but only two cervical vertebrae and seven caudals. The sacrum, pelvis, and right leg were complete. There was a left femur but no forelimb bones, which were "restored in plaster chiefly from Amer. Mus. 970, from Hell Creek, Montana, 200 feet above Fort Pierre, and from other individuals, mostly of larger size than this skeleton, including some guidance from the National Museum mounted skeleton (Nat. Mus. 4842)." Osborn allowed that the mount was composite and that the skull and skeleton probably came from different species, "yet it appears to be correct in proportion." The skeleton as mounted by Charles Lang in 1923 takes up 5.8 m of floor space, but the length of the specimen measured along the curves (i.e., as if the vertebrae were laid on a horizonal surface) is 7.1 m (23 ft 4 in.). The highest point on the skeleton, in the middle of the back, measures 2.3 m (7 ft 7 in.). Unfortunately, the basis for the measurement of the length of the tail is not specified. Thirty-eight vertebrae were restored, for a reasonable though speculative total of forty-five caudals. The hind limb measurements are as follows:

Femur	95 cm
Tibia	61 cm

Unfortunately, the length of metatarsal III was not given by Osborn.[43]

The mount of *Triceratops* in the Science Museum of Minnesota at St. Paul was described in detail by Bruce Erickson. He reported on specimens of *Triceratops prorsus* that he collected in the Hell Creek Formation of Garfield County, Montana, between 1960 and 1964. Two very good partial skeletons were collected. Neither was complete in itself, but together they comprised almost all of a skeleton. The skull is very large, measuring 2.2 m in length and 1.3 m in basal length from the rostrum to the occipital condyle. It appears to be the largest well-documented skull of *Triceratops*. One dorsal vertebra was restored. Many caudal vertebrae were preserved but an unspecified number were restored. The skeleton measures 7.9 m (26.0 ft) in length and stands 2.9 m (9 ft 7 in.) at the highest point over the back. This was an enormous animal, and it seems to give the best indication of the size of *Triceratops* of any mount of which I am aware. For the sake of completeness, the hind limb measurements are as follows:

Femur	114 cm
Tibia	76 cm

Another note. Erickson describes the species as *T. prorsus*, which indicates to us that it was a *Triceratops* of female morphology. He further observes that in the Hell Creek beds of Montana *T. prorsus* is predominant, in fact probably the *only* species present there, based on his sighting of some two hundred specimens in the field.[44] This seems to be an interesting difference from the fossil beds of Wyoming. Does it strain credulity that these specimens are all females? Possibly.

Five-Horned Face and Friends

TRICERATOPS was a great and successful dinosaur. It had a fine friend in Othniel Charles Marsh and his successors. Yet it was rooted in time (68–65 Ma) and place like every mortal creature. As far as we know, it did not venture south of what is now Denver, nor north of what is now southern Alberta. Other horned dinosaurs held sway to the south. *Triceratops* shared its northern world with another mighty horned dinosaur, albeit one far more rare. Its ancestors flourished in Alberta and Texas eight million years earlier. We will continue our historical examination of the horned dinosaurs, and then later examine how they were all related to one another.

TOROSAURUS—A BULL LIZARD

In among the steady stream of *Triceratops* skulls that John Bell Hatcher was sending east to Yale in 1891 were a pair of ringers: two very large skulls that clearly were not the same as the others. The skulls were found less than 2 km apart in southeastern Wyoming. Marsh was on to them right away. By September 1891, he had described them as two species of *Torosaurus*. The name is an interesting one. It is often construed as "the bull lizard" ("El Toro") in reference to the very large size of the skulls. It may have been a clever pun on Marsh's part. The Greek root of the name refers to perforation, presumably of the crest. In 1891 Marsh had just named *Sterrholophus* ("solid crest"), which we regard as simply *Triceratops*. *Torosaurus* with its parietal fenestrae presented a great contrast to *Triceratops*. In the plausible interpretation of classicist Ben Creisler, the name *Torosaurus* ("perforated lizard") simply forms a contrast with solid-crested *Triceratops*.

The first species, the more perfect of the two skulls, was *Torosaurus latus* ("wide"), in which Marsh observed for the first time parietal fenestrae. This feature is familiar to us, with the benefit of one hundred years' experience, but for Marsh, *Triceratops* was the only point of reference, the only other horned dinosaur besides *Torosaurus*. To Marsh, the parietal crest of *Torosaurus* was not complete because of these holes. But to us, with our broader perspective, the reverse is true—*Triceratops* is very unusual in *lacking* parietal fenestrae. He found the crest thin and light, but very broad and flat, lacking any kind of median ridge. The highly distinctive squamosal of *Torosaurus* is strikingly long and rather narrow, with a free border unadorned by scallops, bumps, or epoccipitals. Its length exceeds 1.2 m! The nasal horn core is rather short and compressed, and the postorbital horn cores, though incomplete, seem less robust than those of *Triceratops*. Initially Marsh described *Torosaurus* as "earlier and less specialized" than *Triceratops*, but this is erroneous as the two were contemporaries.

In the same paper, Marsh described a second species, *Torosaurus gladius* ("sword," in reference to the shape of the squamosal, resembling the blade of a sword). The squamosal measured 1.4 m in this specimen, and the postorbital horn cores stood about 70 cm high. The parietal was quite well preserved in the second specimen.[1] In January 1892, Marsh elucidated his description with illustrations of the two skulls, the second of which measured nearly 1.7 m across the parietal crest. A peculiarity of this specimen is that the postorbital horn cores are situated caudal to the eye socket, rather than above it. These erect horns are 68 cm high. The crest measures more than half the length of the skull: nearly 1.6 m in length. These enormous animals had truly long frills.[2] Hatcher gave the total length of the skull of *T. latus* as 2.2 m and estimated the length of *T. gladius* at 2.35 m, but the width of the occipital condyle was a modest 85 mm.[3] It was, however, incompletely fused. It is a rather frightening thought that the gigantic skull of *Torosaurus gladius* was that of an immature animal, with further potential for growth!

Regrettably, *Torosaurus* has remained a rare and rather poorly known horned dinosaur. More than fifty years passed until the next specimen came to light, this time in northwestern South Dakota, roughly 300 km northeast of the *Triceratops* beds of Wyoming. The well-preserved specimen was discovered by E. H. Colbert, leading a field party for the Academy of Natural Sciences of Philadelphia in 1944. Colbert and James Bump described the specimen in 1947. Colbert is the most beloved of American dinosaur paleontologists, the elder statesman of our field. His books on the subject inspired me during my formative years—too long ago to admit. For much of his career he was associated

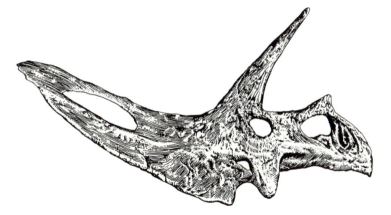

FIG. 4.1. *Torosaurus latus* skull at the Academy of Natural Sciences, Philadelphia. (Robert Walters after Colbert and Bump 1947.)

with the American Museum of Natural History in New York. During the years of World War II, however, he was permitted to supplement his meager salary by spending one day per week curating fossils at the Academy of Natural Sciences in Philadelphia. This arrangement benefited both parties significantly.

The skull that was collected, prepared, and exhibited at the Academy of Natural Sciences is clearly *Torosaurus latus,* as Colbert and Bump recognized. The new specimen is much smaller than the Yale specimens and measures 1.6 m in length, with an occipital condyle of 95 mm in diameter. The postorbital horn cores, at 38 cm in height, are about half as high as those of *T. gladius* (Fig. 4.1). They also are inclined forward at 45°, rather than being erect as in the Yale specimens. The nasal horn core, as in the other two specimens, is low and blunt. Details of the parietal fenestrae that were not accessible in the type specimen of *T. latus* are made clear. The fenestrae are very large, occupying about half of the frill. Marsh thought that the parietal fenestra in *T. latus* contacted the squamosal, but the better preservation of the Philadelphia specimen strongly suggests that Marsh was mistaken. The squamosal in the new specimen measures a paltry 79 cm in length.

Rather than describing the new specimen as yet another species, Colbert and Bump took the opposite tack. Instead, they compared all three skulls and concluded that they represent but a single species, *Torosaurus latus.* They attributed the differences to growth and also hinted at sexual variation but did not come right out with it.[4] I applaud

91

the modernity of their reasoning, and, rushing in where angels fear to tread, I will go one step further. I will suggest that the two Yale specimens are males, with erect, divergent horns, and that the Philadelphia specimen is a female, with procumbent, parallel horns. The analogy with Lehman's interpretation of sex differences in *Triceratops* is clear. In fact, Lehman and I both agree that the Philadelphia specimen is a female and that *Torosaurus gladius* is a male (he believes that the type specimen of *T. latus* is a female). The striking contrast is between the smallest specimen, which has a fused occipital condyle and is close to its maximum size, and the largest specimen, which has an unfused condyle and thus has significant potential for further growth.

Other reports of *Torosaurus* have been sporadic, based on disarticulated material or partial skulls. Fossils from the North Horn Formation of Utah were described in 1946 by Charles Gilmore as *Arrhinoceratops utahensis*.[5] In 1976, Douglas Lawson suggested that these fossils be transferred to *Torosaurus utahensis*. Lawson also claimed, on the basis of his interpretation of certain fragmentary parietals, that *Torosaurus* was present in the Big Bend region of Texas.[6] Today we are inclined to include *T. utahensis* in *Torosaurus latus* unless compelling reasons (i.e., adequate fossil material) emerge. Tim Tokaryk in 1986 described a frill of a *Torosaurus* from Saskatchewan. This specimen measures 1.62 m in breadth across the back of the frill, only about 5 percent smaller than the larger Yale specimen. It has unusually wide squamosals, each of which has an opening in it. Tokaryk's report does an important service in documenting the geographic range of the giant ceratopsid.[7] Parties from the Milwaukee Public Museum working in eastern Montana in 1981 recovered a partial skull of *Torosaurus* and an incomplete skeleton. The skull is of very large size, and the find is enhanced by a superbly preserved right front leg, the significance of which is discussed in Chapter 9.[8]

CHASMOSAURUS—THE CHASM LIZARD

In Alberta, the Red Deer River arises as a babbling, glacier-fed brook in the ice-sculpted Rocky Mountains, crosses aspen parkland, then slashes southeastward across the rolling shortgrass prairies. As it does so, it furrows deeply through sediments that record the last eleven million years of dinosaur time. Its brown waters eventually carry soil the dinosaurs once tread upon into Hudson Bay, half a continent away (Map 4.1). In steep riverine badlands adjacent to the Red Deer, explor-

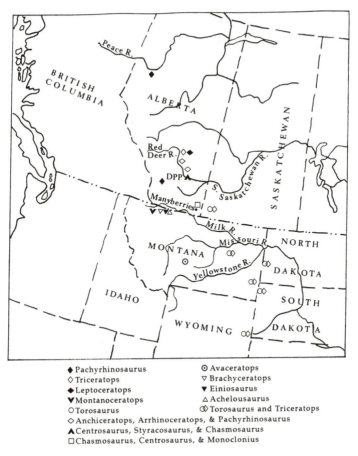

◆ Pachyrhinosaurus
◇ Triceratops
◆ Leptoceratops
▼ Montanoceratops
○ Torosaurus
⊙ Avaceratops
▽ Brachyceratops
▼ Einiosaurus
△ Achelousaurus
⊗ Torosaurus and Triceratops
◇ Anchiceratops, Arrhinoceratops, & Pachyrhinosaurus
▲ Centrosaurus, Styracosaurus, & Chasmosaurus
☐ Chasmosaurus, Centrosaurus, & Monoclonius

Map 4.1. Horned dinosaur localities, northern Great Plains.

ing Canadian geologists and paleontologists from the Geological Survey of Canada in Ottawa uncovered evidence of dinosaurs beginning in the 1870s. George Dawson, Joseph B. Tyrrell, Thomas Chesmer Weston, and Lawrence M. Lambe began to make the first discoveries of Canadian dinosaurs, albeit of a rather fragmentary and tantalizing nature. It is a little ironic that the first important dinosaur fossil found in Alberta, collected by Tyrrell near Drumheller in 1884, was eventually described and named *Albertosaurus sarcophagus* ("Alberta lizard, eater of flesh") by Osborn in New York in 1905. Alerted to the existence of dinosaurs in Alberta, the Survey continued its efforts. In 1888 and again in 1889, Weston explored the Red Deer River by boat, but he lacked the technical skills to collect significant fossils. Lambe, also using a boat, was more successful. He worked the exposures of the "Belly River series" (Judith River Formation) between Berry Creek and Dead Lodge Canyon in 1897, 1898, and 1901. His success in the field was not brilliant, but he was the first Canadian to study dinosaur fossils, initially with Osborn but soon on his own. No complete skulls or skeletons of dinosaurs had yet been found here.

In their diffident, Canadian way, dinosaur studies north of the border did not begin in earnest until 1902, with the publication by Lawrence M. Lambe (1863–1919) of an important monograph entitled "On Vertebrata of the Mid-Cretaceous of the North West Territory." The publication is a well-illustrated presentation of extensive faunal remains from one of the rich fossil localities of the world, the Judith River Formation of Alberta along the Red Deer River. We now know these beds to date from about 74 to 76 Ma, that is some six to eleven million years older than the *Triceratops* beds of Wyoming, Montana, and Alberta. The fossils include dinosaurs and nondinosaurian reptiles. Unfortunately, the dinosaur materials Lambe had at his disposal then were largely disarticulated, no complete skulls yet having been found. Lambe named three species of *Monoclonius*: *M. dawsoni*, *M. belli*, and *M. canadensis*.[9] None has actually turned out to be *Monoclonius*. The first specimen, *M. dawsoni*, belongs to the short-frilled *Centrosaurus apertus*, whom we shall meet in the next chapter. Barnum Brown began collecting skulls and skeletons in Alberta in 1910 and was followed there in 1912 by the Sternbergs, employed in the service of the Geological Survey of Canada. Thus by 1913 Lambe began to have high-quality specimens at his disposal. It was quickly clear to him that *Monoclonius belli*, which he had established on the basis of a fragmentary parietal, was a completely different dinosaur from *Monoclonius*.

By January 1914, Lambe had published a preliminary paper presenting a diagnosis of *Protorosaurus belli*. He expressed it as follows:

> Skull large, broadly triangular in superior aspect, with an abbreviated facial portion and a greatly expanded posterior crest ending squarely behind. Coalesced parietals forming a slender frame-work enclosing large subtriangular fontanelles. Squamosals very long and narrow with a scalloped free border. Epoccipitals present. Supraorbital horn-core small, upright. Orbit small. Supratemporal fossae not greatly developed. Body covered with non-imbricating plate-like, and tubercle-like scales.[10]

This diagnosis is very trenchant. However, the rest of the paper does not follow in the same vein. There are no figures of the skull, but there

are four very interesting plates showing fossilized skin patterns associated with the specimen. No measurements are given. (What does
"large" mean? This dinosaur was certainly not large compared to *Triceratops* or *Torosaurus*.) The paper makes scant reference to a specific
specimen. Lambe's comments deal principally with the relationship of
his animal to Marsh's *Torosaurus*. In 1902, Lambe had expressed the
opinion that his *Monoclonius belli* was "probably ancestral to such later
forms as *Torosaurus latus* and *T. gladius* of Marsh, from the Laramie of
Wyoming."[11] In 1914 he crowed, "This belief is strengthened by the
discovery during the past summer of a skull, with most of the skeleton,
of one individual of this species at the type locality." He clearly and not
unreasonably believed that "its affinities are with *Torosaurus*, Marsh, to
which it apparently leads in a direct line of descent, and from which it
differs by well-marked primitive characters." The name *Protorosaurus*
("before *Torosaurus*") expressed this belief succinctly. He was refreshingly explicit about the characters he considered primitive. These were
smaller size, greater length of the face, retention of a scalloped free
margin of the squamosal, greater size of the parietal fenestrae, smaller
supraorbital horn cores, and erect rather than procumbent horn cores.[12]

A more satisfactory paper followed the next month (February 1914).
It figured the skull with lateral and dorsal photographs, making it
evident that the front of the otherwise exquisite skull was missing
(Fig. 4.2). In addition, an important matter was cleared up. Lambe now
realized that the name *Protorosaurus* was already in use for a Permian
reptile from Europe. He thus substituted the name *Chasmosaurus*
("chasm lizard"), explaining that "The new name has reference to the
openings in the skull, more particularly to the great size of the intra-
parietal fontanelles." He noted that the front part of the skull "had gone
to pieces through weathering," but that the frill was "in a particularly
excellent state of preservation." Lambe was impressed by the length of
the frill, 25 percent longer than the rest of the skull, and the narrowness
of the face. The frill is flat, slightly wider than it is long, and has the
form of a triangle, apex forward, base directed caudally. The squamosals are like long isosceles triangles, broad toward the front of the skull
and tapering backward to reach very nearly the rear edge of the parietal. In this regard, the squamosals of *Chasmosaurus* are very much like
those of *Torosaurus*, and very much unlike those of *Triceratops*. There are
seven epoccipitals on the right squamosal, eight on the left, plus a large
one on the lateral corners of the parietal. The squamosals measure
86 cm in length. The parietal also contrasts greatly with that of *Triceratops*.

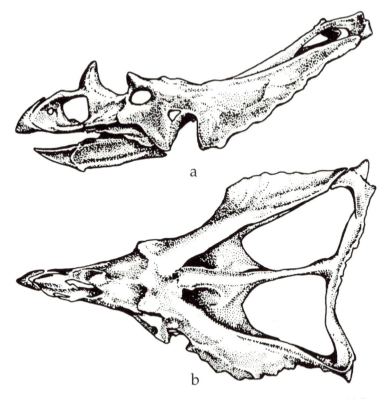

FIG. 4.2. *Chasmosaurus belli,* American Museum of Natural History. (a) Lateral view, (b) dorsal view. (From Dodson and Currie 1990. Donna Sloan. Courtesy of the University of California Press.)

The fenestrae in *Chasmosaurus* are enormous, so that the caudal half of the frill is reduced to a slender but strong T-shaped framework, completely unlike the solid, saddle-shaped parietal of *Triceratops.* The horn cores are not at all impressive. Only the postorbital horn cores are preserved, and these are erect, low, conical structures about 90 mm high, situated immediately above the eye sockets. They are a far cry from the half-meter-long horns of *Triceratops* and *Torosaurus.* Lambe estimated the length of the skull of *Chasmosaurus belli* at 1.65 m. The breadth of the frill is 106 cm.[13]

In 1915 Lambe published an important monograph on horned dinosaurs, the diversity of which had doubled in the few years since the publication of the Hatcher, Marsh, and Lull monument of 1907. The ostensible purpose of this paper was to deal with the third of his three species of 1902, *Monoclonius canadensis* ("from Canada"). This specimen was in fact the best of the three that he named as separate species on that occasion. It consisted of a right squamosal, jugal, postorbital, and dentary of moderate size. In 1902, no mention was made of a nasal, but by 1915 a right nasal had joined the skull. Lambe did not have a new specimen to elucidate his type, but he did have a much better grasp on the structure of ceratopsid skulls. He believed that his species did not fall into any described genus. He thus erected a new genus to receive his species, which he named *Eoceratops canadensis,* the "dawn horned face from Canada." Whereas he believed that *Chasmosaurus* was ancestral to *Torosaurus,* he drew comparisons between *Eoceratops* and *Triceratops,* although allowing the difference that the former had parietal fenestrae.

Eoceratops seems to be small and compact, with a deep skull as Lambe restored it. The squamosal measures 58 cm in length and is rather broad at 36 cm. There are six gentle scallops along the free edge, dying out caudally. The postorbital horn core is erect, a tall 216 mm high, and curves gently backward. The low, blunt nasal horn core is particularly interesting, consisting only of the right half. It thus is clearly formed by separate left and right nasals, fusing later in life. Moreover, there is a scooped, crescentic "gouge" at the base of the right nasal with an apparently separate bone suture into the horn core. Lambe described this as a separate bone, which he named the "epinasal."[14]

Lambe believed the orientation of the horn cores, especially the postorbital horn cores, to be very important in ceratopsian taxonomy. He wrote:

> From our present knowledge of the horn-cores of the Ceratopsidae their curvature, or the direction of their growth, is constant in any species. In individuals of the same species there is remarkably little variation in the curvature or direction of growth of both the supraorbital and the nasal horn-cores. . . . Differences of size occur, no doubt due to age and possibly to sex, but apparently the growth of a horn-core is in a definite direction, forward, upward, or backward, and also with inward or outward curvature in the case of brow-horns, according to the genus and species to which the individual belongs.

Lambe rejected the synonymy of *Eoceratops* with *Chasmosaurus belli* on account of the difference in the postorbital horn cores.

Lambe compared all of the known ceratopsids and recognized three groups, which he formally designated as subfamilies of the Ceratopsidae: the Eoceratopsinae, Centrosaurinae, and Chasmosaurinae. The first subfamily consisted of *Eoceratops, Anchiceratops, Diceratops,* and *Triceratops,* characterized by trends of large brow horns increasing in size; persistently small nasal horn; broadly triangular squamosal; and parietal fontanelle closing. The second subfamily, the Centrosaurinae, consisted of *Centrosaurus, Styracosaurus,* and *Brachyceratops.* These ceratopsids are characterized by persistently small brow horns and persistently large nasal horns; small squamosal; and parietal fontanelle "diminishing." The third group consists of *Chasmosaurus* and *Torosaurus,* which show trends of brow horns increasing; nasal horn decreasing; squamosal lengthening; and parietal fontanelle diminishing.

With regard to the frill, Lambe saw the trend as "a persistent attempt . . . in all three groups to enlarge and strengthen it and to render it a more efficient means of defence by covering more of the neck and shoulders." He evidently regarded the long-frilled lineages as more highly derived, an assessment with which we will not quarrel: "A final result is reached in the *Eoceratops* and *Chasmosaurus* groups in a neck-frill of increased compactness, strength, and resisting power." Lambe also illustrated in the monograph the excellent complete skull of Chasmosaurus belli found by the Sternbergs in 1914 and provided a photograph of a recently found, large and superb skull of *Centrosaurus apertus.*[15]

History has not looked favorably on the genus *Eoceratops* nor on the subfamily Eoceratopsinae. Thomas Lehman, in the most comprehensive study of the Chasmosaurinae, regards *Eoceratops* as a species of *Chasmosaurus, C. canadensis.*[16] The genera contained within the subfamily are similarly regarded as chasmosaurines. We shall return to a discussion of *C. canadensis* later. Lambe did not return to the subject of horned dinosaurs during the few years left in his lifetime.

In 1920, George Sternberg, collecting for the University of Alberta in the Judith River Formation of the Red Deer Valley, found a fairly small ceratopsid skull. Charles W. Gilmore described it in 1923 and referred the specimen to *Eoceratops canadensis.* He estimated the skull to have been about 1.2 m long, although this is possibly a little too generous. It is well preserved, except that it is missing the back half of the frill. The nasal horn core is prominent but blunt, showing no trace of division into left and right halves. The postorbital horn cores are strongly procumbent and measure 170 mm in height. As in Lambe's specimen, they are concave caudally. The squamosal is broad (28 cm), but its length cannot be estimated. It shows epoccipitals, which are lacking in the

Ottawa specimen.[17] Lull in his 1933 monograph was not convinced that the Alberta specimen was distinct from *Chasmosaurus*, although he referred it to a different species, as discussed later in this chapter.[18]

Reconstructing Chasmosaurus

Charles Mortram Sternberg (1885–1981) was one of C. H. Sternberg's three sons. All three lads came to Alberta with their dad in 1912. C. H. Sternberg left Canada in 1916 to resume his career in the United States as a commercial fossil collector. George Sternberg, the oldest of the three brothers, left Canada in 1923. Levi Sternberg joined the technical staff of the Royal Ontario Museum and lived in Toronto until the end of his life (1976). Charlie remained in Ottawa for the rest of his life. Lambe, his mentor in Ottawa, died suddenly in 1919, and Charlie assumed his duties as museum scientist as well as technician.

For a person with a high school degree (from that misty, bygone era when even elementary schools still taught a great deal), Charlie Sternberg had a distinguished career as a paleontologist. His first paper appeared in 1921, supplementing Lambe's study of the ankylosaur *Panoplosaurus,* and his publications continued to appear until 1970, even though he nominally retired in 1950. He described numerous Canadian dinosaurs. As was the style in those days, he did not engage in rapturous flights of speculative fancy, but his papers have stood the test of time very well. His descriptions and interpretations were *reliable,* and this term is intended as high praise. No published study of Canadian dinosaurs is possible to this today without citing one or another of Sternberg's papers. One of the great thrills of my young professional life (or to be honest, my preprofessional life) was meeting C.M., as we called him, when he would visit the old paleo labs of the National Museum of Canada, where I worked as a student during the summer of 1967.

Following Lambe's death in 1919, C. M. Sternberg assumed the role of director of the paleontology enterprise of the Geological Survey of Canada and soon began publishing. One of his early papers was on the description of a patch of skin impression associated with the skeleton of *Chasmosaurus belli* that he had collected in 1913.[19] Two years later he wrote a paper describing a lovely exhibit at the National Museum of Canada of a pair of *C. belli* skeletons mounted side by side. (Fig 4.3). Sternberg noted that although horned dinosaurs were well known to the public, as most museums of the world had the skull of one or another, skeletons were very rare (Fig. 4.4). The two specimens were the

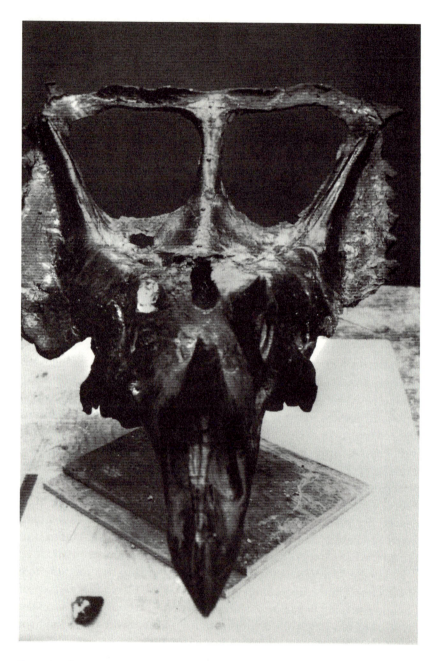

FIG. 4.3. *Chasmosaurus belli,* Canadian Museum of Nature, Ottawa. This exquisite skull belongs to one of the pair mounted by C. M. Sternberg. (Photo by Peter Dodson.)

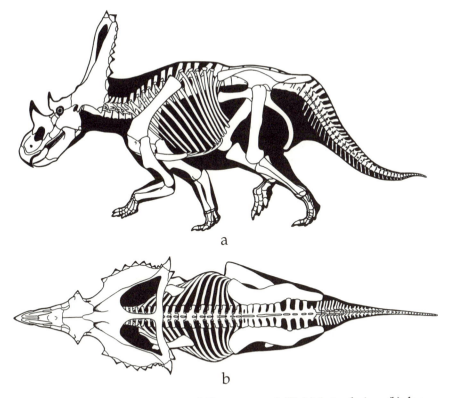

Fɪɢ. 4.4. Skeletal reconstruction of *Chasmosaurus belli*. (a) Lateral view, (b) dorsal view. (Gregory S. Paul.)

first group of horned dinosaurs ever mounted. One (NMC 2245) was collected 5 km downstream from Steveville in 1913, the other (NMC 2280) some 10 km downstream from the first in 1914. In addition to the skull, the first specimen included a vertebral column complete up to the twenty-fourth vertebra of the tail, pectoral and pelvic girdles, humeri, one ulna, femora, one tibia and fibula, and some foot bones. The vertebral column of the second skeleton was complete to the pelvis and included the pelvis and shoulder girdle. Apart from both humeri and a portion of one femur, the limbs were missing. Missing parts were restored in plaster. One striking if unquantified feature of the two skeletons is that the bones of one (NMC 2245) are much lighter in construction than those of the other. Sternberg not only attributed the difference in the otherwise very similar specimens to sexual dimorphism, he also posited that the larger one was the female.

101

Sternberg noted the difficulty he encountered in mounting the shoulder joint, owing to the lateral offset of the head of the humerus. He expressed very clearly the dilemma that every mounter of ceratopsid skeletons has faced: "Consequently the only way the limb could be posed, so the head of the humerus fitted into the glenoid cavity was to place the humerus at almost right angles to the perpendicular. This made the animal very low in front and extremely bow-legged. The humerus placed in this position made a very much better articulation with the ulna and radius than could be gained otherwise." As Sternberg realized, "such a pose does not suggest speed but rather an animal which waddled along with a swaying motion." Our modern aesthetic sensitivity finds this pose somewhat offensive. Today we try to mount ceratopsids with the forelimbs as elevated as possible, but attempts to restore these animals as greyhound-like racers reside more easily in the two-dimensional world of canvas and acrylic than they do in the three-dimensional world of plaster and welder's rod. We will return to this issue in the final chapter.

Sternberg presented an extensive list of measurements of both specimens. As restored, one skeleton (NMC 2245) measures 4.93 m in length, the other (NMC 2280) 4.95 m in length. The skull of the former measures 1.68 m in length, that of the latter 1.65 m in length. NMC 2245 is 1.5 m in height at the pelvis, with a femur length of 75 cm and a tibia length of 53 cm.[20] The two Ottawa specimens of *Chasmosaurus belli* made a striking exhibit. Although the specimens are no longer exhibited in Ottawa, casts of these skeletons may be seen in Philadelphia at the Academy of Natural Sciences and at the Royal Tyrrell Museum of Palaeontology in Drumheller, Alberta.

More Species of Chasmosaurus

For a comparative neophyte in academic paleontology, C. M. Sternberg made a surprisingly cheeky comment that subtly "dissed" his late mentor when in his 1927 paper he stated, "the brow horns vary in length, and it is questionable whether their size or length has any generic or specific significance. The horn itself was probably much longer than the bony core."[21] Although these are words to live by, few chose to live by them, and new specimens soon became new species. Other specimens of *Chasmosaurus belli* were collected from the Steveville badlands of Alberta and were recognized as such. In his compilation of 1933, Lull recognized a number of such specimens in Ottawa,

FIG. 4.5. *Chasmosaurus kaiseni,* American Museum of Natural History. (From Dodson and Currie 1990. Donna Sloan. Courtesy of the University of California Press.)

Toronto, New York, and New Haven. For instance, the beautiful skull of American Museum of Natural History (AMNH) 5402 is photographed, and Yale Peabody Museum 2016 illustrated in ventral view. However, a skull that Barnum Brown collected in 1913 seemed in some respects to stand apart from the normal range of variation. This fine skull, AMNH 5401, measures 1.52 m in length, by no stretch of the imagination a large skull. The frill is only 82 cm wide across the back. However, it has a pair of very tall, procumbent postorbital horn cores that measure 370 mm in length, recurved as in all other specimens of *Chasmosaurus.* The rostrum in front of the external nostril is strikingly long and low. In 1933, Barnum Brown described this specimen as *Chasmosaurus kaiseni,* honoring his associate Peter Kaisen (Fig. 4.5). Brown did consider the possibility that this skull was *Ceratops,* Marsh's genus from the Judith River Formation of Montana that was based on a pair of horn cores and an occipital condyle. However, he wisely decided to refer it to the well-characterized genus *Chasmosaurus.*[22] The University of Alberta specimen of *Eoceratops* is in Lull's judgment better referred to *Chasmosaurus kaiseni* as a sexual variant, presumably a female, according to Lull.

Lull, in his 1933 monograph, named another species, *Chasmosaurus brevirostris* ("short rostrum"), on the basis of a skull at the Royal Ontario Museum collected by the University of Toronto in 1926 (Fig. 4.6). Lull characterized *C. brevirostris* as follows:

This species conforms to *belli,* except that the muzzle is very short and deep, more as in *Monoclonius,* and the nasal horn is large and curves somewhat backward, although inclined forward at the base. The brow

103

Fɪɢ. 4.6. *Chasmosaurus brevirostris,* Royal Ontario Museum. (From Dodson and Currie 1990. Donna Sloan. Courtesy of the University of California Press.)

horns are like the nasal but smaller. The last epoccipital on the squamosal is markedly larger than in any other specimen. . . . There were nine epoccipitals borne on the squamosal.[23]

This is a large skull, approximately 1.8 m in length. Not mentioned in the description is a prominent opening in the left squamosal. This skull really does have a short, deep face, almost antithetical to that described in *C. kaiseni.*

Finally C. M. Sternberg got into the act himself. In 1940 he described *Chasmosaurus russelli* (honoring L. S. Russell, who became one of Canada's most distinguished native sons in paleontology). The skull was collected in 1938 by Russell for the Geological Survey of Canada and came from the Manyberries region of extreme southeastern Alberta, not far from the Montana border. Sternberg designated a second skull as a paratype. This skull, collected in 1928, came from the Red Deer River Valley 5 km south of Steveville. A third skull was collected in 1937 from the Manyberries region, and a fourth specimen consisted of a partial parietal collected in 1935 from a locality along the South Saskatchewan River some 100 km east of Steveville. The type specimen is a large skull, measuring 1.94 m in greatest length. The nasal horn core is described as massive but the postorbital horn cores are lacking. In addition, the squamosal shows poorly developed scallops along the free edge. The caudal bar of the parietal, rather than being T shaped as in *C. belli,* is Y shaped. Sternberg described it as "deeply indented" or "deeply emarginate." The parietal does not bear epoccipitals as it does in *C. belli.* Several limb bones were found with the paratype. These reinforce the idea that *C. russelli* was a large animal. The humerus measures 71 cm and the tibia 61 cm.[24] The comparable measurements in *C. belli* are 51 cm and 53.5 cm. If *C. belli* measures 4.9 m in length, *C. russelli* may measure between 5.7 and 6.9 m in length. Although the latter figure is surely excessive, it does indicate that this animal was unusual for a horned dinosaur of Judithian age.

Matters rested there for many years. As described previously, the trend in more recent years has been to look for ways to consolidate species, not to name more. No one is more sensitive to that trend than Thomas Lehman, who completed his Ph.D. at the University of Texas in 1985 under the supervision of Wann Langston, Jr. Lehman, now a professor of geoscience at Texas Tech University in Lubbock, published an important paper in 1990 in which he demonstrated rational ways to reduce oversplit taxonomy of chasmosaurines and to group species together in meaningful ways. Nonetheless, he too has contributed to the taxonomic history of the genus *Chasmosaurus* by naming a new species. The status of the Texas *Chasmosaurus* somehow seems less ambiguous than that of the multiple species from Alberta, because the geographic separation of more than 2,000 km makes biological sense of a distinct species.

In 1938, a withering depression seared dustbowl-dry landscapes throughout the American West. Franklin D. Roosevelt, determined to get the nation back to work, had established the Works Progress Administration, widely known as the WPA. Labor-intensive projects of staggering variety were carried out. Yet few were as unusual—or took place in such a forbidding (albeit austerely beautiful) landscape—as that supervised by Professor William Strain of the College of Mines and Metallurgy in El Paso. The crew under his supervision excavated three dinosaur bonebeds near the Rio Grande that separates Texas from Mexico, in an area of West Texas that is now Big Bend National Park. The bones came from the Aguja (phonetically the *j* is an *h*) Formation of Late Cretaceous age, roughly equivalent in age to the Judith River Formation of Alberta. Lehman interpreted the sediments as representing deltaic interdistributary marshes, that is to say, swamps. One deposit contained 342 identifiable bones, more than 70 percent of which belonged to the same species of ceratopsid, which Lehman named *Chasmosaurus mariscalensis* ("named for Mariscal Mountain, in the southern part of Big Bend National Park, near the type locality") (Plate II). The specimens reposed unstudied at El Paso for nearly fifty years, until Lehman became interested. The collection is quite marvelous, consisting as it does of a large number of disarticulated specimens of *Chasmosaurus* of small to large size, including both skull and skeletal bones.

Lehman systematically described all identifiable elements of the skeleton. A number of elements are unrepresented in the collection, including the rostral bone and the nasal bone, so the form of the nasal horn core was speculative. Postorbital horn cores, of which there are eleven, are among the most abundantly represented elements in the collection. These range from 140 mm to 350 mm in height and are generally erect. As in all specimens of *Chasmosaurus,* they are recurved, that is, concave backward. Lehman noted subtle differences among them, notably that some are quite erect, whereas others are slightly procumbent and also diverge laterally. He posited that the erect horn cores were those of males and that the less vertical ones were those of females. There are four braincases in the collection. The smallest occipital condyle measures 41 mm in diameter and is unfused, showing all three separate bones that contribute to its formation; the largest measures 70 mm in diameter.

Regrettably, the parietal is poorly represented, but one fragment that includes both the median bar and part of the caudal border shows the Y shape seen in *C. russelli.* Lehman speculated that there were epoccipitals on the parietal, unlike in *C. russelli,* because there are swellings that may represent points of attachment for these elements. Fortunately, squamosals are well represented, although they are somewhat fragmentary in condition. Lehman reported that they are triangular as in the general form of chasmosaurine squamosals, but that they are relatively short and broad, more so than in other forms of *Chasmosaurus*—in fact, more so than in other chasmosaurines except for *Triceratops, Torosaurus utahensis,* and *Anchiceratops.* He believed that the squamosal of *C. mariscalensis* did not change shape or elongate significantly during growth. There are six or seven coarse scallops on the squamosal, expressed as epoccipitals in mature specimens. This number is low compared to other specimens of *Chasmosaurus.* He tentatively suggested that differences in the epoccipitals (broader squamosals with rounder, blunter scallops versus more slender squamosals with more pointed epoccipitals) might have a sexual basis. Five maxillae that range in length from 258 mm to 452 mm show that the number of teeth in the upper jaw increases from twenty to twenty-eight during the phase of growth represented.

Of the postcranial skeletal elements, the best represented are humeri (thirteen), femora (seventeen), and tibiae (fourteen), all robust, well-constructed bones. Humeri range from 352 mm to 545 mm in length, femora from 355 to 713 mm, and tibiae from 389 to 556 mm. The general consistency in the size spread of all elements encourages the interpretation that the deposit contains the disarticulated remains of just a few individuals rather than the water-transported remains of bones gathered some distance upstream and deposited willy-nilly, like driftwood upon a beach. We can go further, using the wonderful reference specimen of *Chasmosaurus, C. belli* (NMC 2245), which is 4.93 m long and has these measurements: humerus 508 mm, femur 749 mm, tibia 533 mm. We can make the following estimates for *C. mariscalensis:*

	Small		*Large*	
	Bone Length (mm)	*Body Length (m)*	*Bone Length (mm)*	*Body Length (m)*
Humerus	352	3.41	545	5.29
Femur	355	2.34	713	4.69
Tibia	389	3.60	556	5.14
		Avg. 3.12		Avg. 5.04

This exercise is not a rigorous one, but it seems to convey a reasonably consistent picture of the size of the animals in the deposit. Note that in our real animal, that is to say, NMC 2245, as in all ceratopsids, the humerus is not the same length as the femur but only about two-thirds the length of the femur. Thus the 355-mm femur of *C. mariscalensis* certainly comes from a smaller animal than the 352-mm humerus.

Lehman concluded his work by diagnosing *Chasmosaurus mariscalensis*. He noted that it is similar to the large-horned *Chasmosaurus canadensis* from Alberta but has an abbreviated snout, a flat maxilla, taller postorbital horns, and shorter squamosal with larger epoccipitals. *C. mariscalensis* is the most highly derived species of *Chasmosaurus,* according to Lehman. He noted similarities with *Pentaceratops* but felt that the latter was still more highly derived and sufficiently distinct to maintain generic separation. He felt that *C. mariscalensis* had a relatively shorter frill than any other species of *Chasmosaurus,* which made it an unlikely ancestor for *Pentaceratops.*[25]

The only major shortcoming of Lehman's project was the lack of a definitive, articulated, adult skull. This was not long in coming. In 1991, the indefatigable Paul Sereno, resting in the United States in between jaunts around the globe, led a small party of University of Chicago

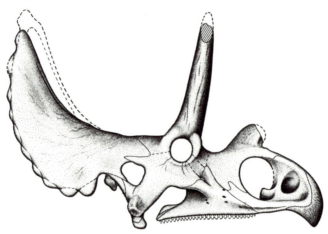

FIG. 4.7. *Chasmosaurus mariscalensis*, Texas Memorial Museum, University of Texas, Austin. (From Forster et al. 1993. Carol Abraczinskas. Courtesy of Catherine Forster. Reproduced courtesy of the Society of Vertebrate Paleontology.)

paleontologists on a brief excursion to Big Bend National Park, and in the process discovered a fine adult skull of *C. mariscalensis* (Fig. 4.7). Fortunately, Cathy Forster was an expedition member, and with her expertise on ceratopsids, a paper quickly resulted. The skull, housed at the Texas Memorial Museum in Austin, measures 152 cm in length, convincingly adult but certainly not of notably large size. Several features were striking. The openings in the skull are rather prominent, especially the external nostrils and the infratemporal fenestra located behind and a little below the eye. In most ceratopsid skulls, this opening is rather small, but in *C. mariscalensis* it is about the size of the orbit itself. The postorbital horns are 43 cm high as preserved, and their tips are missing! There is a blunt but very well-formed nasal horn core. The squamosal is fairly wide but is elongate and has ten epoccipitals, so Lehman was mistaken in his inference about their number. It is also strikingly elevated, a marked departure from the flat frill of *Chasmosaurus belli*. The squamosal measures 89 cm in length, 32 cm in breadth. This is very large for a *Chasmosaurus* squamosal. Forster and associates show that squamosal length in Canadian specimens of *Chasmosaurus* ranges from 70 to 81 cm. In fact, the squamosal of mature *C. mariscalensis* is rather narrow, not at all wide as Lehman inferred. However, the squamosal does not taper backward as in northern specimens of *Chasmosaurus*. Rather, the free border is rounded and convex ventrally, giving the whole blade a wider appearance than in *C. belli.* The new skull regrettably did not provide further information about the parietal.[26]

Although the new specimen does not confirm all aspects of Lehman's analysis, it does confirm the validity of *C. mariscalensis* as a distinct species. The erect postorbital horn cores are unique among ceratopsids. The convex shape of the squamosal is also convincing. The compressed nasal horn may prove significant as well.

Two Canadian paleontologists, Stephen J. Godfrey and Robert Holmes, published in December 1995 a reanalysis of Alberta specimens of *Chasmosaurus.* Their account is superbly illustrated, and, true to the tenor of the times, they too advocate a taxonomy of reduction rather than amplification of species. They recognize *Chasmosaurus belli* and *C. russelli* as the only valid Canadian species of *Chasmosaurus.* Shockingly, however, they posit that Sternberg's male and female of *Chasmosaurus belli* are actually two different species, a conclusion about which I am gravely skeptical.[27]

Chasmosaurus is one of the most important of the horned dinosaurs, though far from the most famous. It is known from many fine specimens, has a broader geographic range than almost any other horned dinosaur, and shows rich and interesting variation in skull morphology. It is of interest that it is the earliest of the chasmosaurine or long-frilled horned dinosaurs but is by no means the most basal representative. In addition, it is not self-evident that it was a precursor of *Triceratops,* which is derived in very different ways from *Chasmosaurus.* Our narrative must continue, but we are not finished with *Chasmosaurus.*

ANCHICERATOPS—ALBERTA AGAIN

Barnum Brown (1873–1963) was one of the most successful dinosaur collectors of the twentieth century. He surely had the knack for finding bone as only a gifted few since have had. His name is indelibly associated with the American Museum of Natural History in New York. The museum had been founded only four years before his birth, but it languished for three decades until Henry Fairfield Osborn joined its staff in 1891. Osborn founded the Department of Vertebrate Paleontology, and by 1897 the museum was sending dinosaur-hunting expeditions to the American West. Young Brown, recently arrived from his native Kansas, cut his teeth as a field paleontologist in the Jurassic Morrison fossil beds of southern Wyoming, first at Como Ridge in 1897, then at the rich fossil site of nearby Bone Cabin Quarry in 1898. The Morrison Formation yielded the great fossil reptiles that inflamed scientific interest in the 1870s and 1880s. However, the ambitious neophyte yearned to explore new territory, two years collecting fossil mammals in Patagonia having whetted his appetite.

In 1902 Brown arrived in Jordan, Montana, to hunt for bones on his own, no longer as an assistant to someone else. Here the badlands near the Missouri River were cut into Late Cretaceous rocks of the Hell Creek Formation. He quickly found a skeleton of a monster predator. By 1905 the monster had been freed from its stony crypt, and Osborn brought it to life under the dramatic name of *Tyrannosaurus rex*, the tyrant reptile king. Brown outdid himself in 1907 by finding an even better skeleton of *Tyrannosaurus*, with a complete skull, 137 cm (4.5 ft) in length. He also collected run-of-the-mill Hell Creek fossils such as *Triceratops* and *Edmontosaurus*. But fresh trophies were harder to come by. The restless and energetic Brown required new challenges. Where to turn?

Brown was aware of paleontological activity in Canada. A serendipitous visit to the American Museum in 1909 by an Alberta rancher convinced him that a visit to western Canada was in order. In 1910, the plucky Brown arrived in Red Deer, Alberta, and took possession of a 9-m flatboat he had ordered. Its 3.5-m-wide deck accommodated a wall tent with stove and left room for plastered specimens and supplies. Casting off into the brown waters, he began the first of six years of exploration that were destined to reveal the Red Deer River Valley as one of the richest dinosaur treasure troves on earth. He found skulls and skeletons with such ease (and with such fanfare) that it quickly became clear that the Canadian government had to do *something*. Fortunately for science, the *something* was to hire Charles Hazelius Sternberg, a fossil collector of great note who had collected for E. D. Cope from Kansas to Montana. By 1912, Sternberg and his three sons, C.M., George, and Levi, were active in the Alberta badlands, achieving equally spectacular results. Initially the museums in New York and Ottawa, and later that in Toronto, filled with the skeletons of horned dinosaurs, duck-bills, armored dinosaurs, and meat-eaters that the world had never seen before. So abundant were they that specimens were exchanged with museums around the world, with the result that Canadian dinosaurs are among the best known in the world today.

Barnum Brown began his work in the Red Deer River Valley by exploring exposures of the so-called Edmonton beds, which we now know as the Horseshoe Canyon Formation. The Horseshoe Canyon Formation (dating from circa 72 to 68 Ma) is several million years younger than the Judith River Formation (76–74 Ma) farther downstream in the Steveville badlands, but several million years older than the Lance and Hell Creek Formations (as well as the Scollard and Frenchman Formations; 68–65 Ma) from which *Triceratops* and *Torosaurus* come. Brown achieved success in the Edmonton beds, but not to the same degree as in the Judithian beds downstream. He stated that remains of horned dinosaurs were comparatively rare in the Edmonton beds and that he had seen only ten specimens in four years.

In 1912 he collected several partial skulls 11 km below Tolman Bridge, and in 1914 he named the ceratopsid *Anchiceratops ornatus* ("almost *Ceratops*, ornate"). The type specimen bore an impressive frill, but the facial part of the skull was missing. He included in his description a paratype that consisted of a pair of long postorbital horns and a beautifully preserved braincase. The large postorbital horn cores are 57 cm high and diverge laterally from their bases for an estimated spread of 95 cm from one tip to the other. They also curve forward. The broad, flat crest is long and has very coarse epoccipitals, rather small parietal fontanelles, and a distinctive pair of short, knob-like processes on top of the midline, at the caudal border of the crest. The crest has a quadrilateral or rectangular form. The squamosals are long and narrow, terminating at the level of the third epoccipital from the rear of the crest. Five epoccipitals on the squamosal are moderate in size, whereas the sixth and final one is much enlarged. There are three pairs of very coarse, almost exaggerated epoccipitals on the caudal edge of the parietal.[28]

In 1924, C. M. Sternberg collected a specimen of *Anchiceratops* 20 km northwest of Morrin, Alberta. In his description, published five years later, he noted "at this horizon the remains of *Anchiceratops* are more numerous than in any other part of the formation. . . . It is below the oyster bed which is to be found throughout most of the region." He thus documented a very important and interesting aspect of the Horseshoe Canyon Formation: the strong influence of the nearby sea, expressed in the estuarine conditions that permitted the development of oyster beds. Brown had alluded both to oysters and to the plesiosaur, *Leurospondylus*. Coals are also an important part of the Horseshoe Canyon Formation, laid down in lower delta, interdistributary swamps of the sort that may be seen in Louisiana today, near the Gulf of Mexico.

Sternberg described how the skull had separated from the skeleton and was preserved upside down, owing to the weight of the heavy

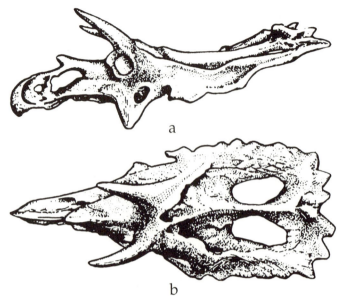

FIG. 4.8. *Anchiceratops longirostris* skull, Canadian Museum of Nature. (a) Lateral view, (b) dorsal view. (From Dodson and Currie 1990. Donna Sloan. Courtesy of the University of California Press.)

horns. Sternberg's skull is 1.66 m long, significantly smaller than the type of *A. ornatus*. He found the crest of his specimen to be "proportionately narrow and thin." The feature that most impressed him was the very long, slender rostrum, for which he named his animal *Anchiceratops longirostris* ("long snout") (Fig. 4.8). Sternberg concurred completely with Brown on the characters that define the genus *Anchiceratops*. He characterized his species as having a skull that was light in construction, with a small, triangular nasal horn core; a long, slender snout; circular postorbital horn cores directed forward, upward, and slightly outward; and a moderately thin crest with fontanelles of moderate size. The postorbital horn cores are 31 cm high, and the tips of the two horns are the same distance apart. The crest is some 86 cm long (measured from the occipital condyle) and 66 cm in breadth. The maximum thickness of the crest is 30 mm, but Sternberg states that the average thickness is half that figure.[29]

In 1929, the interpretation that this gracile specimen was a separate species was consistent with the tenor of the times. In 1933, Lull tentatively accepted the validity of the two species but lamented, "the variations of ceratopsian skulls due to age, sex, or the individual are such that no two ever seem to agree."[30] Today, the most obvious interpretation is that *Anchiceratops longirostris* is the female of *Anchiceratops ornatus* and that only a single, sexually dimorphic species should be recognized. Lehman explicitly expressed this interpretation in 1990.[31]

It was very clear to Sternberg that the affinities of *Anchiceratops* lay with the *Chasmosaurus-Torosaurus* group. He was very explicit in his choice of characters: "large, flat, sub-rectangular crest; long, narrow squamosals, extending almost to the back of the elongated parietals; abbreviated face; long, slender nose or that portion in front of the nasal horn." He also challenged Brown's view that *Anchiceratops* was intermediate between *Monoclonius* and *Triceratops*, but generously allowed, "it is doubtful with our present information if he would hold to this view."[32]

Sternberg collected a skeleton of *Anchiceratops* lacking the skull in 1925, 12 km southwest of Rumsey, Alberta. This he assembled as a panel mount with a cast of his skull of *A. longirostris*. Unfortunately, he never described the skeleton, which is very complete. Our cursory description of it comes from Lull, who anticipated a more complete account of it later. The skeleton, a photo of which appears in Lull's monograph, is robust, but it was smaller in overall size than the skeletons of *Centrosaurus* and *Chasmosaurus* from the Judith River Formation. This was especially due to a short tail, with only thirty-eight abbreviated caudal vertebrae instead of the expected forty-five. The limb bones are distinctly heavier in construction than are those of either of the Judithian genera. Dale Russell kindly supplied me with measurements. With a femur length of 74 cm and a tibia length of 51.5 cm, the limbs are just slightly smaller than those of *Chasmosaurus*.

Several other skulls of *Anchiceratops* have been collected, including two in Toronto, but none of these has ever been described, and *Anchiceratops* remains a relatively poorly known long-frilled ceratopsid. There have been two reports of *Anchiceratops* in southern Alberta outside the Red Deer Valley. A few parietal fragments were recorded in the St. Mary River Formation by Wann Langston from Scabby Butte, near Lethbridge, from deposits of the same age as the Horseshoe Canyon Formation.[33]

A very intriguing report that I confess I find puzzling comes not from a Horseshoe Canyon equivalent but from the Judith River Formation. I find this puzzling because, although there is ample evidence for geographic distribution of ceratopsian species, for the most part they do not pass from one formation into another of a younger age. The degree of turnover in faunal composition between time-successive geological formations is quite high. *Anchiceratops* may be an exception. While he was collecting excellent skulls of *Centrosaurus*, *Monoclonius*, and *Chasmosaurus* in the Manyberries region of extreme southeastern Alberta in 1937, Sternberg picked up skull fragments from two localities 15 km removed from one another. He thought they resembled *Anchiceratops*, which did not "belong" there. He cataloged them in the collections of the National Museum of Canada but, as he had plenty of very interesting specimens to study, did nothing with them.

In 1959, Wann Langston, who had succeeded Sternberg as dinosaur paleontologist at the National Museum following the latter's retirement in 1950, reexamined the specimens and concurred completely. One specimen consists of a chunk of right squamosal and parietal that measures more than 50 cm in length and preserves a series of coarse epoccipitals. It seemed to come from an individual about 15 percent smaller than would be expected based on the skull of *A. longirostris,* and considerably smaller than specimens of *A. ornatus.* The second specimen consists of three epoccipitals and a single "knob" or boss from the pair that is so distinctive for *Anchiceratops.* If this was *Anchiceratops*, what was it doing there, out of its time? To Langston, it was very important that the specimens were found in a region at the very top of the Judith River Formation, where it is grading into the overlying marine Bearpaw Formation. The sediments in this zone are rich in carbonaceous plant material, and the general environment might be described as estuarine. In the Horseshoe Canyon Formation, Langston described a similar situation. Near the Drumheller marine tongue, marked by the oyster beds, *Anchiceratops* again became common. Langston wrote:

> It is apparent from this that *Anchiceratops* occurs mostly where other ceratopsians are uncommon. Its remains are often associated with fine-grained clastic sediments that were deposited near the strand line, probably in quiet but not coal-swamp environmental conditions. The fact that *Anchiceratops* is well differentiated at its first appearance suggests that its appearance may be correlated with a change in the physical environment. The creature evidently found unsuitable the relatively open better-drained situations, evidently preferred by more ubiquitous ceratopsians. In life *Anchiceratops* was evidently restricted to low-lying, even marshy habitats.[34]

PENTACERATOPS—ANOTHER GREAT ONE

C. H. Sternberg was an unabashed entrepreneur. After he left the employ of the Canadian government in 1916, he returned to the United States and continued collecting. He became what is euphemistically called a "paleontological supplier." That is to say, he supplied fossils to those who would pay for them. There is nothing dishonorable about this. He was certainly one of the most accomplished fossil collectors in the world, and one who had dedicated his life and that of his entire family to the science of paleontology. In 1921, he was working in the Late Cretaceous Fruitland Formation in the San Juan Basin in northwestern New Mexico when he began to find ceratopsian fossils. He collected excellent material that year and during the following several years. The fossils he collected in 1921 were purchased by the University of Uppsala, in Sweden. The fossils collected in 1922–1924 were sent to the AMNH. There Osborn described one of them in 1923 as *Pentaceratops sternbergii* ("Sternberg's five-horn face").[35] This was a fitting honor for one with such total devotion to paleontology and whose contributions stretched back nearly half a century at that point. Other museums with *Pentaceratops* material include the Smithsonian Institution (supplied by George Sternberg in 1929), the University of Kansas, the Museum of Northern Arizona in Flagstaff, and the University of New Mexico (UNM).

The type specimen is a very good skull that lacks the back of the crest.[36] There is a remarkable and lamentable dearth of published measurements on *Pentaceratops*, Wiman being the exception.[37] The type skull of *P. sternbergii*, based on the illustration, would have been about 2.2 m long as reconstructed (Fig. 4.9). My measurements of the skull of AMNH 1624, collected in 1923, indicate a skull about 2.3 m long. Lehman reconstructed a skull based on a composite of three specimens and came out with a rough length of 2.7 m. It is clear enough that *Pentaceratops* was an enormous animal, possibly the largest of all ceratopsids. Despite the enormous size, the occipital condyle appears to be rather small. It measures 73 mm in diameter in AMNH 1624, and a width of 57 mm is estimated for the braincase illustrated by Lehman; however, he also stated that this skull is from an individual about two-thirds the size of other specimens.

The nasal horn core is distinct but of modest dimensions. The postorbital horns are long and curved forward, in contrast to those of *Chasmosaurus*. They measure some 56 cm in length in the type, 49 cm in

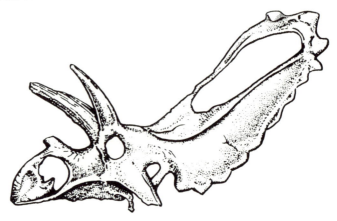

FIG. 4.9. *Pentaceratops sternbergii* skull, American Museum of Natural History. (From Dodson and Currie 1990. Donna Sloan. Courtesy of the University of California Press.)

AMNH 1624, and only 37 cm in a University of New Mexico specimen, which Lehman believes to be a female.[38] Three horn cores form, of course, the normal complement for any horned dinosaur. *Pentaceratops* allegedly has an extra pair, formed by the epijugals at the tips of the jugals. In all ceratopsids, the ventral tip of the jugal flares laterally to some degree, often accentuated by an epijugal, a bony scute attached to the tip. In *Pentaceratops*, the thickness of the epijugal is carried to an extreme. Whereas in a typical skull of *Chasmosaurus* the lower end of the jugal might measure 30 mm in thickness without the epijugal and 40 or even 50 mm with one, the skull of *Pentaceratops* measures a whopping 144 mm with the conical epijugal in position. It is not a true horn core in any sense, yet it certainly is a robust, bony projection. A blow received from the swing of a muscular neck, delivered by an enormous skull, would certainly get the "point" across to any sensible opponent. We can well imagine a pair of bull *Pentaceratops* facing each other with heads dipped, snorting with rage, glaring, stepping forward, advancing past each other in close contact, swinging heads against each other, shoving, throwing their weight around, and generally behaving in a boorish, thoroughly masculine, testosterone-plagued sort of way.

The frill of *Pentaceratops* is highly distinctive, being very long and open, with enormous parietal fenestrae reminiscent of those of *Chasmosaurus*. It is comparatively narrow, evidently less than a meter in breadth across the back despite its great length,[39] and is decorated by prominent epoccipitals. The squamosal is very long and tapering, much as in *Chasmosaurus*. The dimensions, however, are those of *Torosaurus*, ranging from 1.2 to 1.5 m in length! Lehman documents a range of eight to twelve scallops on the squamosal. The last epoccipital on the squamosal is the largest on that bone, about equal in size to the adjacent epoccipital on the parietal. As is typical of all chasmosaurines, the squamosal bends upward along its length, resulting in an erect frill, clearly visible from the front with the head horizontal, but even more visible with the head dipped forward.

Restoration of the parietal of *Pentaceratops* has proven a little more difficult, as most specimens are damaged at the rear. A very important specimen for clarifying the parietal was collected by the Museum of Northern Arizona in 1977 and described by Tim Rowe, E. H. Colbert, and Dale Nations in 1981. This specimen confirms the enormous size of the parietal fenestrae, amounting to 60 percent of the length of the frill. The rear border of the parietal is deeply emarginate, with a U-shaped notch on the midline somewhat reminiscent of that of *Chasmosaurus russelli*. The midline length of the parietal is some 33 cm shorter than the length of the parietal measured to the corners. On the inside border of the notch is a large triangular epoccipital that measures 21 cm at the base.[40] There is on the upper surface of the back edge of the parietal a pair of midline structures that Lehman interprets as a pair of upturned epoccipitals. These resemble the pair of knobs on the frill of *Anchiceratops* and are yet one more distinctive feature of *Pentaceratops*.

The Swedish scientist Carl Wiman was the paleontologist who had the privilege of examining the *Pentaceratops* material that came to Sweden. The specimen was from the Kirtland Formation, which may be slightly younger than the Fruitland Formation. The skull is a fair one, generally similar in nature of preservation to the ones at the AMNH that Osborn described, but perhaps a little more distorted. The skull measures 2.16 m in length and differs in minor details from the type of *Pentaceratops sternbergii*. Wiman named the new material *Pentaceratops fenestratus* ("fenestrated"). One notable feature was a fenestra in the left squamosal. This seems a dubious feature, as squamosal fenestrae, generally only on one side, are known from a variety of chasmosaurines and do not seem to have taxonomic significance. Rather, they have been cited as evidence for aggressive behavior within species, although this claim is itself controversial. The differences that Wiman cited—including number of epoccipitals, position of the postorbital horn core, size of the infratemporal fenestra, nature of the epijugal, and length of the

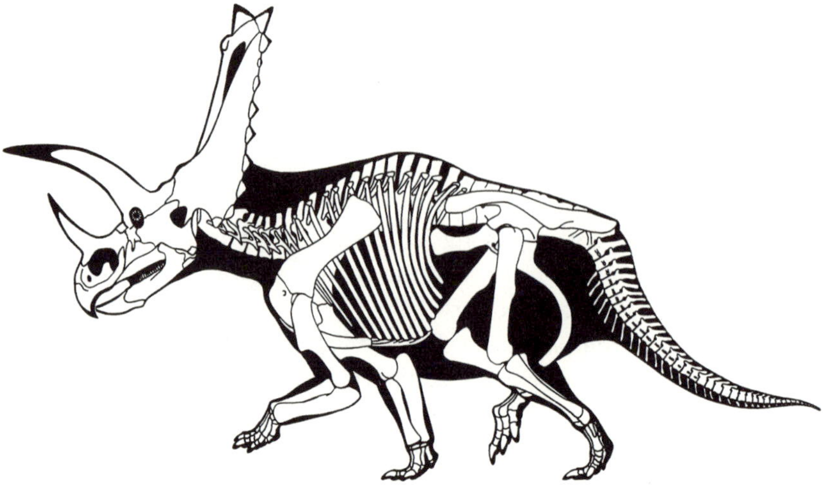

FIG. 4.10. *Pentaceratops* skeleton. (Gregory S. Paul.)

nasal horn core—do not seem to be convincing characters, and there has been little support for the idea that *P. fenestratus* is a valid species.

Nonetheless, Wiman's report has value, because he also described a skeleton. The skeleton was not, unfortunately, associated with the skull, although he did add the skull to his skeletal reconstruction (Fig. 4.10).[41] Critics have carped that there is no evidence that the skeleton even belonged to *Pentaceratops*, but the obvious answer to this complaint is that there is no evidence for any other ceratopsid from the Fruitland-Kirtland beds. Though his thesis has not been proven, it is highly probable that Wiman was correct, and I feel comfortable in assuming that he was. The length of the vertebral column, complete to the twenty-eighth caudal, was 4.1 m, or about 5 m including the skull. Apparently the tail was quite short in *Pentaceratops*. The length of the femur was 88 cm and that of the tibia 65.5 cm. A comparison with our reference specimen of *Chasmosaurus belli* would suggest that if *Pentaceratops* had similar body proportions, it would have been 6.0 m long and not 5.0. Of course a simple comparison of skulls would yield an animal 7.0 m long or longer. Evidently the body proportions of *Pentaceratops* are rather significantly different from those of *Chasmosaurus*, seemingly being very "stubby," although we do not understand the significance of this difference. We also do not know if the single skeleton known is that of a small individual like that from which the UNM skull to which Lehman referred was taken, or from an individual of

FIG. 4.11. *Pentaceratops* reconstruction. (Robert Walters.)

large size. How large are isolated ceratopsid limb bones from the San Juan Basin? This information is unreported (Fig. 4.11).

The San Juan Basin of New Mexico is far removed from the famous horned dinosaur beds of Wyoming, Montana, and Alberta. The precise stratigraphic correlation has proven to be somewhat difficult to establish, in part because there are intervals of missing time in the basin, in which sediments range in age from Late Cretaceous to Eocene. With the use of powerful stratigraphic tools such as paleomagnetism, it seems that the correlation is now secure. Lehman, in his 1993 paper, summarizes the evidence. He concludes, "The Fruitland Formation and the Kirtland Shale up to the base of the Naashoibito Member probably span the Judithian and Edmontonian Land Mammal Ages. . . . Hence, *Pentaceratops* is of Judithian to Edmontonian age, but not Lancian."[42]

ARRHINOCERATOPS—NO NOSE HORN

W. A. Parks (1868–1936) at the University of Toronto already had solid credentials in invertebrate paleontology, when, late in his career, he had the opportunity to study dinosaurs as well. Barnum Brown had already finished his collecting in Alberta. The Sternberg team had broken up, and C. H. Sternberg had returned to the United States. C. M. Sternberg in Ottawa had replaced Lawrence Lambe, who was now dead. Fortu-

119

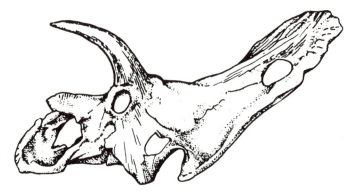

FIG. 4.12. *Arrhinoceratops brachyops* skull, Royal Ontario Museum. (From Dodson and Currie 1990. Donna Sloan. Courtesy of the University of California Press.)

nately for the University of Toronto, Levi Sternberg came to work for them. With Parks participating, expeditions to Alberta were launched in 1920, and they came back successful. Despite the plethora of fossils described by Brown and Sternberg, there were still more to go around. For fifteen years (1920–1935), Parks made significant contributions to the study of duck-billed dinosaurs, armored dinosaurs, and meat-eating dinosaurs. Probably his greatest contribution was the description of the trombone-crested duck-bill, which he named *Parasaurolophus walkeri* in 1922.

In 1925, Parks described a new horned dinosaur from the Horseshoe Canyon Formation of the Red Deer River Valley near Blériot Ferry, not far upstream from Drumheller. Little did he, or anyone else, dream that this would be the last new genus of the long-frilled Ceratopsia to be named from that day to this. (New short-frilled ceratopsians continued to be named as recently as 1995.) Parks named his skull *Arrhinoceratops brachyops* ("without a nose horn face, short face") (Fig. 4.12). This name, though technically well composed, certainly wins no popularity contests. I have yet to be approached by a bright-eyed, eager young lad bursting to share his trove of *Arrhinoceratops brachyops* lore with me. Nor have I seen *Arrhinoceratops* worn with pride on belt buckles, tie clasps, or other produce of the dinosaur knickknack industry. *Arrhinoceratops* remains a rare entity, only a single skull ever having been described, and only two significant papers appearing in print. Nonetheless, no one has ever questioned the validity of the taxon. Parks was terse in his description:

Supraorbital horn cores large, directed outwards and forward; nasal horn core absent; facial region short; crest relatively large, subquadrate, flat; squamosals long; parietals with oval fontanelles of moderate size; anterior process of jugal unusually long.

For Parks, the defining feature that plays so prominently in his name comes from the snout:

The nasal horn core is apparently absent, but the nasal bone is sharp above and somewhat rugose, suggesting that it may have carried a horny sheath. The nasal bone rises very abruptly, posterior to the suture with the rostral, suggesting the condition that maintains in *Triceratops prorsus*, but there is no trace of a horn core nor of an epinasal; neither does the surface of the bone indicate that a structure of this kind has been lost.[43]

He also felt that the facial region in front of the orbits is very short. The skull is 1.5 m long and exceeds a meter in breadth across the frill (1.08 m).[44] The postorbital horn cores curve forward and diverge strongly laterally. The estimated height of the postorbital horn cores is about 41 cm. The frill is broad, with a straight rear border. The parietal fenestrae have an unusual oval form, with the long axis parallel to the central axis of the frill. Several features are important. The frill is very thin, several thickened portions attaining 30 mm in thickness, but most of it being 5–10 mm thick, according to Parks. The other feature is that *both* the dorsal and ventral surfaces are covered with vascular grooves in a reticular pattern, except for a smooth channel that extends from the parietal fenestra to the supratemporal fenestra.[45] The squamosal is 72 cm long and nearly half that wide. Ornamentation on the free edge of the squamosal is modest. The left squamosal has a prominent fenestra in it. The epijugal is quite prominent and measures 10 cm in thickness.[46]

Parks presented very little analysis and unfortunately is remembered more for the errors he made. Lull expressed skepticism concerning some of the anatomical details, but more than fifty years passed before *Arrhinoceratops* was restudied, by a master's candidate at the University of Alberta, Helen Tyson. In Tyson's cogent analysis, Parks's failure to accept the existence of a true nasal horn core followed from his failure to see evidence for a separate center of bone growth, e.g., an epinasal. This despite the fact that a horn-like growth is as obvious as—dare we say it?—the noses on our faces. As Tyson put it, "To deny the presence of a horn core in *Arrhinoceratops*, which lacks such a suture but possesses a distinct horn-like organ, contributes neither to the problem of the homology of this structure nor to an accurate characterization of the genus." The emperor has no clothes! Furthermore, Tyson, benefiting from the progress made in understanding ceratopsian morphology over the decades since Parks's time, corrected some egregious errors. Parks failed to discern the dorsal limit of the rostral bone, which he thought contacted the nasal in front of his nonexistent horn.[47] Similarly, he thought that the jugal reached far forward, almost contacting the premaxilla and cutting off the maxilla from the lacrimal, another unprecedented and erroneous interpretation. The sutures in the specimen are difficult to interpret, and he simply read too much into the cracks and fissures of a mature specimen.

Tyson performed a significant service by setting the record straight on these matters. Moreover, she proceeded to carry out some phylogenetic analysis to place the position of *Arrhinoceratops* in a needed context. She unequivocally recognized *Arrhinoceratops* as a long-frilled ceratopsid. Moreover she saw evidence for a close relationship between *Arrhinoceratops* and *Torosaurus*. Both have thin parietal frills and squamosals with modest or absent epoccipitals. She cautioned against the assignment of isolated parietals or squamosals of thin construction, simple pattern, and prominent fenestrae to one genus or another without further confirming evidence. Accordingly, both genera remain rare.[48]

Thus it is that the most recently described genus of chasmosaurine was described in 1925. The long-frilled horned dinosaurs remain of interest. New specimens continue to be described, and a new species, *Chasmosaurus mariscalensis*, was named by Lehman in 1989. Can it be that our knowledge of chasmosaurines is actually *complete*? That view is a little optimistic. For starters, a complete skeleton of even *Triceratops* would be nice. And I would be pleased to add skeletons of *Torosaurus* and *Arrhinoceratops* to the list. A population of *Anchiceratops* would also be nice. Chasmosaurines lived in what is now northern Mexico and on the North Slope of Alaska, but we do not yet know the generic identity of these animals. Finally, convincing intermediates among known chasmosaurines would clear up some mysteries, foremost among them the ancestry of *Triceratops*. *Chasmosaurus* is the earliest chasmosaurine, but not the most generalized by any means. Where is its ancestor and what did it look like? The job is certainly not finished.

Color Plates

PLATE I. *Triceratops horridus* encounters *Tyrannosaurus rex*. The meek shall inherit the earth. Late Cretaceous, Montana, Hell Creek Formation, circa 65 Ma. Chapter 3.

PLATE II. A lone *Chasmosaurus mariscalensis*, wading in the swamps in search of fodder, tenses at the scent of danger. Late Cretaceous, Texas, Aguja Formation, circa 70 Ma. Chapter 4.

PLATE III. *Styracosaurus albertensis*, bellies filled, tranquilly survey the land-scape, oblivious of nearby crocodile and *Euoplocephalus* in background. Late Cretaceous, Alberta, Judith River Formation, circa 74 Ma. Chapter 5.

PLATE IV. Two bull *Pachyrhinosaurus canadensis* joust. Boys will be boys. Late Cretaceous, north slope of Alaska, Prince Creek Formation, circa 69 Ma. Chapter 6.

PLATE V. Two *Leptoceratops gracilis* refresh themselves, the young one not certain what to make of the turtle (*Aspideretes*) in the water. Late Cretaceous, Alberta, Scollard Formation, circa 65 Ma. Chapter 7.

PLATE VI. Two *Psittacosaurus mongoliensis* warily eye marauding troodontids. Early Cretaceous, Mongolia, Ondai Sair Formation, circa 100 Ma. Chapter 7.

A Big One on the Nose

THE SHORT-FRILLED HORNED DINOSAURS

WHEN horned dinosaur remains began to be discovered in what is now Montana, they were those of the short-frilled types we know today as centrosaurines. This group includes the exotic *Styracosaurus,* the enigmatic *Pachyrhinosaurus,* and the recently discovered *Einiosaurus* that shows an unexpected state of horn development. I confess to a special fondness for centrosaurines. My own work in Alberta and Montana has often involved centrosaurines. In fact, part of my slender claim to paleontological legitimacy stems from my discovery and description of a new centrosaurine from Montana, which I named *Avaceratops* in 1986, and which occupies a place of honor in the wonderful dinosaur hall (safely away from the glare of the *Tyrannosaurus*) at the Academy of Natural Sciences in Philadelphia. Besides all this, the first ceratopsid dinosaur to be identified and that we still recognize today was the centrosaurine *Monoclonius,* named by E. D. Cope in 1876.

When Barnum Brown named *Leptoceratops,* the first protoceratopsid, in 1914, there was a ready yardstick of comparison because ceratopsids, especially *Triceratops,* were by that time well known. It is hard to imagine that there was a time when no one knew that there were dinosaurs that sported horns on their faces. It had been suspected, for instance, that the two-legged plant-eater *Iguanodon* had a spike on its nose. With the discovery of skeletons of *Iguanodon* in the coal mines of Bernissart, Belgium, in 1878, however, the 15-cm conical spike was removed from the nose and placed in its proper, if surprising, position on the thumb. There were no horned ornithopods. Clearly, having horns on the head seemed like such a good idea that, if nothing else, someone would just have to invent such an animal.

Early dinosaur discoveries in England and Europe were rather fragmentary. When Richard Owen coined the name *dinosaur* in 1842, no accompanying visual image was presented, because to do so would

123

FIG. 5.1. Joseph Leidy (1823–1891). (Robert Walters.)

simply entail too much speculation. The first discoveries of dinosaurs in the New World were little more auspicious. Not surprisingly, the discoveries were made in Cretaceous beds of Montana, in what we now recognize as the Judith River Formation. The geological explorer Ferdinand V. Hayden picked up a small collection of teeth in 1855 and sent them to Professor Joseph Leidy (1823–1891) at the Academy of Natural Sciences in Philadelphia (Fig. 5.1).[1] Leidy, a brilliant young physician and anatomist, was one of the first scientists to make extensive use of microscopy in his research and made important contributions in many fields of natural science in addition to paleontology. He recognized Hayden's teeth as dinosaurian and in 1856 named the first American dinosaurs: *Trachodon, Troodon, Deinodon,* and *Paleoscincus.* Unfortunately, these names have proven troublesome ever since, because the plain fact is that teeth are not a very good basis for naming dinosaurs.[2] Only the name of *Troodon,* a nasty little meat-eater with a large brain, is still in use today. Interestingly, when we examine the vial containing the type specimen of the dubious duck-bill *Trachodon mirabilis* at the Academy of Natural Sciences, there are several teeth. One of these teeth is clearly that of a

FIG. 5.2. Edward Drinker Cope (1840–1897). (Robert Walters.)

duck-bill, although which duck-bill we cannot say. Another tooth, less well preserved, is equally clearly that of a ceratopsid, split root and all! Thus we can trace the earliest evidence of ceratopsids to 1856.

Dinosaur science was greatly advanced with the discovery of the first good partial skeleton of a large dinosaur in 1858. The discovery came not from Europe, or from the great fossil beds of the western United States or Canada, but from Haddonfield, New Jersey, barely 10 km from Philadelphia. In Joseph Leidy's able hands, the duck-bill *Hadrosaurus foulkii* took form as an animal that walked on its hind legs, no longer a monster lizard but a kangaroo-like biped. Ten years later, the skeletal reconstruction of *Hadrosaurus* that went on display at the Academy of Natural Sciences was the first dinosaur skeleton ever exhibited anywhere in the world. Young Edward Drinker Cope (1840–1897) was enthralled by the fossil treasures of the Academy (Fig. 5.2). As a young lad of eight years, he used to spend free hours there, sketching fossil ichthyosaurs and plesiosaurs from Lyme Regis, England. Cope went abroad to study zoology and paleontology in Europe then returned to

125

Philadelphia to pursue his professional career. He was a brilliant man who published more than 1,400 papers in his career (and he died young!). Many of these papers were mere notes, and others contained embarrassing errors. A substantial number of his papers, however, represent fundamental insights for which paleontologists and vertebrate and invertebrate zoologists today are indebted to him. He also possessed, shall we say, an energetic personality. Many found him hard to get along with, truly an *enfant terrible*. He was impulsive, impetuous, and sometimes forgetful of social graces. He did not have a successful record of institutional employment, although he taught for one year at Haverford College. He also had a stormy relationship with the Academy of Natural Sciences. He was at his best when on his own; his family's wealth permitted him this luxury.

Cope began his career in fossil reptiles by describing Cretaceous specimens from New Jersey, the most important of which was an intriguing meat-eater, which he named *Laelaps aquilunguis* ("eagle-clawed hunting dog") in 1866. Unfortunately, the name *Laelaps* was already in use, and O. C. Marsh at Yale was only too happy to rename the dinosaur *Dryptosaurus* ("tearing lizard"), a less poetic but suitably rapacious name. In 1868, Cope described a large marine reptile, sent to Philadelphia from Kansas, as *Elasmosaurus*. However, in his haste, he blundered by reconstructing the head on the end of the tail—an error indelicately pointed out to him by both Marsh and Leidy, to Cope's everlasting chagrin. Soon his restless energy had brought him out west to work in the fossil beds of Kansas, Wyoming, and Colorado.

In 1872, F. B. Meek discovered bones in southwestern Wyoming, 80 km east of Green River. Cope was soon there and excavated a partial skeleton, consisting of sixteen vertebrae from the tail, sacrum, and back; two large ilia from the pelvis measuring 1.2 m in length; and some ribs. He named his specimen *Agathaumas sylvestris* ("great wonder of the forest") in 1872, noting that "the measurements . . . of the present animal exceed those yet described from North America." He was impressed with the proximity of the bones to coal seams: "It appears that the forests that intervene between the swamps of epochs during which the coal was formed were inhabited by these huge monsters; that one of them lay down to die near the shore of probably a brackish-water inlet, and was soon covered by the thickly fallen leaves of the wood."

Cope was at a loss to explain what kind of animal this dinosaur was. He compared it to *Hadrosaurus*, to *Dryptosaurus*, and even to *Cetiosaurus*, concluding that "it is evidently new to our system."[3] Cope was undoubtedly correct: today we can recognize the remains, which are in the collection of the American Museum of Natural History, as those of a large ceratopsid, but no skull material has ever been associated with *Agathaumas*, nor have further specimens of this or any other dinosaur been found in the region. The "great wonder of the forest" remains only of historical interest, as the first ceratopsid to receive its own name, but a name that falls into the dreaded category of "nomen dubium." We may guess that the animal in question was either *Triceratops* or *Torosaurus*. These are only guesses, but they are consistent with its large size.

Cope again visited the West in 1873, and somewhere in northeastern Colorado he made a small collection of bones, which are cataloged today in the collection of the American Museum of Natural History as "fragments of horn cores, vertebrae, etc." As we have already discussed, dinosaurs were not then known to have possessed horn cores. When Cope described these materials in 1874, he failed to recognize fragments of horns as such. He did not shrink, however, from applying a name to his find, which he called *Polyonax mortuarius* ("master over many, dead").[4] Whereas *Agathaumas* tantalizes, *Polyonax* simply clutters the literature. When John Bell Hatcher reviewed the species thirty-three years later, his scorn was little disguised:

> Cope's description and figures demonstrate conclusively the extremely fragmentary and totally inadequate nature of the material upon which the genus and species were based. The fragments supposed by Cope to pertain to the ischia are now known to have been portions of the frontal horn cores. The "paleontological wastebasket" would be a fit receptacle for what remains of the type material, while the name should be dropped from the paleontological literature. It was perhaps a premonition of this which suggested to Professor Cope the specific appellation *mortuarius*. Unfortunately vertebrate paleontology is burdened with too many genera and species founded, as in the present instance, on fragmentary and insufficient material.[5]

Touché!

MONOCLONIUS—THE FIRST CERATOPSID

Twice Cope failed in his quest to name a valid ceratopsid. The third time he achieved some degree of success, though it was hard won and even today does not elude controversy. Cope's first forays out west were under the escort of parties from the U.S. Geological and Geo-graphical Survey of the Territories. By 1876, he had begun to use his favorable personal circumstances to hire his own workers. Soon he made contact with Charles Hazelius Sternberg (1850–1943), who was destined to become one of the great American dinosaur hunters of all time. The senior Sternberg, then a student at Manhattan State College in Kansas, attempted to sign on with Benjamin Mudge's expedition to the Cretaceous chalk deposits of western Kansas. Mudge had been sent to the chalk on behalf of Yale's O. C. Marsh. No positions were available when Sternberg applied, so in desperation he wrote to Cope in Phila-delphia. We have the account in Sternberg's own words:

> Almost with despair, I turned for help to Professor E. D. Cope, of Phila-delphia, who was becoming so well known that a report of his fame had reached me at Manhattan.
>
> I put my soul into the letter I wrote him, for this was my last chance. I told him of my love for science, and of my earnest longing to enter the chalk of western Kansas and make a collection of its wonderful fossils, no matter what it might cost me in comfort or danger. I said, however, that I was too poor to go at my own expense, and asked him to send me three hundred dollars to buy a team of ponies, a wagon, and a camp outfit. . . .
>
> I was in a terrible state of suspense when I had despatched the letter, but, fortunately, the Professor responded promptly, and when I opened the envelope, a draft for three hundred dollars fell at my feet. The note which accompanied it said: "I like the style of your letter. Enclose draft. Go to work," or words to the same effect.
>
> That letter bound me to Cope for four long years, and enabled me to endure immeasurable hardships and privations in the barren fossil fields of the West; and it has always been one of the greatest joys of my life to have known intimately in the field and shop the greatest naturalist America has produced.[6]

Sternberg greatly repaid the trust Cope showed in him, and the resulting collection of Late Cretaceous marine fossils provided Cope with specimens of fishes, reptiles, and birds to describe for several years. For the purpose of our narrative, however, it is events later in the summer of 1876 that hold our attention. Cope came west, arriving by rail on August 1 in Omaha, where he was met by Sternberg. Cope desired to investigate the location from which Hayden had collected the first American dinosaurs in 1855. An arduous and circuitous jour-ney ensued, to north-central Montana via Nebraska, Utah, and Idaho. A momentous event in American history had occurred in southern Montana just a few weeks earlier, on June 25, 1876—the Battle of Little Bighorn. Nerves were greatly ajitter, but Cope, a devout Quaker and pacifist, refused to take heed. In Fort Benton on the Missouri River, horses, a wagon, and equipment were purchased, and a cook and crew hired. Heading overland, the small band reached the badlands near the mouths of the Judith River and Dog Creek, 80 km east of Fort Benton, on August 27. Surrounding the Missouri River here were steep and forbidding riverine exposures, the likes of which Cope had never seen. Adventures included a visit by 2,000 peaceful Crow Indians and the subsequent desertion of the cook and scout; hair-raising nocturnal forays through the badlands; and a life-threatening slip down a steep slope described by Sternberg. The work was physically exhausting. Fossil remains of many animals—of both dinosaurs and other animals, including turtles, garfishes, and freshwater rays—were found. However, these were predominantly fragments or "spare parts," rather than major portions of skeletons. Sternberg recounts evidence of Cope's vivid imagination:

> Every night when we returned to camp, we found that the cook had spent the whole day in cooking. Exhausted and thirsty,—we had no water to drink during the day (all the water in the Bad Lands being like a dense solution of Epsom salts),—we sat down to a supper of cakes and pies and other palatable, but indigestible, food. Then, when we went to bed, the Professor would soon have a severe attack of nightmare. Every animal of which we had found traces during the day played with him at night, tossing him into the air, kicking him, trampling upon him.
>
> When I waked him, he would thank me cordially and lie down to another attack. Sometimes he would lose half the night in this exhausting slumber. But the next morning he would lead the party, and be the last to give up at night. I have never known a more wonderful example of the will's power over the body.[7]

The results of their efforts were less than spectacular. Cope left from Cow Island, 50 km downstream, on a river steamer bound for Omaha around October 15. He carried 1,700 pounds of fossils with him. Sternberg stayed on until a heavy snowfall on November 1 and then left. However, before he left, he found a significant dinosaur fossil near Cow Island that Cope described in 1889 as another species of *Monoclonius*.

Complete articulated dinosaur material, or even associated partial skulls and skeletons, were still a rarity in 1876. The best fossils yet known were exquisite small specimens from lithographic limestones of Germany, notably single specimens of *Compsognathus* and of the proto-bird *Archaeopteryx*. *Hadrosaurus* and *Laelaps* from New Jersey were far more complete than anything yet seen in the American West. Cope was undeterred by the quality of his specimens, and he set out quickly to describe some of his finds. When I say quickly, I mean that by *October 31, 1876*, scarcely two weeks after he began steaming down the Missouri, he published a fourteen-page paper in which he described four new genera and thirteen species of dinosaurs, plus seven species of turtles and fish. It may well be suspected that such an outpouring of taxonomic activity was not accompanied by mature reflection, insightful description or comparison, or even minimal illustrations—of which there were none. Furthermore, the specimens were not numbered with an accession number or museum collection number, nor were any specific locality data noted, other than that conveyed in the title of the paper: "Descriptions of some vertebrate remains from the Fort Union beds of Montana."[8] These facts being the case, the reader will not be surprised to learn that the majority of taxa therein described have since been relegated to oblivion.[9] One dinosaur, however, commands our attention: *Monoclonius crassus*.

Monoclonius crassus ("thick single-sprout") was not merely a tooth-genus. Although teeth played prominently in Cope's concept, he also included vertebrae, limb bones, and other materials in his description. Historians have complimented Cope on the elegance of his names: replete with classical allusions, grammatically well formed, and euphonious. It is all the more impressive to realize that these names were not necessarily composed in his wainscotted study—surrounded by shelves of scholarly tomes, reference works, and dictionaries—but, at least in the present instance, by the flickering light of a campfire or on the decks of a river steamer that plied the waters of a muddy river somewhere in the middle of the western wilderness of Montana.

Cope's names were not explained to his classically educated contemporaries, but we may presume that they were understood. Their meanings have suffered through the years, and the name *Monoclonius* is a case in point. David Lambert, for instance, in his *The Dinosaur Data Book* (1990), defines the name as "one horned." The late Helen Roney Sattler, in her *The New Illustrated Dinosaur Dictionary* (1990), carefully and correctly translates the name as "single stem," but then adds "referring to the single horn on its nose." This view has flourished since Osborn wrote Cope's biography in 1931.[10] Recently, however, Benjamin S. Creisler, a classicist who is also interested in dinosaur taxonomy, did a neat and convincing job of unraveling Cope's probable intention. He interprets the name *Monoclonius* as a contrast with *Diclonius*, named immediately before *Monoclonius* in the same publication. *Diclonius* consists only of teeth, and Cope was attempting to contrast the method of tooth replacement (i.e., of sprouting) in the one dinosaur with that in the other. He compared tooth replacement in *Monoclonius* with that in *Hadrosaurus*.[11] The sad fact is that he simply did not have enough information upon which to base his conclusions. His *Monoclonius* tooth was probably that of a duck-bill, not a horned dinosaur. Were this all there were to *Monoclonius*, it would by now simply be another of his forgotten dinosaurs, such as *Dysganus* or *Diclonius*.

Fortunately, however, Cope added further details to his description. He described the sacrum as having ten vertebrae, measuring 27 in. (69 cm) in length. Three anterior dorsal vertebrae are fused together, with a deep cup facing forward. The limb bones are said to be robust, with a femur of 22 in. (56 cm) and a tibia of 20 in. (51 cm). The forelimb was not measured but is said to be robust in contrast to that of *Hadrosaurus*, which certainly makes sense. The most intriguing bone is a large "episternum" or breastbone, for which he gives a length of 21 in. (53 cm). These twenty-seven lines of text conclude his description. There is no attempt either to interpret the bones or to determine what manner of animal his *Monoclonius* was.[12]

The following year, Cope expanded on his subject and included some figures for the first time. The descriptions are wordy and uninformative, and the figures of several skull elements are hard to interpret. Indeed, he wrote, "Positive determination of these elements is impracticable, as they do not resemble the corresponding bones in any animal known to me." Despite having figured a nasal horn core, Cope did not recognize that it actually was a horn! In many words he described it as L shaped (i.e., including the nasal base), massive at the base, narrowing gradually to the extremity, and roughened with grooves for blood vessels.[13] It is clear enough that Cope had no concept of what kind of herbivore *Monoclonius* was.

The next major event in the history of horned dinosaurs was the discovery of *Triceratops*, whose history we discussed in Chapter 3. In 1889 Marsh named and figured a skull of *Triceratops*, which has since become one of the best-known dinosaurs in the world. It now became clear for the first time that horned dinosaurs were large herbivores with horns on their faces and expansive bony frills. Moreover, these were important animals in the Late Cretaceous communities of Wyoming and Montana. Cope was for the first time in a position to make some sense of his collection of *Monoclonius* bones from 1876. He reviewed his type species, *Monoclonius crassus,* and was now quite definite about it. He states that Marsh's figure "enables me to determine more exactly the affinities of several species of the family which have been in my possession for many years. The most complete specimen in my collection is that of *Monoclonius crassus* Cope. This includes representatives of all elements excepting the bones of the feet."

He now characterized *Monoclonius* as having small horns over the eyes and having an "enormously expanded parietal" with huge openings. Although he did not admit his mistake, his "episternum," or breastbone, of 1876 now had become a skull element, the parietal, which is the correct anatomical assignment. Proper placement of this element gave the skull such an improbable, even outlandish, appearance that we may forgive Cope his error. The parietal is now figured, and it is a very important specimen indeed, as we shall see presently. It resides in a basement treasure room of the American Museum of Natural History, where I have pored over it on many occasions. Cope referred to a squamosal, which he described in very general terms, but unfortunately he did not figure it. He compared pelvic and sacral bones to similar bones in *Agathaumas* and therefore preferred to use his name, Agathaumidae, for the family of horned herbivorous dinosaurs, consisting of *Agathaumas, Polyonax,* and *Monoclonius.* "This family is called by Marsh the Ceratopsidae; but as it is not certain that *Ceratops,* Marsh, is distinct from one of the genera previously named, I shall call it the Agathaumidae."[14]

Whatever the scientific merits of his case (and it is true that *Ceratops* is a poorly founded genus), it is Cope's misfortune that *Agathaumas* and *Polyonax* are forgotten genera, and his name has enjoyed no currency whatsoever. It is also Cope's misfortune that Marsh's ceratopsian genera, *Triceratops* and *Torosaurus,* are known from far more complete material than any specimen of *Monoclonius* discovered before 1937. Perhaps most important of all, the definitive 1907 monograph on the horned dinosaurs was a Yale production.[15] As was ever the case, the victors wrote the history, and the name *Ceratopsidae* has served for the family of horned dinosaurs since Marsh proposed it in 1889. It is a felicitous name, and I for one am glad that it has triumphed over Agathaumidae.

Cope named three further species of *Monoclonius* in his 1889 paper. *M. recurvicornis* specimens had been figured in his 1877 paper, but he states that, "suspecting they might belong to some of the species already known," he did not name them. The "recurved horn" species consists of a pair of moderately prominent (210 mm high) orbital horns and a low, blunt nasal horn about half that height. Again Cope mentions a squamosal but does not figure it. The second species, *M. sphenocerus* ("wedge horn"), was collected by Sternberg at Cow Island and is said to consist of "numerous parts of the skeleton, including parts of the skull." The most striking aspect of *M. sphenocerus,* and really the only one that Cope expanded on, is a very tall, straight nasal horn. It was 325 mm tall. It was as clear to Cope then as it is to us today that this nose horn is imposing. He wrote without hyperbole: "The *Monoclonius sphenocerus* is an animal of large size, exceeding a rhinoceros in height, and the nasal horn is the most formidable weapon I have observed in a reptile." The third species, *M. fissus,* has caused the fewest problems. It is based only on a single bone, which Cope misidentified as a squamosal. No wonder it did not resemble the squamosal of *M. crassus*—it was a pterygoid, and a broken one at that. It was not figured, and the species was soon ignored, as we shall do.[16]

Thus was born *Monoclonius,* now clearly a horned dinosaur of the Judith River Formation. Its Late Cretaceous age was established. Its name was destined to be conspicuous and important. Late in Cope's life, Osborn arranged for major parts of Cope's personal collection to be purchased by the American Museum of Natural History, which in 1895 had not yet come to prominence in dinosaur paleontology (or in any another branch of paleontology, for that matter).[17] Hence, the type specimens of *Monoclonius* species came to reside at the American Museum, where they may be studied today. *Monoclonius* remained the preferred name used by Barnum Brown in his writings on horned dinosaurs from beds of Judith River age, including those collected in Alberta. (I will argue later that this usage is incorrect.)

On the other hand, the Yale camp, represented by the trio of John Bell Hatcher, O. C. Marsh, and Richard Swann Lull, had no vested interested whatsoever in preserving Cope's name. In 1907, there appeared a large, brown, folio-sized volume entitled *The Ceratopsia* under the authorship of these three scholars.[18] As we have seen, neither Marsh (died 1899) nor Hatcher (died 1904) lived to see his triumph. The task of completing the monograph was turned over to Lull, a young professor at Yale, who did an admirable job of completing the text, modestly listing himself as editor. The book, published in 1907, is one of the most beautiful and significant tomes on dinosaurs ever published, as useful today as the day it was published.

Hatcher did not beat around the bush. He made it clear that there were serious problems with *Monoclonius*. The problems stem from the failure of Cope to document his finds with identifying marks (e.g., field numbers), illustrations, or locality data. Hatcher suspected, and I absolutely agree, that Cope's species are based on composite specimens rather than single individuals. That is to say, there is no evidence that he ever succeeded in finding a single specimen representing the greater part of the skeleton of one individual; rather, it is probable that the specimens he attributed to *Monoclonius crassus* represent a synoptic collection of elements picked up from the Missouri River badlands over the six-week period that he was there. If we knew for certain that there were only a single genus and species of horned dinosaur living at that time and place, any remains that were collected from there might justifiably be referred to that animal. However, we have no such confidence, as Marsh had in 1888 described *Ceratops montanus* from the same region. Although this genus is poorly founded on only two horns and an occipital condyle, the prominent horns suggest that *Ceratops* was a chasmosaurine, a completely different subfamily of horned dinosaurs from *Monoclonius*. Thus a dinosaur assembled from a collection of skeletal elements of unrelated specimens may end up being a chimaera, an animal that never existed.

Specimen numbers were applied to Cope's specimens after the fact, that is, when the bones were placed in the collections of the American Museum. Bones attributed to *M. crassus* are now cataloged under the number AMNH 3998. Hatcher carefully and critically described, analyzed, and figured all of the material associated with Cope's type specimen of *M. crassus*. He specifically included only the material that he could positively identify as pertaining to the type specimen. He was unable, for instance, to attribute any squamosal (and there were several in the collection) to *M. crassus*. This is unfortunate, as the squamosal is one of the most important bones in diagnosing ceratopsids.

The parietal, which forms the major expanse of the bony frill, is a beautiful bone (Fig. 5.3). Part of the right side is missing in the type of *Monoclonius crassus*, but as the bone is symmetrical most of the information is preserved. It is a large, fan-shaped bone with huge parietal fenestrae. It is about half a meter in length along the midline, and its width may be estimated at 830 mm. For its size, it is somewhat delicate

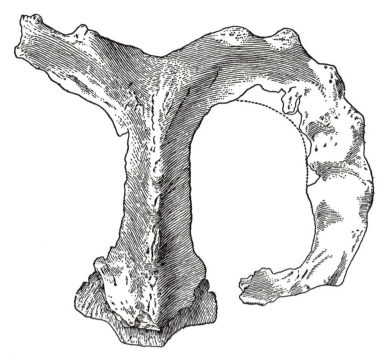

FIG. 5.3. Parietal frill: type specimen of *Monoclonius crassus* Cope, American Museum of Natural History. (From Hatcher et al. 1907.)

in construction, with five rather poorly defined scallops or ornaments along the edge. There are also several more or less prominent bumps on the median parietal bar between the parietal fenestrae. The back edge of the parietal is not very thick, measuring only 18 mm on the midline and 22 mm nearby. The significance of these details will become apparent later. The orbital horn described by Cope was cataloged under a separate number, and Hatcher was firmly of the opinion that it did not pertain to the type.[19] I agree. Thus this important detail too is wanting for *Monoclonius*. Hatcher described no further skull elements but proceeded to the sacrum, fused cervical vertebrae, and isolated miscellaneous vertebrae. Of the sacrum, he noted the following: "although Cope in his description has referred it to the same skeleton with the parietals and the postfrontal [*sic*] described above, there would seem from such general characters as size, color, degree of petrifaction, etc., to be little doubt that all three pertained to as many different individuals."[20] The alleged skeleton includes an ilium, an ischium, a shoulder girdle, a humerus, two femora, and a tibia and fibula. We know these to be ceratopsian, and all may potentially come from *Monoclonius*, but as there is no documented association, it is dangerous to infer one.

The material pertaining to *M. recurvicornis* had not been amplified since being figured by Cope in 1877 and thus still consisted only of an occipital condyle, two orbital horn cores, and a nasal horn core. Hatcher compared the orbital horn cores with those of Marsh's *Ceratops montanus* and advocated the transfer of Cope's species to Marsh's genus, as *Ceratops recurvicornis*. This step might be justified if *Ceratops* were regarded as a valid genus, but it too has spent most of its years in the taxonomic limbo of "nomen dubium." Thus *M. recurvicornis* and *C. recurvicornis* are equally forgotten.

M. sphenocerus is, in effect, based only on the nasal horn core and associate premaxilla. There is no sign whatever of the "numerous parts of the skeleton" to which Cope referred. It is certainly an interesting animal, and certainly a centrosaurine, but there is no compelling reason to associate it with *Monoclonius*. It too is in limbo. At some future date, a new find may associate a nasal horn like this one with a skull showing a full complement of diagnostic characters, including orbital horns, squamosal, and parietal. Until then, we can only speculate—an activity in which we paleontologists are not constitutionally averse to indulging from time to time.

MONOCLONIUS AND THE BIRTH OF CENTROSAURUS

This completes a review of Cope's understanding of *Monoclonius*. But as time passed the confusion only got worse. To paraphrase what my good friend Philippe Taquet wrote of iguanodonts from Europe, the murky swamps in which *Monoclonius* waded in the Cretaceous were nothing compared to the taxonomic swamp in which it wallows today! For a while there was a tendency to name everything that came from Judithian beds *Monoclonius*. For instance, the first dinosaur species described in Canada were horned dinosaurs described by Lawrence M. Lambe in 1902. He described three fragmentary skulls as—what else?—species of *Monoclonius*!

Dinosaur finds had come steadily from the American West since 1855, but dinosaur discoveries were a little slower in coming from western Canada. The exploration of the Canadian West by Geological Survey explorers had begun in the 1870s. Initially fossil finds were sent to Cope for identification. As we saw in Chapter 4, Lawrence Lambe was the first Canadian scientist to collect fossils in Alberta, to recognize their significance, and then to describe them. His first major monograph was published in 1902. Many vertebrate specimens, principally reptilian, were found; these were described and handsomely illustrated in an important monograph. Among the fishes, turtles, and crocodiles were dinosaurs, including horned dinosaurs.

Without drawing specific comparisons with Cope's *Monoclonius crassus,* Lambe erected three new species: *Monoclonius dawsoni, Monoclonius canadensis,* and *Monoclonius belli.* Subsequent finds of good specimens further helped to elucidate the latter two species. These turned out to be the chasmosaurine *Chasmosaurus belli,* as we discussed in Chapter 4. *Monoclonius dawsoni* has been harder to resolve, but it is to this species that our attention now must turn. Lambe referred two specimens to *Monoclonius dawsoni* (honoring George Dawson, who had found the first dinosaur remains in Canada in 1874): a poorly preserved skull of one specimen and *the posterior crest of another,* both of which he collected in 1901. The badly crushed skull lay on its right side; its poorly preserved crest could provide little information. Lambe drew instead from the second specimen, which he illustrated separately—an error that he soon realized. There was a prominent nasal horn core 33 cm high that was recurved; that is, it curved backward toward the eyes rather than forward toward the beak. There was an upper jaw, a quadrate, an occipital condyle that measured 58 mm in diameter, and a set of bones from both sides of the skull that define the eye socket. The orbital horn cores were very low; in fact, Lambe recorded that there were none.[21] He was very impressed with a large, saddle-shaped bone, which he called the "posterior crest." He also included descriptions of some skeletal material, including a shoulder girdle and a sacrum that he "provisionally" associated with the species. This provisional association scares me away from considering them.[22]

By 1904, Lambe had had an important insight concerning the separate frill and horn core that greatly clarified his interpretation of *Monoclonius dawsoni:*

> These remains were thought to belong to the same species, but it has since become evident that the separate posterior crest and horn core belong to a distinct species and probably also to a different genus. . . . For the form represented by the posterior crest the writer has proposed the generic name of *Centrosaurus* in allusion to the remarkable inwardly directed hook-shaped processes springing from the posterior border of

the frill. The species has been designated by the name *apertus* in reference to the very large openings or fontanelles lying wholly within the parietal expansion.[23]

Lambe was very impressed by the parietal, which he illustrated twice in dorsal view and once in lateral view. Curiously, it is about as complete as the parietal of *Monoclonius crassus*, with which *Centrosaurus apertus* may be directly and most satisfactorily compared. Both are well preserved and both are about two-thirds complete, *Monoclonius* lacking a portion of the right side, *Centrosaurus* a portion of the left side. In neither case was a squamosal found. Lambe was impressed not only by the inward-directed hooks on the caudal border of the parietal, but also by the thickness of this bone, 6 cm (2.5 in.) thick along the midline at the back of the frill. The similarity in size of the frills of the name-bearing specimens of *Centrosaurus* and *Monoclonius* is uncanny. Their lengths are identical within 10 mm (485 mm for the former versus 495 mm for the latter, measured along the midline); the dimensions of the large parietal fenestrae are also very close. The frill of *Centrosaurus*, however, seems significantly wider (each half being nearly as wide as the bone is long), giving it a rounder appearance in dorsal view. The crest of *Monoclonius* is of simpler construction. Its caudal border is quite thin, 18 mm at the caudal margin where *Centrosaurus* is 60 mm thick, and there are no hooks or excrescences along the caudal border, only the familiar pattern of gentle scallops or ornaments.

Lambe made an interesting error in his description. He described what he believed was a somewhat distorted, compressed, and incomplete horn core 30 cm in length. It was broken at the base, which is 40 mm thick and deeply fluted longitudinally. He took it to be a nasal horn, but he was mistaken. Barnum Brown visited Lambe in Ottawa in the summer of 1909, in preparation for his explorations in Alberta the following year. When Brown examined the parietal of *Centrosaurus apertus*, he noticed that there was a fresh break on the forward face of the thick bone on the back of the parietal. Brown happened to notice that the distorted horn had a fresh break, too. When the two broken surfaces were opposed, they fitted together perfectly! As Lambe expressed it, "The defensive frill or crest of *Centrosaurus*, so singular in its general form and contour, has lately been found to be even more grotesque than it appeared at the time of its discovery."[24]

He thus recognized that *Centrosaurus* bore a long, strong spur that points forward and overhangs the entire parietal fenestra, but only on

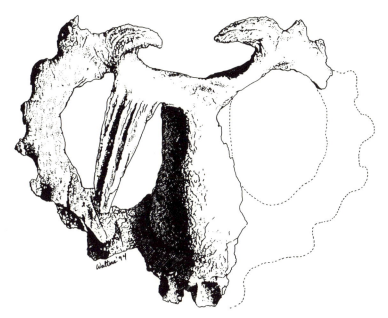

FIG. 5.4. Parietal frill: type specimen of *Centrosaurus apertus* Lambe, Canadian Museum of Nature. (Robert Walters after Lambe 1910.)

the right side (Fig. 5.4). On the left side, the bone is smooth, showing no indication of such a structure. Lambe believed that the caudal hooks that he documented earlier were "probably of some use in a protective sense," but the forward processes were puzzling. Presumably they did not show above the surface of the frill, and he could not attribute any defensive function to them. He attempted to relate them to a trend toward closure of the parietal fenestrae that culminated with *Triceratops*, but he was no more convinced of this than we are. How could anyone mistake such a distinctive animal for *Monoclonius*?

Another of Lambe's errors is incorporated into the titles of both of his 1904 papers, each beginning "On the squamoso-parietal crest. . . ." He believed that the squamosal was fused with the parietal, but Hatcher pointed out this error to him, presumably on the basis of a careful reading of the 1902 paper.[25]

The magnificent Hatcher, Marsh, and Lull monograph was published in 1907. The scope of that monograph is remarkable, in that it summarizes the entire fossil record of ceratopsians known at the time of completion of the manuscript. Included are both transcriptions of the original descriptions and reproductions of the original figures (these are particularly welcome features, as papers published prior to 1900 are frequently not readily accessible to researchers today). The treatment is confusing and slightly ambiguous with regard to *Centrosaurus,* which is acknowledged in a lengthy footnote written by Lull. It seems there should have been ample opportunity between 1904 and 1907 to accord it full status as a genus in the index and headings, but this was not done. However, Hatcher wrote the following of the parietal that Lambe described in 1902: "The peculiar parietal . . . associated by Lambe with the type of *M. dawsoni* I regard as pertaining to a distinct species and perhaps also to a distinct genus." Lull in his footnote opined that Lambe's decision to name *Centrosaurus* was "mainly due to Hatcher's suggestion." Hatcher also pointed out that the sacrum Lambe included was not ceratopsian, but was probably hadrosaurian.[26] Lambe, in his second *Centrosaurus* paper, published July 7, 1904, wrote:

> The writer's attention was drawn . . . to the wrong interpretation in the original description of the nature of the side extensions, by Mr. J. B. Hatcher, Curator of the Department of Vertebrate Palaeontology of the Carnegie Museum, Pittsburgh, who, with his intimate knowledge of the *Ceratopsidae,* is justly regarded as one of the foremost authorities on this interesting family.[27]

Lambe was a good scientist, a careful observer, and a diligent recorder of facts. His papers are far more satisfactory than Cope's, because he documents specimens with numbers, localities, and excellent figures. He was not, however, a distinguished collector. The quality of the specimens available to him for study improved radically when C. H. Sternberg and sons began in 1912 to collect for the Geological Survey of Canada in Ottawa. For the first time, complete skulls of Canadian dinosaurs were available, which Lambe was pleased to study. He described the splendid *Styracosaurus albertensis* in 1913, *Chasmosaurus belli* in 1914, and *Eoceratops canadensis* in 1915. A fine skull of *Centrosaurus* was collected in 1913 and another in 1914. The latter was figured in Lambe's 1915 monograph.[28] The new specimens confirmed Lambe's interpretation of *Centrosaurus* as a distinct, valid animal. For the first time, centrosaurine fossils showed parietal, squamosal, nasal and orbital horn cores, and jugal in single, articulated, unambiguous specimens. In short, all of the diagnostic characters (synapomorphies in modern parlance) were there.

Matters might have rested and clarity might have prevailed, but Barnum Brown had other plans. The view at the American Museum continued to favor Cope, and Brown's new discoveries in Alberta were interpreted as new finds of *Monoclonius*. Brown entitled a 1914 paper, "A complete skull of *Monoclonius*, from the Belly River Cretaceous of Alberta." In beginning it, he wrote:

> A rare specimen secured by the American Museum Expedition of 1912 is a complete skull of *Monoclonius* from the Judith River (Belly River) exposures on the Red Deer River, one mile below the mouth of Berry Creek. It is unusually perfect, lacking only the vomers, and the sutures are for the most part still well defined, a condition that enables us to understand more clearly the structure of the primitive Ceratopsian skull. Compared with other known skull material it shows the range of variation in horns, and in the peculiar outgrowths on the back of the crest in this genus.

Brown began his paper by reviewing Cope's species of *Monoclonius*, quickly dismissing *M. fissus*. He agreed with Hatcher that *M. recurvicornis* should be transferred to *Ceratops*. He believed that his Alberta skull, with its tall nose horn, "leaves little doubt" that *M. sphenocerus* is synonymous with *M. crassus*. Thus, "This leaves only one identifiable species of the genus from the Judith River beds of Montana, *M. crassus*, although future discoveries will probably disclose as great a variety of horned dinosaurs there as further north in Canada." Unfortunately, this prediction has yet to be realized, and more than one hundred years after Cope's discoveries, we are still none the wiser about *Monoclonius* from Montana.

Brown set to the task of analysis. He had visited Ottawa and with Lambe's cooperation examined the type specimens of *Monoclonius dawsoni* and *Centrosaurus apertus*. He concluded, "With the complete skull as a guide I can see no characters that distinguish *Centrosaurus apertus* from *M. dawsoni* and consider the former a synonym of the latter." He then proposed a formal diagnosis of *Monoclonius* Cope:

> *Generic characters:* Skull small to medium sized with three horns; nasal horn large, curved or straight, rising from middle of nasals immediately above the posterior border of the nares; supraorbital horns small or incipient and flattened on the outer surface. Nasals large; nares nearly separated by osseus septum formed by premaxillaries and nasals. Premaxillaries deep with vertical plate forming septum non-fenestrated. Crest composed of short, broad squamosals and extension of elongate coössified postfrontals ("parietals")[29] perforated by large fenestrae; each fenestra wholly within the boundary of the postfrontal. Margin of crest

crenulated, each prominence bearing a separate ossification. A pair of long curved hook-like processes on posterior border of postfrontals.

Brown named his new species *Monoclonius flexus* and listed its specific characters: "Skull medium sized. Nasal horn long and curved forward. Supraorbital horns short." He described the skull carefully. He was very impressed by the nasal horn, which he believed was longer than that of any described species of the family. (It is barely longer than that of *M. sphenocerus,* by 15 mm, but is about 33 percent shorter than that of *Styracosaurus,* described by Lambe the year before.) The orbital horns are prominent and asymmetrical, the left horn being roughly conical and 107 mm above the rim of the orbit, the right one being lower, only about three-quarters the height of the former, and more rounded and blunt. He noted the similarity of the crest with that of *Centrosaurus apertus,* which he insisted on referring to as *Monoclonius dawsoni.* He described the squamosal as a "thin irregular quadrilateral plate" with a long axis oblique to the long axis of the skull. There are five "projections" (variously termed scallops, ornaments, or epoccipitals) on the free edge of the squamosal on the left side, and four on the right side, another example of asymmetry in the same specimen. As had been inferred from the parietals alone, the squamosals are much shorter than the parietal, which forms the caudal two-thirds of the crest. As in *Centrosaurus apertus,* the parietal thickens to 6 cm at the back and supports the distinctive hook-like processes at the rear that curve toward the midline. Also as in *Centrosaurus apertus,* there is the sturdy "pseudohorn" that projects forward over the parietal fenestra, this time only on the *left* side. Brown was very concerned with the structure of this process and thought its texture was like that of the bony tendons along the vertebral column of many dinosaurs. Thus he speculated that this intriguing structure anchored huge jaw-closing muscles. There are about thirty-five vertical rows of teeth in the maxilla. The premaxilla lacks teeth. This character had already been firmly established by Marsh in *Triceratops* and is a general character of all ceratopsids. Brown provided several measurements that convey the sense that this was a good-sized animal. The total length of the skull is 1.57 m, and it is in fact one of the longest of the Alberta centrosaurine skulls (Fig. 5.5).

In his brief analysis, Brown made a very important statement:

Now the type of *Monoclonius crassus* is the posterior half of a crest or frill of an old individual, and it had *evidently been subjected to considerable abrasion during fossilization* so that the sutural borders of the epoccipitals

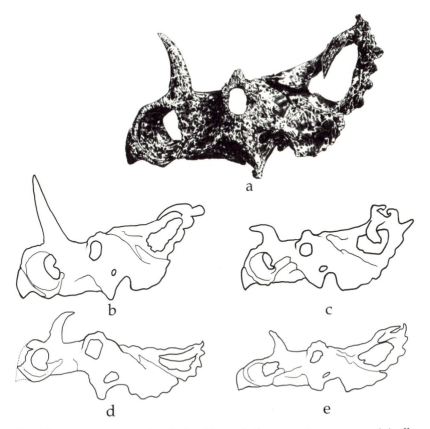

FIG. 5.5. *Centrosaurus apertus* skulls. Although these specimens were originally referred to several different species, today only a single species, *Centrosaurus apertus*, is recognized. (a) *Centrosaurus "flexus,"* American Museum of Natural History; (b) *Centrosaurus "nasicornus,"* American Museum of Natural History; (c) *Centrosaurus "flexus,"* Canadian Museum of Nature; (d) *Centrosaurus "dawsoni,"* Canadian Museum of Nature; (e) *Centrosaurus "longirostris,"* Canadian Museum of Nature. (Robert Walters after Lambe 1915; Brown 1914b, 1917; Sternberg 1940; Dodson 1990a.)

are indistinct. Moreover, the two posterior prominences, one on either side of the central concave border, *are the remains* of the posterior hook-like processes complete on the type of *M. dawsoni* (*Centrosaurus apertus*) and on the present specimen. (emphasis mine)[30]

This interpretation is crucial. Brown assumed that *Monoclonius crassus* originally had been just like *Centrosaurus apertus*, only damaged by erosion. Specifically, a thickness of some 45 mm of bone had been stripped from the rear margin of Cope's specimen, neatly and symmet-rically removing the caudal hooks and the forward processes, but somehow not leaving evidence of this abrasion on the bone surface, which appears to the casual observer as having a natural, undamaged outer surface. Could this have been wishful thinking on Brown's part? Are there other undamaged specimens that show the thin, simple frill of *Monoclonius crassus* reinforcing the morphology described by Cope? Ironically, Brown may have collected just such a specimen himself, in Alberta in 1913. AMNH 5442 is a fine, slightly dainty skull with a good, simple frill and a squamosal, but unfortunately missing the nasal horn and snout. Not perhaps of exhibition quality, the skull has never been figured or described and so has played no role in discussions about the validity of *Monoclonius*. In a basement storage room in the bowels of that great museum, it keeps company with thousands of pounds of dinosaur wealth. It was twenty-five years before any new specimens came to light to help elucidate the status of *Monoclonius*.

It may well be surmised that Lambe was not pleased by Brown's treatment of his genus. He was too genteel to object that by Brown's diagnosis of *Monoclonius, Monoclonius crassus* itself is eliminated from its own genus, because it did not bear "a pair of long curved hook-like processes on posterior border of postfrontals." In 1915, Lambe made his most important synthetic statement on ceratopsians, having now named four himself (*Centrosaurus*, 1904; *Styracosaurus*, 1913; *Chasmo-saurus*, 1914; *Eoceratops*, 1915). In this paper, he named three subfamilies of the Ceratopsidae: the Centrosaurinae, the Chasmosaurinae, and the Eoceratopsinae (the latter no longer considered valid). Lambe's Centro-saurinae included *Centrosaurus, Styracosaurus,* and *Brachyceratops,* which had been described by Gilmore in 1914. He emphasized the large nasal horn, small orbital horns, and parietal extending much farther back than the squamosals. The centrosaurines have come to be known as the short-frilled or short-squamosaled ceratopsids. Not surprisingly, his view of generic distinctions was a trifle simplistic in terms of what we know today. He regarded the shape of the nasal horn as diagnostic of genera: *Centrosaurus* had a horn that curved forward, *Styracosaurus* had a horn that was straight, and *Brachyceratops* had a horn that curved backward toward the eyes (said to be recurved). Would that it were so simple!

With regard to *Monoclonius*, Lambe did an interesting thing. He did not rebut Brown directly. Rather, he transferred his *Monoclonius dawsoni* to Gilmore's new genus, *Brachyceratops,* because in both the horn was recurved. Cope's *Monoclonius sphenocerus* was transferred to *Styracosaurus* because of its tall, straight horn. With regard to *Monoclonius crassus*, Lambe reviewed Cope's material and repeated Hatcher's reservations. He concluded: "With *Monoclonius* its non-employment as a generic term is considered advisable on account of the composite nature of the material described which lacks data both as regards localities and the association of the skeletal elements in the field."

To my mind, Lambe missed the argument he might have made—that the parietal of *Monoclonius crassus* is actually well preserved, not damaged, and qualitatively different from that of *Centrosaurus apertus*. Instead, he in effect argued that without a nasal horn, one could not be sure. There are no reliable species of *Monoclonius*, which is thus a doubtful name. He therefore addressed this eyebrow-elevating statement to Barnum Brown: "Reference may here be made to the opinion expressed by Mr. Brown in a recent paper that *Centrosaurus apertus* is a synonym of *Brachyceratops dawsoni*." Brown may have been surprised to have such a statement attributed to him, as he did not cite *Brachyceratops* at all in his 1914 paper. In any case, Lambe (correctly in my opinion) regarded Brown's description of *"Monoclonius flexus"* as supporting in every way his original description of *Centrosaurus apertus*. He thus regarded Brown's *Monoclonius flexus* as another specimen of *Centrosaurus apertus*.[31]

Was Brown chastened? Not at all! In 1917 he wrote his second paper on centrosaurines from Alberta. This is a valuable paper, as it describes a complete skull and skeleton collected in 1914 (Fig. 5.6). Brown pointed out that at the time of publication of the Hatcher, Marsh, and Lull monograph in 1907, not a single complete skeleton of a ceratopsian was known, and that the structure of the feet and length of the tail were simply unknown. He repeated and emended the diagnosis of *Monoclonius* he had written three years earlier, adding details about the skeleton for the first time. Among them were the following:

Vertebral column composed of 77 vertebrae comprising 21 presacrals, 10 sacrals and 46 caudals.

Carpus with 2 ossified carpals.

Manus with five digits, the inner three bearing hoofs, phalangeal formula, I = 2, II = 3, III = 4, IV = 5, V = 2.

Tarsus with 4 ossified tarsals in two rows.

Pes with four functional digits bearing hoofs and the fifth vestigial, represented by reduced metatarsal. Phalangeal formula, I = 2, II = 3, III = 4, IV = 5.

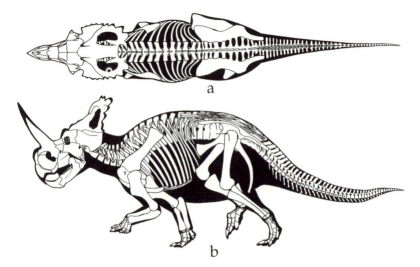

FIG. 5.6. *Centrosaurus* skeleton. (a) Dorsal view, (b) lateral view. (Gregory S. Paul.)

Brown's descriptions are quite good. His account of the vertebral column included numerous useful details. For example, in the neck vertebrae, there is a shortening of the centra back to the eighth, which is the shortest. The tail has transverse processes back to the twenty-third centrum. Brown believed that the general structure of the tail indicated that it was used neither for balance nor for swimming. He described a mass of ossified tendons along the vertebral column, particularly over the sacrum. They extended forward as far as the sixth vertebra of the dorsal series but did not extend backward beyond the sacrum. They did not show the regular, lattice-like organization of those in hadrosaurs. The explicit comparison throughout was with *Triceratops,* and the similarity is considerable.

It is fair to say the ceratopsids do not differ greatly among themselves in the structure of the postcranial skeleton. There are differences in size and robustness among genera and species, with relatively minor differences in proportions. Brown sniffed that his 1906 paper on the structure of the sternum in *Triceratops* had been overlooked in the 1907 monograph, but this is not too surprising in view of the fact that two of the three authors were dead by 1906 and that Lull clearly had had his hands full in successfully preparing the monograph for publication. He noted that the tibia of *Monoclonius* was only 140 mm shorter than the femur, whereas in *Triceratops* the tibia is half the length of the femur. Brown showed definitively that the foot of ceratopsids was four-toed, not three-toed as the Ceratopsia monograph and a mounted skeleton of *Triceratops* at the Smithsonian showed.

Brown did an admirable job of presenting measurements, which I heartily applaud.[32] Some salient ones:

Total length, snout to tail	5.16 m
Femur length	74 cm
Tibia length	60 cm
Metatarsal III length	21.5 cm

Thus Brown's specimen was of an animal about 5.2 m (17 ft) long and 1.5 m (5 ft) at the hips.

Let us turn now to the skull to learn more about this animal. Brown started with an encouraging statement of philosophy, which resonates well with modern views about the need to incorporate biological variation into definitions of fossil species. He editorialized:

> Most writers in describing species of the Ceratopsia have attached greater importance to the development of horns and accessory frill growths than to me seems warranted. These parts are subject to great individual variation and too much stress should not be laid on such characters when they are not corroborated by differences of the fundamental skull elements.

His comparison now was not with *Triceratops* but with his *"Monoclonius" flexus*. For Brown, the new specimen was strikingly different in the facial region. The two skulls were of similar width and height, but the new skull was 15 cm shorter, at 1.42 m total length. It has a long, straight, massive, laterally compressed nasal horn that resembles the nasal horn of Cope's *Monoclonius sphenocerus*. The orbital horns are low and blunt, less well developed than in *"Monoclonius" flexus*. The excrescences of the frill, so typical of *Centrosaurus apertus*, are in evidence, both the hook-like processes and the free spurs (as he called them), for the first time two in number, one projecting forward over each parietal fenestra. There are five rather coarse epoccipitals on each squamosal, in contrast to the dainty ones of *"Monoclonius" flexus*, and four on each side of the parietal. There is even evidence of a partial sclerotic ring, the bones that support the eyeball. There are twenty-nine tooth positions in the maxilla and twenty-nine (right) or thirty-one (left) positions in the lower jaw. Unfortunately, Brown did not figure the skull, other than to present a photo of the entire panel-mounted specimen. The left side of the skull is unavailable for inspection and undocumented. Panel mounts are not a paleontologist's best friends!

Was Brown convinced that *Monoclonius* was not a reliable genus? Let us say that Cope's ghost was still strong at the American Museum. Brown did not even cite Lambe's paper. In fact, he did not cite any literature other than the Ceratopsia monograph of 1907 and his own paper of 1914. Citing his paper, he wrote:

> I expressed the opinion that *M. sphenocerus* was a synonym of *Monoclonius crassus,* a view I continue to hold, but in the light of material recently collected it seems advisable to await more complete material from the type locality of the Judith River species before referring doubtful Belly River Ceratopsians to Judith River species. I shall therefore describe this complete skeleton as a new species.

Accordingly, he named *Monoclonius nasicornus,* in reference to its large nasal horn. My personal reaction is one of incredulity that the solution to the taxonomic problem is to *increase* the number of species. Is this the same writer who wisely counseled his colleagues not to be misled by minor variants? However, this is my modern overprint on history. It is fair to say that the importance of biological variation was not at that time fully appreciated, either by paleontologists or by the majority of biologists. Just to make his position clear, Brown proceeded to describe yet another species, *M. cutleri.*[33]

The second specimen described in the 1917 paper was the back half of a skeleton, the front having been destroyed by erosion, so that only a few skull fragments remained. There were forty-nine vertebrae, a few bones from the forelimb, the pelvis, and much of the hind limb. He referred the skeleton to *Monoclonius* on the basis of "coössified anterior ends of nasals and premaxillaries which are diagnostic." Oh? Brown again made a careful disclaimer: "Sufficient material is not yet available in this family to determine what species characters in the skeleton are constant nor are we able to correlate skull characters with those of the skeleton in many instances." He then proceeded to list characters that he believed differed significantly from those of *"Monoclonius" nasicornis.* The sternal plate is longer, narrower in front, and wider behind. The ilium is narrower and more elongate, the post-pubis is relatively longer, and the ischium is most distinct from that of *"Monoclonius" nasicornis,* being long and fairly straight, with a sharp downward curve at the tip. The hind limb is, I admit, interesting, in that the tibia is very much shorter than the femur, barely 60 percent of the length of the latter. In this regard, it resembles *Triceratops* and contrasts with *"Monoclonius" nasicornis*. This was a good-sized animal, with a femur 80 cm in length but a short tibia of only 50 cm. With a metatarsal III measuring 23 cm, the height at the hips was 1.53 m, about the same as that of *"Monoclonius" nasicornis*. Brown also described a section of skin impression around the knee, consisting of small polygonal (five- or six-sided) tubercles and large round tubercles. The tubercles were low and were "disposed in rows over a part, probably the ventral surface of the body." History has not looked kindly on *Monoclonius cutleri,* and no subsequent finds have elucidated it. Its referral to either *Monoclonius* or *Centrosaurus* seems none too secure. It is usually ignored, but could in principle be resurrected if a skeleton with an interesting skull were to show these hind limb proportions.[34]

The next contribution on the subject of *Centrosaurus* came from Toronto in 1921. With exciting scientific results being achieved by the Geological Survey of Canada and the American Museum of Natural History, it was opportune for Canada's leading city to become involved as well. William A. Parks (1868–1936) was a geologist and invertebrate paleontologist at the University of Toronto who made his first expedition to the dinosaur beds of Alberta in 1918 at age fifty. He succeeded in collecting a good skeleton of the duck-bill *Kritosaurus* for the Palaeontology Museum (which later joined with the Zoology Museum and the Museum of Archaeology to become the Royal Ontario Museum, often simply known as the ROM). The following year, Levi Sternberg, the youngest of the Sternberg brothers, was hired by the museum, and thoroughly professional collections and exhibits were mounted for many years thereafter. Parks continued to enjoy forays into the dinosaur beds, and many significant papers resulted until his death. An early paper was on the head and forelimb of a specimen of *Centrosaurus apertus.*

Parks disavowed the intention of taking a firm position on the controversy of *Monoclonius* versus *Centrosaurus,* although he wryly observed that *Monoclonius crassus* failed to conform to the definition of *Centrosaurus apertus* established by Lambe. His inclination was to support Lambe's position until evidence to the contrary was found. The Toronto specimen was found in 1919, in the extensive badlands south of the Red Deer River within several kilometers of where Lambe's and Brown's specimens had been discovered. The Toronto skull is one of the smallest skulls of *Centrosaurus* known, although at 1.24 m in length it is by no stretch of the imagination small in any absolute sense. However, the sutures are very clear. The specimen comprises, in addition to the skull, a forelimb and bones of the hyoid apparatus that supported the tongue and larynx. Parks regretted that Lambe had not described the Ottawa skulls of *Centrosaurus*, a regret that I continue to share today, as this has never been done. Parks wrote that such differences as could be found from the Ottawa and New York skulls

> may easily be accounted for by the less advanced age of our skeleton. In fact, there is sufficient difference in the two sides of the frill to equal in importance the differences observed above. The parietal part of the frill on the left side shows a peculiar overlap in the bone outside the fontanelle and this orifice is itself much smaller than on the other side.

There are superb, high-fidelity ink-wash illustrations, but the skull is only seen on its right side, both in illustration and in exhibition mount today at the ROM. The asymmetry of the left side is not shown. The nasal horn is erect and heavy. The two orbital horns are essentially symmetrical and are of moderate size. The two hyoid bones are slender bony rods 168 mm long that fit in the throat region immediately behind the jaws.[35]

Richard Swann Lull (1867–1957) trained in entomology but became a professor of paleontology at Yale University in 1906, the scientific heir to the great dinosaur collection compiled by Marsh, with whom he authored a monograph but whom he never met. Lull returned to the subject of ceratopsians in 1933 when he published an important monograph, "A Revision of the Ceratopsia or Horned Dinosaurs." Like the 1907 monograph, this was an attempt to summarize the entire fossil record of ceratopsians known as of that date. It was a noble undertaking, as many new genera and species had been described by that time. Lull provided the rationale and paid tribute with these words:

> Of the older forms, where fragmentary material only was known, we now have perfect skulls, and in several instances, more or less complete skeletons with occasional impressions of the external covering. Perhaps the most outstanding explorer of all is Barnum Brown of the American Museum, whose remarkable skill both in discovery and exhumation has enriched our collections amazingly, especially by forms derived from the Belly River beds of Alberta. Then that noted family of collectors, the Sternbergs, should be mentioned, headed by the veteran Charles H., and ably seconded by his several sons.[36]

Unlike the 1907 classic, however, this monograph was rather minimally illustrated with simple line drawings, and with few of them at that. The old standards were slipping. In consequence, it never was an entirely satisfactory publication, and it does not hold a candle to the earlier book in terms of quality of description, illustration, or general level of insight.[37] I am perhaps a little harsh. One feature that I have always found useful is the listing of specimens collected, with date of collection and museum specimen number. It would be very useful to see this list brought up to date.

For our purposes, however, two points are noteworthy: a description of another skull and skeleton from Alberta, and yet more opinions on the same old problem. Lull was not a field man himself. The Yale specimen was collected by Barnum Brown from the Red Deer River in 1914 and subsequently purchased by Yale from the American Museum. It is exhibited today in the Great Hall of the Peabody Museum of Natural History, underneath a vast and majestic mural by Rudolph Zallinger, who in 1949 received a Pulitzer fellowship in art. The mural shows a timeline of vertebrate and plant life on earth from the Late Devonian to the end of the Cretaceous. The *Centrosaurus* skeletal mount is, shall we say, unique. The left side presents the skeleton, but the right side is a reconstructed fleshy body.

As far as taxonomy was concerned, Lull sat magnificently, squarely, securely on the fence. In discussing *Monoclonius,* he opined, "On the basis of present evidence, I am inclined to consider *Centrosaurus* a local, Belly River phase of *Monoclonius,* possibly of sub-generic rank." But he quickly added, "The nasal horn presents little difficulty, as it is not demonstrably absent from Cope's type, but the processes of the crest are, to my mind, a serious obstacle to the acceptance of generic identity, as they are demonstrably absent in the type of the genus. Geographically, the two genera come from areas at least 250 miles apart."

He thus decided to designate *Centrosaurus* as *Monoclonius (Centrosaurus) flexus* (Fig. 5.7). In this practice he has been followed by no one. The Yale specimen, YPM 2015, is not large. In fact, it is comparable to the Toronto skull, with an overall length of 1.35 m. Lull described this skull as very similar to Brown's type specimen of *"Monoclonius" flexus,* but it has two forwardly directed processes over the parietal, as in *"Monoclonius" nasicornis,* rather than one, as in *"Monoclonius" flexus.* The nasal horn curves forward, and the orbital horns are asymmetrical. The left orbital horn is well formed, but the right is rudimentary. The squamosal has five prominences on the free border. The parietal is of typical form.

FIG. 5.7. Reconstruction of *Centrosaurus*. (Robert Walters.)

After a good description of the skull, illustrated by a lateral view and one of the underside and palate, Lull proceeded to a thorough description of the skeleton, with an extensive set of illustrations of limb bones and representative vertebrae back to the sacrum, but with no tail vertebrae illustrated, as only a few were preserved. Despite the small size of the skull, the skeleton is that of a good-sized animal. The femur, at 79 cm long, is slightly greater in length than that of *"Monoclonius" nasicornis*. The tibia has a length of 55 cm and MT III is about 21 cm long, so the height at the hips is similar, indicating an animal about 1.5 m at the hips. Lull estimated the length of his specimen at "17 ft. 8 inches" (5.4 m), slightly longer than the AMNH skeleton of *"Monoclonius" nasicornis*. The limb proportions of the Yale specimen are intermediate between those of *"Monoclonius" nasicornis* and *M. cutleri*. For instance, the ratio of tibia to femur length in *"Monoclonius" nasicornis* is 0.81, and in *M. cutleri* the ratio is 0.63. In the Yale specimen the value is 0.70. This intermediate value suggested to Lull that *M. cutleri* was not well founded. He wryly observed, after comparing the Yale specimen (which he comfortably referred to as *"Monoclonius" flexus*) with *M. cutleri*,

> These are distinctive features which characterize the Yale skeleton; but without the skull, one would hardly be justified in erecting a new species for its inclusion. The feeling is quite strong, however, that were the skull of *cutleri* known, it would prove to resemble one of those we have included under the species *flexus*.[38]

BRACHYCERATOPS—A PROBLEM THEN AND NOW

Before moving ahead with the narrative of *Monoclonius* and *Centrosaurus*, we will consider another small dinosaur. It must have been exciting to be a student of horned dinosaurs during the second decade of the twentieth century. New kinds of dinosaurs were coming out of the ground from Montana and Alberta at a remarkable pace never since equaled. Charles Whitney Gilmore (1874–1945) had a long and productive career as a vertebrate paleontologist. Gilmore worked first at the Carnegie Museum, but his name is usually associated with the Smithsonian, for it was there that he spent most of his career. He is particularly noted for his contributions to the study of Jurassic giants, although by no means did he neglect the Cretaceous. His name is not associated with pioneer exploration of virgin fossil beds of North America or of some more exotic terrain, nor is he indissolubly linked with every kid's favorite dinosaur. (There did, however, appear posthumously in 1946 a description of the exquisite skull of an animal that he named *Gorgosaurus lancensis*, which has since attracted considerable attention as the recently rechristened *Nanotyrannus*.) Gilmore seems to have been a fine and upright gentleman. He was not involved in any celebrated quarrels or tainted by any personal scandals. I would not go so far as to say he was *boring*, but we may say that no particular colors or hues of his personality are well known to my generation. Or we might say that all of his skeletons are on full view. It has been charged that he invented the "lead-footed" dinosaur. It is true that the skeletal mounts whose erection he supervised at the Smithsonian certainly do have all four feet planted firmly on the ground. Yet I hardly feel that this is a just criticism of the man.

For our purposes, we turn to Gilmore's expedition to the Blackfoot Indian Reservation in north-central Montana, near the gorgeous, ice-sculpted peaks of Glacier National Park. In the summer of 1913, he discovered in the badlands adjacent to the Milk River the partial skeletons of five small ceratopsids, in his account "the smallest known representative among the Ceratopsian dinosaurs." Gilmore did not yet know of *Leptoceratops* or *Protoceratops*. In 1914, he named his new dinosaur *Brachyceratops montanensis* ("Montana short horn face"). His 1914 paper was preliminary, establishing the name and illustrating the nasal horn, the rostral, and the dentary. Most importantly, he showed a reconstruction of the skull in both side view and top views. He estimated the length of the skull as 565 mm. This is not a tiny animal by any means,

153

certainly not one just out of the nest, but its skull is about one-third the size of the skulls of *Centrosaurus* that were soon to become known, and a quarter or less the size of Marsh's giant horned dinosaurs, *Triceratops* and *Torosaurus*. It certainly seemed very small. Gilmore acknowledged that these might well be the remains of young animals. He noted that the open sutures of the skull and vertebrae indicated immaturity. Nonetheless, he was very impressed by the shortness of the face and the division of the nasal horn cores into left and right halves. *Brachyceratops* seemed demonstrably different from *Triceratops*, in which the nasal horn core may have grown from a separate bony center. The orbital horns are low but well formed. Gilmore noted the possibility that *Brachyceratops* might have something to do with *Monoclonius*. This possibility, given the obvious juvenility of the specimens in question, has perpetually clouded the history of *Brachyceratops*.[39]

Three years later, in 1917, Gilmore published his definitive monograph on *Brachyceratops*, a slender, folio-sized, beautifully illustrated paper, including a 45-cm foldout reconstruction of the complete skeleton.[40] The 600-m-thick, Late Cretaceous formation in which *Brachyceratops* had been discovered was unnamed in 1914, but by 1917 the U.S. Geological Survey had named it the Two Medicine Formation. The Two Medicine Formation was stated by Gilmore to be equivalent in time to the upper part of the "Belly River" (i.e., Judith River) Formation of Alberta.

The five specimens of *Brachyceratops* were found together in a small area measuring about 2 m in diameter. It did not occur to Gilmore that this was a nest, but the thought certainly arises now. He collected eggshell fragments but did not comment on them. (Sixty-five years later, paleontologist Jack Horner, then at Princeton University, earned paleontological immortality by finding eggs, nests, and babies of several kinds of dinosaurs, including the duck-bill *Maiasaura*, elsewhere in the Two Medicine Formation, some 125 km southeast.) The bones of the five *Brachyceratops* skeletons of about the same size were thoroughly commingled, three tails still remaining in natural articulation, of which one was attached to a pelvis and hind limb. The number five was arrived at by the count of nine ischia.

The bones of the skull of the type specimen of *Brachyceratops*, USNM 7951, were exquisitely preserved and entirely disarticulated. This condition permitted thorough analysis of anatomical details, only a few of which we will examine (Fig. 5.8). The nasal horn core is well developed, laterally compressed, and gently recurved. It is divided into left and right halves. The split nasal horn core is a fine character. It undoubtedly

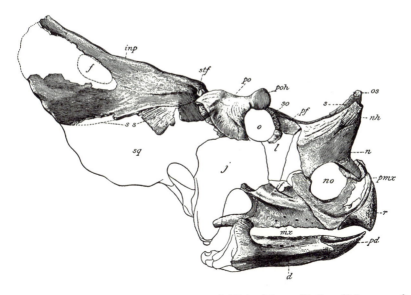

FIG. 5.8. Skull of *Brachyceratops montanensis*, United States National Museum. d, dentary; f, parietal fenestra; if, antorbital foramen; inp, parietal; j, jugal; l, lacrimal; mx, maxilla; n, nasal; nh, nasal horn core; no, nostril or external naris; o, orbit; os, ossicle on top of nasal horn core; pd, predentary; pf, prefrontal; pmx, premaxilla; po, postorbital; poh, postorbital horn core; r, rostral; s, internasal suture; so, suture on prefrontal for supraorbital; sq, squamosal; ss, articular surface for squamosal; stf, supratemporal fenestra. (From Gilmore 1917.)

is a juvenile trait, but one that provides important insight into the nature of the nasal horn core in centrosaurines, which do not show the split owing to normal fusion in adulthood. The postorbitals bear well-formed small horn cores 31 mm high above the dorsal rim of the orbits. These show no sign of separate centers of bone growth for the horn core. As we will see, there are centrosaurines for which such a center is suspected. Regrettably, there is no squamosal. The parietal[41] figured by Gilmore is complete only along the midline and broken on both sides. It nonetheless shows interesting features. The bone is 325 mm long and has a sharp ridge along the midline, which contrasts with the broad, low, rounded shape of the median region of the frill in *Centrosaurus* and *Monoclonius*. The caudal border of the frill is thin, and no ornaments are seen. Gilmore *assumed* that there were parietal fenestrae, but I am not convinced that this is so. I find it curious that no specimen shows any fenestrae unambiguously. The maxilla shows sockets for about twenty teeth, and for the dentary the tooth count appears to be seventeen.

155

Many vertebrae are preserved and illustrated, including representative types from the neck and trunk, a sacrum, and a complete tail. There are fifty vertebrae in the tail of *Brachyceratops*, the highest count in any ceratopsian. Also illustrated are ribs, pelvis, a scapula, an ulna, a radius, and a few phalanges. The ulna shows a strong process at the elbow, the olecranon, which is prominent in ceratopsids but not so in protoceratopsids. There is a nearly complete hind limb. The hoof-like unguals are small and blunt miniatures of the similar bones of the feet of *Centrosaurus*. A complete restoration was presented, indicating an animal 6 ft 9 in. (2.06 m) in total length and 2 ft 4 in. (71 cm) in height at the hips. For comparison with the *Centrosaurus* specimens we have previously considered, we present the following measurements:

Total length, snout to tail	2.06 m
Femur length	33.7 cm
Tibia length	26.8 cm
Metatarsal III length	9.7 cm

It is worth noting that the ratio of tibia to femur length is 0.80, comparable to that of *Centrosaurus* and unlike those, exceeding 1.00, of *Protoceratops* and *Leptoceratops*, which had elongated lower legs. This was an animal the size of a protoceratopsid, but one that did not show the legs of fleet-footed smaller animals but rather the proportions found in larger, presumably slower-moving, animals.

In his analysis of relationships, Gilmore continued to believe that relationship to *Monoclonius* was most likely. He noted that Brown had redefined *Monoclonius* to include forms with hook-like processes on the midline of the parietal. He thought that the split nasal horn "would represent a most important structural difference, provided that such a difference can be shown to exist, but I am of the opinion that when juvenile specimens of *Monoclonius* are found they will also show a similar development of the nasal horn core."[42] This observation was insightful and almost certainly correct, as we know now, not from juvenile specimens of *Monoclonius* but from other ceratopsids.[43]

Gilmore returned to the Blackfoot Reservation to collect in the Two Medicine Formation in 1928 (we will discuss those results later) and for a third time in 1935. He published an intriguing paper in 1939 entitled "Ceratopsian Dinosaurs from the Two Medicine Formation, Upper Cretaceous of Montana." In this paper, Gilmore described a partial specimen of *Leptoceratops gracilis*, which is of some interest (Chapter 7). But of special interest is a disarticulated partial skull that he regarded as the first undoubted *Brachyceratops* specimen since the discovery of the type in 1913. Furthermore, the new specimen, USNM 14765, is nearly twice the size of the type material, apparently confirming the juvenile nature of the type and showing adult characters of *Brachyceratops*. The locality of the new find is "about a mile distant from the spot where the type was discovered, and on the south side of Milk River at approximately the same level in the formation."

Gilmore reported the discovery of several bones not present in the original that enabled him to make a new reconstruction of the skull. Specifically, a complete circumorbital series was found, including a lovely, erect jugal and a lacrimal, postorbital, and prefrontal that define the eye socket completely. The length of the socket was typical of those of mature *Centrosaurus* specimens, 98 mm. The length of the jugal, from the ventral rim of the socket to its tip, was significantly smaller than that of typical adult *Centrosaurus*, in fact about 3 cm shorter than the jugal in the small ROM specimen of *Centrosaurus apertus*. The maxilla is damaged, but Gilmore estimated that the number of tooth sockets was about twenty. If this is an accurate estimate, it is something of an eye-opener, because we might expect that the number of tooth positions would increase with age as it does in many reptiles, including some dinosaurs. An animal the size of *Centrosaurus* would be expected to have about thirty tooth sockets in the maxilla, and an individual midway between the size of the juvenile *Brachyceratops* and an adult *Centrosaurus* might be expected to have twenty-five tooth sockets. If USNM 14765 really only had twenty positions, this count would be important evidence about the unique status of *Brachyceratops*.

The character that really arrests our attention is the parietal. It is *very* large. The bone is imperfect (isn't that always the case?), with much of the left half missing and the thin central portion of the right half gone. Nevertheless there is much information preserved. The bone measures a whopping 57.5 cm in length along the midline, which is perfectly intact and complete. This frill is not a juvenile of anything. My notes record only three other specimens of centrosaurines that show longer measurements along the midline, the longest measuring 64.5 cm. Very striking in this adult parietal is the pattern of coarse ornaments along the edge, perhaps seven in number on each side, contrasting with the smooth margin of the juveniles. There is little thickening of the caudal border as there is in *Centrosaurus apertus*. The thickness of 23 mm on the midline in USNM 14765 compares with 18 mm in the type of *Mono-*

clonius crassus and contrasts starkly with the value of 60 mm in the type of *Centrosaurus apertus*. The type of *Centrosaurus apertus* has gentle scallops along the margin and very large parietal fenestrae measuring 31 cm by 22.5 cm. By contrast, there is no way that USNM 14765 could have accommodated such large parietal fenestrae. If any were present, they were small and widely separated. I believe it is entirely possible that the frill of *Brachyceratops* was solid and not fenestrated at all.

A few other skeletal bones were mentioned but not described, among them a pair of femora, each measuring 58 cm in length. This is unexpectedly short for a mature centrosaurine, whose femora, as we have seen, typically measure 74–80 cm in length.[44]

No further specimens of *Brachyceratops* have been claimed. The status of *Brachyceratops* has been controversial at best. C. M. Sternberg accepted the idea that *Brachyceratops* is a juvenile of *Monoclonius*. Offering his assessment in 1949, Sternberg quotes a letter to him from Gilmore: "Enjoyed your article appearing in the last number of the Journal [*of Paleontology*]. It makes me wonder if my genus *Brachyceratops* is going to remain valid much longer. I am very dubious, looks to me now as though the type may be a juvenile *Monoclonius*."[45]

The trend has been to dismiss *Brachyceratops* as a juvenile *Monoclonius*, a position accepted by Romer in his authoritative textbook of 1966 and maintained by Carroll in 1988. There is nothing improbable about the claim. Certainly the type material of *Brachyceratops* is a juvenile *something*. The two animals lived at the same time in the same general region, although known specimens are separated by some 300 km and arguably lived in different ecosystems. The Two Medicine beds are part of an upland community, members of which include the duck-bills *Maiasaura* and *Hypacrosaurus;* the hypsilophodontid *Orodromeus;* the primitive ceratopsians *Leptoceratops* and *Montanoceratops;* the ankylosaur *Edmontonia;* and the small raptor *Troodon.* The lowland fauna of the Judith River Formation is far more diverse, and faunal exchange between the two communities is more limited than might be expected. My own analysis suggests that *Brachyceratops* is real, but the situation remains complicated.[46] I assume that Gilmore's juveniles and the adult found a mile away belong to the same taxon. If they do not, two things follow: my argument is without merit and *Brachyceratops* is not a valid taxon, if it can be established that the type specimens are juveniles of a taxon described prior to 1914; and Gilmore's adult specimen is something else entirely different and must be renamed. It is clear to me that *Brachyceratops* as I understand it, including the juvenile and adult specimens, has nothing to do with *Monoclonius.* The adult *Brachyceratops* shows no resemblance whatsoever to *Monoclonius,* either the Montana or the Alberta variety. It is a distinct type.

We rejoice in the collection of new specimens, but sometimes it has the effect of complicating what we know, or what we think we know. Such is the case with the remarkable finds made by Museum of the Rockies parties from Montana State University, led by Jack Horner. In 1987 they prospected in Two Medicine exposures near Glacier Park and encountered several ceratopsian bonebeds not far from Gilmore's localities. These bonebeds contain disarticulated remains of juvenile and adult centrosaurines that tell an interesting but complicated story. This we will consider in due course, but not just yet.

MONOCLONIUS IN ALBERTA

Now our narrative moves forward in time beyond the coverage of the two comprehensive monographs on horned dinosaurs. I would be delighted if a monograph existed today that brought the first two up to date, but I am very much afraid I would have to write it myself. I have neither the resources nor the inclination to devote the time that would be required. Monography, like monogamy, is no longer in vogue—yet one more victim of the pace of modern life. Charles Darwin would never have gotten tenure today. Publication is an expensive undertaking, especially because of the meticulous program of illustration required to do it satisfactorily. Yet none of this is to say that monographs are not appreciated today—they most assuredly are. They are treasures upon the bookshelf, and books that we turn to again and again because they never become dated. Morphology does not change.

Alberta is a province gifted by nature. Oil and oil sand lie beneath the skin of chlorophyll at the surface. Although the winter winds blow cold, in summer long hours of sunshine nurture amber waves of grain, pale aqua fields of flax, and sulfur-colored swatches of rape, a seed crop from which the euphemistically named canola oil is pressed. There is a surprising softness to the open rolling land; traveling from south to north, one passes from irrigated cropland to aspen parkland and eventually to boreal forest. In 1937, C. M. Sternberg left the well-watered valley of the Red Deer River and traveled to the vicinity of Manyberries, 175 km to the southeast, in a remote corner of the province. This region, only 50 km north of the Montana border and 100 km from

Rudyard, Montana, resembles more and more the dry wildness of eastern Montana rather than the cultivated civility that one observes from the Trans-Canada Highway in Alberta.

Sternberg had significant success there, collecting several ceratopsid skulls and a skeleton of the splendid duck-bill *Lambeosaurus magnicristatus*. One of the ceratopsids pertained to *Monoclonius*—not the ersatz Alberta kind but the genuine Montana kind. Sternberg quickly produced a preliminary report with no figures in 1938, then issued an important paper in 1940. In the earlier paper, he expressed himself as follows:

> While collecting vertebrate fossils from the uppermost strata of the Belly River series in the southeastern corner of Alberta last summer, my ambition of the last twenty years was realized in the discovery of two skulls which can be identified as *Monoclonius*. . . .
>
> The coalesced parietals so closely resemble those of Cope's type as to leave no doubt of their generic identity. The apex of the median bar carries undulations, and the posterior portion is broad and moderately thin and shows no evidence of any kind of the hornlike processes so diagnostic of *Centrosaurus*.

He was particularly struck by the shape of the nasal horn, which closely resembles that of *Brachyceratops*, being laterally compressed, divided into left and right halves, and recurved. There was no evidence of orbital horn cores, not even a roughened area on the postorbital bone. There are seven thin scallops on each side of the parietal. The parietal fenestrae are damaged; the left one is incomplete, and the right one may have been injured in life and then covered over with a thin sheet of bone.[47]

In preparing his expanded treatment of the new skulls, Sternberg visited the American Museum of Natural History and examined the type of Cope's *Monoclonius crassus*. He was convinced that, contrary to Brown's assertion, the parietal had not been altered by erosion, a conclusion with which I am in complete agreement. After examining the Cope material, Sternberg stated, "It is not possible to demonstrate that our new specimen is distinct from *M. crassus*, though it certainly is distinct from *M. sphenocerus*, which Brown referred to *M. crassus*."

Notwithstanding this observation, Sternberg proceeded to designate a new species, *Monoclonius lowei*, named in honor of a field assistant, Harold Lowe (Fig. 5.9). He described the skull as that of a young adult, with widely open sutures. The skull was crushed dorsoventrally and

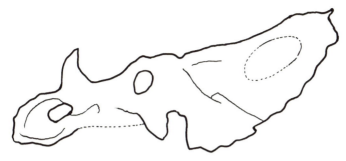

FIG. 5.9. Skull of *Monoclonius lowei*, Canadian Museum of Nature. (After Sternberg 1940.)

has a somewhat peculiar look to it. The right maxilla is displaced toward the midline and forced upward. The orbits are partially collapsed. The damage to the frill has been mentioned. Whatever the shortcomings of the specimen, it is a welcome one indeed. It is the first articulated skull of *Monoclonius* to show the nature of the nasal horn, orbital horns, parietal, and squamosal. On the basis of this specimen and the type of *Monoclonius crassus*, Sternberg offered this emended diagnosis of *Monoclonius*:

> Generic characters (emended): Nasal horncore outgrowth of nasals laterally compressed; facial part of skull moderately deep; no interpremaxillary fontanelle; anterior edge of premaxilla almost at right angles to lower border; frontal fossa shallow; postorbitals well developed, forming a broad rectangular surface behind the orbits; squamosals short and well forward, with proximal part standing nearly erect; coalesced parietals arched anteriorly but flattening out posteriorly; posterior edge broad and moderately thin, no hooklike processes at back, no process over fenestrae; fenestrae wholly within parietals; no epoccipitals.

For characters defining the species, Sternberg seemed to choose some unfortunate ones, including the shapes of the external nostril and of the orbit, both of which were affected by crushing. He was impressed with the sharp-edged, recurved nasal horn core, of a sort never before seen in Canada, and also by the complete absence of orbital horn cores. The comparison of the nasal horn core of *Monoclonius* with that of *Brachyceratops* is obvious. He noted that the scallops on the parietal were relatively coarse. The parietal is thin on the midline, measuring only 16 mm in thickness at the rear margin, contrasting as always with the greatly thickened rear border in *Centrosaurus*. The squamosal is de-

161

scribed as being short and almost vertical in orientation, a rather un-usual position. There are five very thin scallops on the squamosal. The maxilla showed twenty-nine or thirty vertical rows of teeth, a number that recurs in these discussions. Sternberg included some useful mea-surements that provide an indication of the size of the animal:

Total length of skull	1.65 m
Basal length of skull	80 cm

This was not a small animal. Thus small size cannot account for the thinness of the bones. Several other centrosaurines exceed this skull in basal length—the longest skull known to me (Brown's "*Monoclonius*" *flexus*) measuring 83.5 cm in basal length—but only the stunning skull of *Styracosaurus* clearly exceeds *Monoclonius lowei* in total length. (See note 36, Chapter 7, for a discussion of basal length.) Thus *Monoclonius lowei* has a very long frill. In summary, Sternberg had finally succeeded in demonstrating beyond reasonable doubt that dinosaurs of the aspect of *Monoclonius crassus* lived in Canada and were demonstrably different from dinosaurs of the classic *Centrosaurus apertus*.[48]

Regrettably, *Monoclonius* remains rare, although there is evidence of it in Dinosaur Provincial Park in Alberta, where *Centrosaurus* was a com-mon dinosaur. No unequivocal skulls of *Monoclonius* have come from Montana since the days of Cope's discovery in 1876. Such a discovery is long overdue and would be of considerable value. The Judith River of Montana has been stingy to would-be collectors over the years: bones are to be found there, but articulated materials are rare. Although Jack Horner cut his dinosaur-collecting teeth (so to speak) in exposures of the Judith River Formation north of Rudyard and Havre, it is no coinci-dence that he has enjoyed his greatest triumphs and has spent most of his time in the Two Medicine Formation farther west.

No one has seriously questioned the distinctiveness of *Monoclonius* and *Centrosaurus* since Sternberg's time, but the inertia of past usage continues. Cope's ghost and Brown's long shadow have hung over the American Museum of Natural History for many years, and the old signage over the skulls of "*Monoclonius*" *flexus* and "*Monoclonius*" *nasi-cornis* remained intact until as recently as 1992, when the old dinosaur halls were dismantled. In the new ones that opened in June 1995, at long last the name *Centrosaurus* is finally established.

In this same 1940 paper, Sternberg described a second skull, not from Manyberries but from the classic Red Deer River site. This skull, with

FIG. 5.10. C. M. Sternberg and G. Lindblad collect a *Centrosaurus* skull, Steveville Badlands, Red Deer River Valley, Alberta, 1917. (Courtesy of the Canadian Museum of Nature.)

jaws, had been collected in 1917 but remained undescribed until 1940 (Fig. 5.10). He was impressed that this skull, a classic *Centrosaurus,* was rather long and low in front of the orbits. He decided to designate it a new species, which he called *Centrosaurus longirostris,* the "long-snouted *Centrosaurus."* The nasal horn core curved forward but was somewhat wasted in appearance, with the surface facing the orbits eroded or excavated. The orbital horns were rather rugose but low, and the frill was comparatively short. The caudal process of the premaxilla failed to reach the lacrimal bone, and in consequence the nasal was in contact with the maxilla, a feature unusual in *Centrosaurus,* but one that also occurred in *"Monoclonius" nasicornis* and *Monoclonius lowei.* Sternberg gave the total length as 1.45 m. He did not give the basal length, despite a beautifully exposed and prepared palate, but my own measurement is 81 cm. It is not small, but it has a rather short frill, to be sure.[49]

Historically, *Centrosaurus longirostris* is the last-named species in the *Centrosaurus-Monoclonius* complex. Fortunately, the vogue to name

163

CENTROSAURUS A VALID NAME?

Centrosaurus has had a beleaguered history. Not only has it had to fend off the assaults of the Cope crowd, so to speak, but another "threat" to its status has arisen. The name *Centrosaurus* is not quite pure. It has been known for a number of years that the name had been used in 1843 by L. J. Fitzinger, a German herpetologist, in connection with a modern lizard. Although this usage was technically "forgotten"—that is, not used in a paper for fifty years following publication, and therefore unlikely to cause confusion—the possibility was enough to suggest to some workers that perhaps a name change was required. Accordingly, Chure and McIntosh obliged in 1989 by coining the name *Eucentrosaurus*. Ben Creisler has delved more deeply into the matter and has uncovered some interesting information. It seems that Fitzinger never proposed *Centrosaurus* as a senior synonym, but rather as a junior synonym for *Phrynosoma*, the horned lizards of western North America. The name *Centrosaurus* appeared in small print, without description. Nothing about its original usage invalidates its application to Lambe's horned dinosaur. Creisler speculates that *Centrosaurus* may have been a manuscript name that Fitzinger never published, having decided that *Phrynosoma* was a valid name to be used.

There is yet another problem to comment upon. The correct pronunciation of *Centrosaurus* is "SEN-trow-SORE-us," with a soft C. In general, I prefer not to insist upon one pronunciation over another, believing as I do in a certain amount of linguistic flexibility, as in "toe-MAY-toes" versus "toe-MAH-toes." In this case, however, the pronunciation affects another dinosaur, *Kentrosaurus*, a stegosaur from East Africa. If the name of the horned dinosaur is pronounced with a hard C, it will have the same sound as that of the stegosaur, and homonyms are not permitted; differing spelling does not suffice. Hennig proposed the name *Kentrurosaurus* in 1916, thinking that *Kentrosaurus* was preoccupied by homonymy. However, the use of both a different spelling *and* a different pronunciation for Lambe's dinosaur renders this change unnecessary.

each new specimen as a new species began to pass into oblivion. Less fortunately, new specimens ceased to be collected, and when collecting resumed thirty or forty years later, new specimens ceased to be described. Trust me: they are still there in the rocks, stacked up and waiting to see the light of day.

STYRACOSAURUS—THE SPEAR-BEARER

Recall that when Lawrence Lambe began to describe Canadian dinosaurs in 1902, he only worked with fragments. Barnum Brown was collecting magnificent skulls and skeletons of horned dinosaurs in Alberta by 1912, but the Sternbergs were not far behind. To Lambe fell the honor of describing in 1913 the first complete skull of a Canadian horned dinosaur, which remains one of the most spectacular skulls ever found (Fig. 5.11; Plate III). The name of the animal is *Styracosaurus albertensis,* meaning "Alberta spike lizard." In fact, the spike is not a garden variety spike, or even a railroad spike (the ancient Greeks not being very big on railroads). Instead the Greek word *styrak* or *styrax* denotes the spike at the lower end of the blade of a fancy spear—you know, the nasty one that sticks out sideways. Lambe explained his choice of name in this way: "The name selected for this genus has reference to the shape of the large processes on the frill, which resemble spikes, and must have made this bristling reptile in life a veritable moving chevaux [*sic*] de frise.[50″]

For an inoffensive plant-eater, this animal had remarkable equipment. The skull, including its spikes, is 1.84 m long (just over 6 ft), by far the longest skull of the Judith River centrosaurines. Three pairs of long spikes radiate out from the back of the parietal, and the nasal horn core is erect and tall, an estimated 50 cm high. However, when the spikes are not taken into consideration, the skull dimensions of *Styracosaurus* are only average, with a basal length of 76 cm and a frill length of 52 cm. Though further evidence of *Styracosaurus* has been found, both in Alberta and in Montana, never again has a magnificent complete skull like this one been found.

The straight nasal horn points slightly forward of the vertical. Its base measures 17 cm in length and nearly 11 cm in width. Unfortunately it broke before it was collected, and the upper half was lost. The tall, straight nature of this horn recalls those of *Monoclonius sphenocerus* and *"Monoclonius" nasicornis,* but it exceeds both of those in size. Combined with this tallest of nasal horns was a remarkable lack of horn on top of

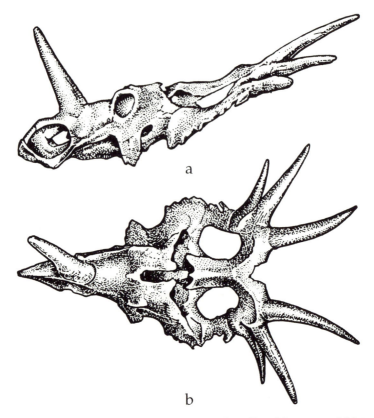

FIG. 5.11. Skull of *Styracosaurus albertensis*, Canadian Museum of Nature. (a) Lateral view, (b) dorsal view. (From Dodson and Currie 1990. Donna Sloan. Courtesy of the University of California Press.)

the orbits. In fact, there is a depression or pit in the position where an orbital horn core would be expected. This character suggests that the orbital horn core actually has a separate center of bone growth. If so, this would be a unique feature found neither in adults nor in immature specimens of *Centrosaurus, Monoclonius,* or *Brachyceratops*. Scott Sampson suggests that this feature indicates that the horns were actually resorbed in certain adult specimens, for reasons that are difficult to fathom. With the short squamosal having five scallops, the frill may be recognized as that of a typical centrosaurine. The parietal is the most luxuriant aspect of the cranial anatomy of *Styracosaurus*. It has ample parietal fenestrae and a caudal bar that measures 5 cm in thickness on the midline and 8 cm laterally. The right side is not well preserved,

166

despite elegant reconstruction. The better-preserved left side shows three remarkable spikes. None of the three is complete, although all are nearly so. Lull estimated the lengths of the three spikes as 55 cm, 50 cm, and 39 cm, respectively. Each is thickest at its base and tapers distally to a point. The largest is closest to the midline, is 11 cm wide and 7.5 cm thick at its base, and points almost directly backward, bending laterally with a graceful curve. The next horn is less prominent in all its dimensions, projecting from the corner of the frill. The short one is overlapped by the second spike and projects almost completely laterally. This is a most distinctive skull.[51]

There have been no further Canadian contributions to our knowledge of *S. albertensis* since 1913, although Lambe provided several comparative illustrations in his 1915 monograph on *Eoceratops*. As a footnote, in 1935 the University of Toronto investigated the site where C. M. Sternberg had collected *S. albertensis* 22 years earlier. Some bone was exposed at the site, and further excavation resulted in the recovery of the missing lower jaws and a good portion of the skeleton. These bones have never been described but were eventually traded to Ottawa to accompany the type skeleton. Dale Russell kindly supplied me with measurements of the skeleton. The lengths of the femur and tibia, 83 cm and 61 cm respectively, suggest that *Styracosaurus* was a little larger than the *Centrosaurus* skeletons in New York and New Haven. It was perhaps 5.5–5.8 m in total length and 1.65 m high at the hips.

In 1937 Barnum Brown and E. M. Schlaikjer described a specimen of *Styracosaurus* that Brown had collected on the Red Deer River in 1915. The specimen was prepared as a striking, two-sided panel mount. This was exhibited in the Cretaceous Hall of the American Museum of Natural History in the company of other fine specimens of horned dinosaurs from Alberta, including *"Monoclonius" flexus*, *"Monoclonius" nasicornis*, and specimens of *Chasmosaurus*. It was even described as a new species, *Styracosaurus parksi*, in honor of W. A. Parks, who had died the year before. There are problems, however. The skeleton is restored as a fine specimen, some 5.4 m (17.6 ft) long and perhaps 1.75 m (5.75 ft) at the hips. The skull is very impressive. The problem is that there is too much plaster; little of the skull is real. As was the style in those days, the exhibit fails to show the distinction between plaster and bone, and the investigating scientist today can have little confidence in what he or she sees. The actual bones that went into the reconstruction were not figured in the paper. Preserved skull bones include the squamosal, quadrate, quadratojugal, and lower jaw and a partial maxilla, all from the left side. Broken fragments of the skull showered down the hillside. These were collected and utilized where possible. The skeleton was much better preserved: it was essentially complete back to the twentieth caudal vertebra, and fifteen other caudal vertebrae were found.

A few fragments of nasal show that the horn core was prominent. Another fragment shows that the postorbital presented "a rather large and roughened dorsal knoblike elevation, which suggests that it probably was capped by some sort of epidermal padlike development." If so, it is a striking departure from *S. albertensis*. Could it represent the mature counterpart of the separate center of bone growth inferred in *S. albertensis*? However, as reconstructed, the horn core bears a remarkable resemblance to that of *"Monoclonius" nasicornis*. Was this actually the model on which a reconstruction was based? The shortcomings of Brown and Schlaikjer's paper leave this an open question. The squamosal evidently was well preserved, with interesting features. It shows three coarse scallops rather than the five fine ones in *S. albertensis*. The parietal is poorly preserved. Only the bases of several of the spikes are extant. However, a very interesting detail is preserved. Brown and Schlaikjer noted that a short rostral process originates on the thick caudal border of the parietal, just medial to the base of the first spike; the process "probably extended only a short distance over the frill fenestra." I find this process fascinating, as it is clearly homologous with the much better-developed process seen in all specimens of *Centrosaurus*. I have seen this structure in fragmentary specimens of *S. albertensis* at the ROM and in Alberta at the Royal Tyrrell Museum of Palaeontology in Drumheller. Its presence clearly establishes a strong similarity between *Centrosaurus* and *Styracosaurus*. I see the rostral processes in *Styracosaurus* as abortive, as if they were present in the developing embryo and then switched off by some regulatory gene before growth continued.

Brown and Schlaikjer noted a few distinctive features of the skeleton. There was fusion of cervical vertebrae four and five, similar to the fusion of cervicals five and six in the Yale specimen of *"Monoclonius" flexus*. They reported differences from *Monoclonius* in the sacrum, ribs, and pelvis, but it is not clear to me what the value of such differences may be. For example, they determined that the tenth sacral vertebra was incompletely fused to the ninth, and thus that "the sacrum of *Styracosaurus* is more primitive than in *Monoclonius*, yet more advanced than in *Brachyceratops* which has eight sacrals." The authors provided a series of measurements of the bones of the hands and feet, but, unaccountably, not of the hind limb. The length of the humerus is given as 618 mm and that of the radius as 375 mm. This humerus compares with the humeri of *"Monoclonius" flexus* at Yale, measuring 585 m, and *"Monoclonius" nasicornis* at the American Museum of Natural History, measuring 600 mm. This length is consistent with the estimate of an animal that slightly exceeded the length of the aforementioned specimens.[52]

There is one further report of *Styracosaurus*. Although it was published in 1930 by Gilmore, I mention it last because it is based on the least material and comes not from the Judith River Formation of Alberta but from the Two Medicine Formation of Montana. Gilmore figured a piece of the caudal border of a ceratopsid parietal. The specimen is somewhat distressed, but it shows two or possibly three spikes projecting caudally from each side. The longest of the spikes is 295 mm long, significantly shorter than the spikes of *S. albertensis*. Furthermore, the spikes closest to the midline converge toward the midline, rather than away from it as in *S. albertensis*. The second pair is directed caudolaterally as in *S. albertensis*. That is all there is. Gilmore named a new species, *Styracosaurus ovatus*, apparently in reference to the oval cross sections of the spikes.[53] This material is particularly intriguing and badly needs elucidation.

Newer Developments and Modern Studies

WITH THE END of World War II, an era in North American dinosaur paleon-tology had passed. During the war, nonstrategic scientific activity natu-rally took a backseat to more pressing concerns. By the war's end, Lull had retired, Gilmore had died, Brown's wonderful career in the exploration and description of dinosaurs had come to an end, and those who had known Marsh and Cope were themselves rapidly becoming extinct.

In 1946, a 61-year-old C. M. Sternberg might have had thoughts of retirement on his mind. If so, this would not have readily been guessed by those who tracked his field itinerary or read his papers, which continued to appear until 1955. Among other locations Sternberg vis-ited that summer was the picturesque spot of Scabby Butte in south-western Alberta, 250 km southwest of the famous Red Deer River sites of the Judith River Formation, and 30 km northwest of the city of Lethbridge. A second locality in the vicinity was on the Little Bow River, 30 km north of Scabby Butte. The beds in this region were not of the Judith River Formation, but of what was subsequently described as the St. Mary River Formation—beds equivalent to the Horseshoe Can-yon Formation in the Red Deer River Valley near Drumheller. These beds are several million years younger than those of the Judith River Formation. At both sites, Sternberg located remains of three incomplete skulls of an unusual ceratopsid that puzzled him greatly. In 1950, he described the animal as *Pachyrhinosaurus canadensis* ("Canadian thick-nosed reptile"), a name whose appropriateness is undoubted. As he stated in his introduction, "The specialized development, the large massive head and the great thickness of bone is suggestive of the freakish development that took place among some of the dinosaurs near the very close of the Cretaceous." Although I would hesitate to apply the term *freakish* to any of God's creatures, there is no hiding the fact that "thick nose" is a strange creature.

Sternberg recognized the overall ceratopsid form of the skull, with a narrow beak, a broad skull, a round occipital condyle, and at least a short crest. There were, however, no horns, but instead a broad, very thick, flat, or even concave dorsal surface. The texture of the bone suggested to him that the structure was covered with "a chitinous sheath" (by this I presume he meant horn, which is a vertebrate epidermal material, not the chitin of insects) that formed a "battering-ram." The nasals that form the battering ram are 18–25 cm thick and composed of extraordinary, massive bone. It is not an easy skull to interpret, as fusion of sutures among bones is extensive. The maxilla, jugal, and quadrate are described as ceratopsid-like but shorter and heavier. The squamosal and parietal are both present but incomplete; both are characterized as short and thin. There are thirty-five dentary tooth positions, with five teeth in each tooth position. The teeth show the characteristic split roots of ceratopsids.

Sternberg was so amazed by this animal that, though clearly recognizing its ceratopsid affinities, he placed it in a new family within the Ceratopsia, the Pachyrhinosauridae.[1] There has been little enthusiasm for this proposal, because it is apparent that *Pachyrhinosaurus* is nested phylogenetically within the Centrosaurinae.

Sternberg officially retired in 1950 at age 65, though he remained active. By now, vertebrate paleontology had been transferred from the Geological Survey of Canada to the independent National Museum of Canada (NMC). He was replaced at the museum by Wann Langston, Jr., who had completed his Ph.D. on fossil crocodiles from Colombia at the University of California. This crusty Texan greatly enjoyed the company of his predecessor, who willingly passed on his knowledge and experience. Langston was intrigued by *Pachyrhinosaurus* and in 1957 visited Scabby Butte, where he collected more material for the NMC. His first paper on the subject of *Pachyrhinosaurus* was published in 1967, after he had returned to Texas, where he became curator of fossil reptiles at the Texas Memorial Museum of the University of Texas. The subject of his enquiry was not material from Scabby Butte, but rather a new specimen that had been collected from near Drumheller in the lower part of the Edmonton Formation, now called the Horseshoe Canyon Formation of the Edmonton Group. The new skull was collected and preserved by a small natural history museum, the Drumheller and District Museum.

The new specimen was better preserved than the first specimens from southern Alberta but nonetheless was still incomplete. Langston expressed his surprise that after a half-century of exploration of the productive fossil beds in the area no evidence of this strange dinosaur had previously come to light. The specimen, found only 15 km north-west of the town of Drumheller, was located in a bonebed populated largely by remains of duck-billed dinosaurs, and it was the only ceratopsid in the deposit. Parts of the skull were densely mineralized, with geode-like infillings of chalcedony and "tiny quartz stalactites" in open spaces within the skull. The Drumheller skull is a little larger than Sternberg's skulls. In fact, Langston correctly observed that *Pachyrhinosaurus* is among the largest of ceratopsids, "being surpassed in size only by species of *Triceratops* and *Torosaurus*." With this statement I am in complete agreement, except that I would add *Pentaceratops* to this list of very large ceratopsids.

Langston lavished much description on the unique thickening and flattening of the top of the skull. He coined the term *nasofrontal boss* to describe it. The structure attains a width of 35 cm in front of the orbits, and he judged it to consist principally of the nasals, with lesser contributions from prefrontals, frontals, and even postorbitals. The sides of the boss are vertically corrugated, giving it a "palisaded" appearance. The dorsal surface of the boss has a hummocky aspect, irregular pits giving it a cinder-like texture. In dorsal view the boss is triangular, narrowest in front and expanding caudally toward the orbits. Langston speculated on the growth of this extraordinary structure:

> It seems clear the nasofrontal boss of *Pachyrhinosaurus* developed from the concerted upward growth of many closely spaced, finger-like spongy excrescencies [*sic*] that arose from the superior surface of the nasal and associated roofing bones. As the individual grew the excrescences became longer and coalesced, while laterally new columns were being added to grow upward on the sides of the mass. Vital fluids nourished the substance of the boss, and whatever tissues it supported, carried upward along the numerous peripheral grooves and through foramina at the base of the mass. With increasing size the older nutrient passages tended to become filled with spongy bone, and growth in the central area of the boss slowed. This permitted the more peripheral area to attain a greater elevation and produced the depressed central area.

The postorbital bears "an extremely rugose exostosis," or irregular bony growth that abuts the caudolateral corner of the boss, but is separate from it. There is also a low, conical, inconspicuous horn core somewhat reminiscent of that in some specimens of *Centrosaurus*. The structure of the nostril seemed very important to Langston. The beak of the Drumheller skull is complete, showing a rostral bone that is fused to the premaxilla, which is short, deep, rounded, and solid. The beak in *Pachyrhinosaurus* strongly resembles the same region in *Centrosaurus* and contrasts with the long, low, and perforated premaxilla of *Triceratops* and *Chasmosaurus*. A process of the nasal projects into the caudal part of the external nostril, as in *Centrosaurus*.

A few fragments offer some insight into the nature of the frill. The median bar of the parietal is intact for a distance of 32 cm and is 43 mm thick on the midline, thinning to less than half that figure away from the midline. The existence of parietal fenestrae is logically inferred.

The Drumheller skull as preserved is about 140 cm long, but this is with a very short frill. From the diagram Langston provided, the basal length of the skull may be roughly estimated at 95 cm, which is indeed very long. By comparison, the longest basal length in *Centrosaurus* is 83.5 cm in the American Museum of Natural History specimen of *C. flexus*. The width of the occipital condyle is 9 cm, compared with 7.8 cm for the largest condyle in *Centrosaurus*, in the *Centrosaurus flexus* skull in Ottawa.

Langston set about to rehabilitate *Pachyrhinosaurus* as a ceratopsid. He observed:

> The Drumheller *Pachyrhinosaurus* now shows that the genus, though aberrant, was by no means as unusual as first supposed.
>
> The unique nasofrontal boss of *Pachyrhinosaurus* seems qualitatively similar to the horn-like structures and other excrescences present in ceratopsians generally. . . . If one ignores the nasofrontal boss and other unusual features obviously related to it, all else known about the general proportions of the *Pachyrhinosaurus* skull conforms fairly well to the pattern of the short-faced genus *Centrosaurus*.

Langston was quite definite that *Pachyrhinosaurus* was a member of the same phylum as *Centrosaurus*, *Styracosaurus*, and *Monoclonius*. He curiously declined to accept Lambe's subfamily Centrosaurinae as a name for this lineage, while admitting that "the time may be close when a formal recognition of the long-acknowledged division of the Ceratopsidae . . . will be found useful." He considered whether the boss might represent the base of a deciduous nasal horn, which had been lost in all specimens. He thought a seemingly identical break in all known specimens was unlikely, and he knew of no evidence for a strikingly large horn in bonebeds. He thus supported Sternberg's interpretation that the boss was covered with a thick horny covering. Because of a faulty interpretation of the fragmentary squamosal, Langston was not certain if the Drumheller specimen pertained to the same species as Stern-berg's, *Pachyrhinosaurus canadensis,* or not.[2] The following year, Lang-ston reinterpreted the squamosal and was satisfied with the referral to *P. canadensis.*[3]

In 1975, Langston published an important paper on the bonebed fauna of Scabby Butte, focusing especially on the ceratopsids. The bone-bed contained the remains of a wide variety of lower vertebrates,[4] including sharks, rays, bony fishes, salamanders, turtles, champsosaurs, mosasaurs, crocodiles, and dinosaurs, the most common of which were duck-bills (usually called hadrosaurs by paleontologists), presumably *Edmontosaurus.* Mammals were also found here. Second in abundance to hadrosaurs were ceratopsids. Most of these remains may be identified as those of *Pachyrhinosaurus,* of which evidence of fourteen specimens was found. Parts of an *Anchiceratops,* a long-frilled ceratopsid, were also detected. This evidence includes a characteristic fragment of frill and two pieces of robust supraorbital horn core, conformable with those of *Anchiceratops* but for which there is no evidence in *Pachyrhinosaurus.*

A fine skull of *Pachyrhinosaurus,* NMC 9485, excavated in 1957, was figured for the first time. It was essentially complete except for the frill, which remained elusive. This skull elucidated further details, for in-stance the nature of the bone above the external nostril. Langston referred to this distinctive region, composed of the premaxillae and nasal bones, as the "supranarial bridge." Here the bone forms "flat-tened tumescent bulges" that are bilaterally symmetrical, forming supranasal bosses that prefigure the immense nasofrontal boss. The size of the supranasal bosses is variable from specimen to specimen; they were too small for Sternberg to comment on in the type specimen, but they are 14 cm long in NMC 9485. The postorbital bosses too are better developed in NMC 9485 than in other specimens, and they resemble "large, flattened sponges appressed to the sides of the skull." On the right side of this skull is a structure that might be taken for the broken base of a very small postorbital horn core, but on the left side is a shallow pit, as in *Styracosaurus,* that suggests the loss of a horn core.

The spherical occipital condyle of NMC 9485 measures a whopping 95 mm in diameter, the size of a large Washington State Delicious apple. This is a respectable size for the condyle of a *Triceratops,* a very large ceratopsid indeed, and far larger than that of any previously reported centrosaurine.

Also found in the Scabby Butte bonebed are ceratopsid skeletal bones, presumably those of *Pachyrhinosaurus.* These include vertebrae, ribs, two scapulae, a humerus, a femur, a tibia, and a fibula. The bones are generally massive. The femur is 93 cm long, and the tibia 70 cm. The femur is 13 cm longer than the femur of Brown's *"Monoclonius" cutleri,* the largest described centrosaurine specimen exclusive of *Pachyrhinosaurus.* Yet again, we are left with the conclusion that *Pachyrhinosaurus* was a very large ceratopsid, probably close to 2 m high at the hips, and perhaps 6.5 m long (compared with 7 or 8 m long for a *Triceratops*).

An intriguing find that Langston discussed with some reticence at the back of his lengthy paper was parts of the parietal frill of "unusual form" that were reminiscent of those of *Styracosaurus* and of *Centrosaurus.* There is a 23-cm-long spike projecting caudolaterally as in *Styracosaurus,* and a small, hooked process directed toward the midline, as in *Centrosaurus.* The spike is 7.4 cm thick at its base. The adjacent parietal bar is 6.9 cm thick. The curvature of the parietal suggested that large parietal fenestrae were present. He also noted three epoccipital scallops of normal centrosaurine aspect located in front of the single spike. As Langston noted, there is a passing resemblance to Gilmore's *Styracosaurus ovatus,* except that the medial spike is more abbreviated in the present specimen, which is less massive than the parietal of *Styracosaurus ovatus.* He considered the resemblance of the skull of *Pachyrhinosaurus* to those of *Centrosaurus* and *Styracosaurus,* the similarity of the frill to those of the same genera, and also the physical proximity of the frill to two skulls of *Pachyrhinosaurus.* He thus, somewhat boldly, offered the first visual reconstruction of the complete skull of *Pachyrhinosaurus canadensis* as a centrosaurine (Fig. 6.1). This reconstruction remains extremely insightful and valuable, the only one yet to appear in the scientific literature.[5]

INTERLUDE—IN PRAISE OF VERDANT PASTURES

It will not have escaped the perceptive reader that Alberta is a very special land. It has provided more skeletons of dinosaurs than any other place on earth. Besides horned dinosaurs, Alberta is also rich in hadrosaurs, ankylosaurs, pachycephalosaurs, and theropods. The United States has more kinds of dinosaurs than Alberta, in fact more kinds of dinosaurs than any other land on earth, but these skeletons tend to be less complete. These come from rocks spanning essentially the entire 165-million-year Mesozoic reign of dinosaurs. The United States also has larger skeletons, the giant sauropods. No giant sauro-

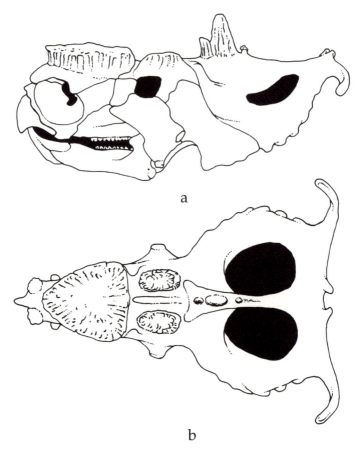

a

b

FIG. 6.1. Reconstructed skull of *Pachyrhinosaurus canadensis* based on remains found at Scabby Butte, Drumheller, and Pipestone Creek, Alberta. (a) Lateral view, (b) dorsal view. (Robert Walters after Philip J. Currie.)

pods are found in Alberta, or anywhere else in Canada. Only a small slice of dinosaur time is represented in Alberta, perhaps the past 20 million years at most. What is remarkable is that dinosaurs in Alberta come from relatively small areas of outcrop, principally less than 100 km of linear exposure of riverine badlands along the Red Deer River. However, as we have seen, other parts of the province have even smaller areas of exposure that contain dinosaur fossils. The younger beds of the Edmonton Group extend for 60 km northwest of Drumheller; the older beds of the Judith River Formation begin at Steveville, 100 km southeast of Drumheller, and extend for 25 km through exposures of the Judith River Formation in Dinosaur Provincial Park. Arguably, this park is the richest dinosaur burial ground on earth. More than three hundred skeletons have been collected here since 1910, and field parties from the Royal Tyrrell Museum of Palaeontology in Drumheller locate on average six complete skeletons there in a typical summer. Another feature of the Red Deer fossil beds is that this is the only location on earth where three time-successive, dinosaur-bearing geological formations are located in proximity to each other, in stratigraphic superposition as geologists say. This allows an unprecedented opportunity to study rates and patterns of evolution in dinosaurs.

The richness of the Alberta fossil beds has been unquestioned for most of this century. We have described little else for the past 85 pages. What is striking, however, is that although the studies described in the previous pages have used Alberta fossils, they have all been based at great eastern institutions, principally in New York, New Haven, Ottawa, and Toronto. In the early 1920s, the University of Alberta hired George F. Sternberg to collect fossils for several years. He met with significant success, and among his finds was the most complete specimen of a pachycephalosaur, the little dome-headed dinosaur *Stegoceras validus*, ever found. Yet the fossils were described by Charles Gilmore from the Smithsonian. As the Province of Alberta grew in population, wealth, and sophistication, particularly following the discovery of oil at Leduc, south of Edmonton, in 1947, a climate slowly developed that encouraged the development of an Alberta-based vertebrate paleontology.

In 1955, Dinosaur Provincial Park was established, preserving forever the fossil legacy of the Judith River (Belly River) exposures below the ghost town of Steveville, made famous by Lawrence Lambe, Barnum Brown, and C. M. Sternberg. Included in the 90-km² park are the famous sites of Steveville, Berry Creek, Deadlodge Canyon, Sandhill Creek, and Happy Jack's Ferry. In 1966, Richard C. Fox, a new paleontologist from the University of Kansas, was hired by the University of Alberta in Edmonton. Fox began his work in Alberta with great vigor, but he targeted an even rarer treasure: fossil mammals of dinosaur age. Over the intervening thirty years, Fox has amassed a remarkable collection of fossil vertebrates of all kinds, including dinosaurs, and has trained some fine students in dinosaur paleontology. Among Fox's students are Hans-Dieter Sues, now at the Royal Ontario Museum, and Peter Dodson, who seems to have ended up in a school of veterinary medicine in Philadelphia, and who, at last report, had a smile on his face as well as a mustache. There are many others as well, but for the most part they, like their mentor, do not specialize in the study of dinosaurs (although David Krause, at State University of New York at Stony Brook, has recently seen the light!).

A most important development occurred in 1976, when the Provincial Museum and Archives of Alberta hired a lanky, long-legged, curly-haired young man who had been engrossed in a study of Triassic marine reptiles from Madagascar at McGill University in Montreal. Toronto-born Philip Currie had not yet finished his Ph.D., but he packed up his papers and moved to Edmonton (finishing his degree in due course). His first undertaking was a salvage project, to study Early Cretaceous dinosaur footprints exposed in the Peace River Canyon of northeastern British Columbia before they disappeared beneath the rising waters of the newly dammed river. This endeavor was a marvel of ingenuity and endurance under truly trying conditions. In successfully prosecuting this work, Currie assembled a first-class team of dedicated technicians. It wasn't long before Currie and crew turned to the more clement locale of Dinosaur Provincial Park, where his success at finding specimens brought attention very quickly. By 1979, major study of dinosaurs was in progress at the park, and it has continued ever since.

So great was the success of the Provincial Museum's crew, and so favorable their timing, that they were to leave a legacy few would have predicted. The year 1980 was the seventy-fifth anniversary of the Province of Alberta, and plans for celebration were underway. Furthermore, world oil prices were at an all-time high, and the public coffers were overflowing. The province established a Heritage Trust Fund to serve the public interest, and it was decided to use some of the fund to establish a museum for the study and display of Alberta's fossil heritage. The museum was established in the badlands of the Horseshoe Canyon Formation in Midland Provincial Park, just outside of Drumheller. It was christened the Tyrrell Museum of Palaeontology (becoming the *Royal* Tyrrell Museum of Palaeontology [RTMP] in 1990 on the occasion of Her Majesty's visit to Alberta) in honor of Joseph Burr Tyrrell, whose claim to paleontological immortality was the discovery in 1884 of the skull of *Albertosaurus* near Drumheller. In 1982, the paleontological staff of the Provincial Museum, now administratively transferred to the new facility, moved to Drumheller to continue collecting and to begin the preparation of exhibits for the new museum. The Tyrrell Museum opened to rave reviews in 1985, and it now attracts half a million visitors per year to the sleepy little town of Drumheller, nestled in a crevice in the rolling prairies beside the Red Deer River 100 km east of Calgary.

The marvelous exhibits of the RTMP include more skeletons of more kinds of dinosaurs than in any other museum on earth. Behind the scenes are state-of-the-art specimen preparation laboratories and cavernous collection rooms. When the visitor tires of indoor activities, pathways through the badlands have the habit of regularly disgorging fresh specimens of dinosaur bone, seeing the light of day for the first time in seventy million years. One day in 1983, Montana paleontologist Jack Horner visited Currie at the construction site of the new museum. As the two famous scientists hiked through the very badlands now visible from Currie's corner-office window, Horner bent over and picked up the jaw of a small meat-eating dinosaur bristling with teeth. When Currie analyzed the jaw, it turned out to be the key to the identity of four different genera of dinosaurs: *Troodon* and three others. Thanks to their serendipitous walk, we now know that *Stenonychosaurus, Polyodontosaurus,* and *Pectinodon* are all synonyms of the small theropod *Troodon.* One of the great charms of paleontology is that one never knows what will turn up where, or who will find it.

Thus it is that dinosaur paleontology has taken root in the prairie soil of Alberta and has flowered magnificently.[6] Dinosaur Provincial Park continues to be a focus of RTMP activities; the C. M. Sternberg Field Station was established there in 1987 to aid the museum's research program at the park. The park provides a congenial locale in which to live and work. The cottonwood trees beside the Red Deer River provide welcome shade at the end of a hot working day, and Sandhill Creek, or the sometimes treacherous waters of the Red Deer itself, provides a cool if turbid opportunity for a dip. The badlands exposures offer a seemingly endless supply of dinosaur fossils. The Tyrrell parties and armies of volunteers have collected thousands of specimens of Judithian dinosaurs, including several scores of skulls and skeletons, most commonly hadrosaurs and ceratopsians. A number of important bonebeds have been investigated. One particularly notable bonebed is a major *Centrosaurus* mass grave. Disarticulated remains of hundreds of specimens —small, medium, and large—may represent a herd that perished during a river crossing. Carcasses scavenged and trampled, bones broken and scattered, predator teeth shed—all give silent witness to an ancient catastrophe sealed in stone.

However, Tyrrell research crews are by no means limited in the geographic focus of their activities. Small parties have ranged widely across the province, and in 1986 a major cooperative initiative with China was begun, the Canada-China Dinosaur Project, which entailed major expeditions in each country. As no ceratopsids were found in China, their expeditions were of less than major interest from my painfully myopic viewpoint. My heart remains in Alberta, the site of yet another find, one that now tugs our narrative back on course.

PACHYRHINOSAURUS—AGAIN AND AGAIN

Darren Tanke is a technician at the Tyrrell who, like me, suffers from an affliction not listed in medical dictionaries: *ceratophilia* (that is, an unreasonable love of horns, preferably old horns). When I first lectured at the Provincial Museum in Edmonton during the seventy-fifth anniversary celebrations in 1980, young Mr. Tanke sternly took me to task publicly on the correct grounds that I had made loose with the identification of a ceratopsid skull I had shown in a slide during my talk. This was in a previous incarnation, when my knowledge of ceratopsids was second hand, my hands-on dinosaur experience being restricted to hadrosaurs and *Protoceratops*. I was duly chastened, and never again have I misidentified *Anchiceratops*, least of all in Darren's presence.

Tanke has been deeply involved in the excavation of the *Centrosaurus* bonebed at Dinosaur Provincial Park, but he is by no means opposed to perambulations either. Following a report of a possible occurrence of dinosaur bone in northwestern Alberta in 1987, he came upon a rich bonebed at Pipestone Creek, more than 600 km from Drumheller. This site is in the Wapiti Formation, which is about the same age as the Horseshoe Canyon Formation, but in a region not previously known to produce dinosaur fossils. Tanke estimated in 1988 that the remains of "thousands of individuals" of *Pachyrhinosaurus* had been entombed by a local catastrophe. Included are specimens spanning a wide size range, from "20% smaller than the type of *Brachyceratops*" to the size of *Centrosaurus*. Thus we may infer a size range of roughly 1.5–5.4 m in length. The maximum size appears to be significantly smaller than that of *Pachyrhinosaurus* from Scabby Butte. The specimens are completely disarticulated and commingled, but the quality of bone preservation is superb, affording detailed insight into the anatomical structures of "thick nose." Furthermore, bones of the skeleton, indeed essentially all bones in the body, are well represented by multiple specimens spanning a wide size range.[7] It is a fossil deposit of the greatest significance.

Analysis is still in progress, and nothing substantial has yet appeared in print, apart from Tanke's abstracts in 1988 and 1989.

Several details are noteworthy. Tanke surveyed bone injuries on more than one thousand bones and reported a total of only ten lesions, a low frequency. Two were on the parietal frill, another elsewhere on the head, three involved ribs, two involved vertebrae, and one each involved a scapula and a toe bone. Most pathologies probably relate to accident or disease, indicating that injuries due to aggressive behavior are uncommon. This suggested to Tanke that the nasofrontal boss was a blunt weapon that did not inflict injury.

Further insight into the frill has also come forward. The Pipestone Creek *Pachyrhinosaurus* shows unicorn horns, a prominent set of conical prominences on the midline of the parietal bar. This feature is unique and very distinctive. As there is no indication of such unicorns in the specimens from Scabby Butte or Drumheller, this finding, along with the smaller size, may confirm the existence of a distinct, not-yet-named species of *Pachyrhinosaurus*. In addition, the spikes at the back corners of the frill curve forward. Otherwise, Langston's reconstruction of the frill holds up very well indeed. In the collection are small parietals said to resemble those of *Brachyceratops*. The tentative conclusion is that the extraordinary characters (such as parietal thickening, bosses, spikes, and unicorns) that make adult *Pachyrhinosaurus* so distinctive are expressed late in the growth trajectory, and that younger individuals resemble the most generalized or basal centrosaurines. This conclusion has not yet been documented, and I confess my skepticism; however, I hereby record it in the event that it proves to be correct. (The corollary being, of course, that *Brachyceratops* bites the dust as a valid taxon.) Tanke also reports concave (typical) and convex nasal bosses, in equal numbers. Thus, sexual dimorphism is a distinct possibility.[8] Whatever may prove to be the case, the Pipestone Creek bonebed is of extraordinary importance, and I await further information with great anticipation.

PACHYRHINOSAURUS IN THE COLD AND DARK

Most dinosaurs that we know—and ceratopsids are no exception—come from the midlatitudes. However, recent finds in both hemispheres have extended the geographic scope of dinosaur discoveries. Notable from the southern hemisphere are ornithopods in Antarctica and on the south coast of Australia, which was within polar latitudes in the Early Cretaceous. Quite recently, an Early Jurassic crested meat-eater, *Cryolophosaurus,* was described from Antarctica. In the northern hemisphere, dinosaurs have come to light on the North Slope of Alaska, along the Colville River at 70° north latitude. Deposits of the Late Cretaceous terrestrial Prince Creek Formation are dated at about 69 Ma, about the same age as the Horseshoe Canyon Formation 3,000 km to the south. Not surprisingly, the first reports of dinosaur bone came from an oil company exploration geologist (who else would be up there?) named R. L. Lipscomb as early as 1961, although these finds were not described until 1987. The fossil deposits are bonebed deposits containing several species of dinosaurs, dominated by the hadrosaur *Edmontosaurus,* including juvenile specimens. Nondinosaurian components of the fauna include a shark, several kinds of bony fish, and mammals, including marsupials, placentals, and multituberculates. A striking difference from typical Late Cretaceous deposits of Alberta and Montana is the lack of amphibians and nondinosaurian reptiles. In addition to the hadrosaur are teeth indicating the presence of a small theropod (either a dromaeosaurid or a troodontid), hypsilophodontids, and . . . *Pachyrhinosaurus!*[9] The specimen is only a partial skull, but the identity is unmistakable. The find has not yet been described, and there is no way of knowing whether it is from the same species as the Scabby Butte or the Pipestone Creek *Pachyrhinosaurus.* Nonetheless, it constitutes evidence for an extraordinary geographic range for *Pachyrhinosaurus:* about 3,500 km from the Colville River to Scabby Butte (Plate IV)!

This discovery sets up the possibility that *Pachyrhinosaurus* was a migratory dinosaur that summered on the North Slope at an estimated Late Cretaceous paleolatitude of 85° north, amid lush, deciduous, large-leafed plant growth that thrived under conditions of 24-hour sunlight. In the autumn, as the sun receded toward the horizon and plants ceased to put forth fresh leaves and new growth and dropped the leaves they bore in preparation for the dark of the boreal night, the ceratopsids and hadrosaurs may have begun a stately southward shuffle in lowlands on the western shore of an inland sea. At a steady rate of 50 km per day, they could have reached what is now southern Alberta in sixty days.[10] This is merely a possible scenario. We do not know for certain that *Pachyrhinosaurus* herds *were* migratory, but the facts as we know them are compatible with this inference. Because the winters were relatively mild, with an average temperature above freezing, there is the possibility that smaller populations of dinosaurs may have overwintered.

MONTANA ONCE AGAIN—AVA'S LITTLE HORNED FACE

With the description of *Pachyrhinosaurus* in 1950, one could believe that ceratopsid diversity in North America had been plumbed to its depths. After World War II, a whole generation of paleontologists entered the lists, passed their entire working careers, and retired, without a new ceratopsid being described. Montana had been brilliantly productive of dinosaurs of latest Cretaceous age. The Hell Creek Formation in eastern Montana produced (and continues to produce) spectacular skulls and skeletons of *Tyrannosaurus, Triceratops,* and *Edmontosaurus;* as we have observed before, the Judith River Formation of Montana was not particularly kind to collectors one hundred years ago, and it still is not today. Jack Horner's magnificent discoveries of hadrosaurs beginning in 1978 stimulated a renaissance of dinosaur paleontology in the state, but neither the Judith River Formation nor horned dinosaurs played a significant role. Yet, happily for our story, the Judith River Formation has not finished yielding treasure.

The Missouri River is one of the great rivers of the United States and of the world. It arises among mountain spruce trees on the sunset side of the Bridger Mountains not far from Bozeman, courses clear on its northerly run to Great Falls, and gradually acquires its brown load of Cretaceous mud (this is not merely *common* mud, but actually a particularly oozy and treacherous kind named bentonite)[11] as it passes Fort Benton and then bends east across a vast expanse of yellow-brown prairies. Whether water falls as unwanted slushy snow on the tent of a shivering summer camper beside Yellowstone Lake in northwestern Wyoming, or as a glutinous coating on the puffy, white bear grass flowers or purple lupine in a meadow beside Two Medicine Lake in Glacier National Park in northwestern Montana,[12] the meltwater ends up in the Missouri, bound for the Mississippi River and then the Gulf of Mexico, 4,000 km away.

Much of Montana is dry. Range lands in southern and eastern regions of the state make do with as little as 25–30 cm of rain annually. If the rain falls at the right time of year with proper spacing between showers, ranchers can get by. But when the clouds refuse to part with their element, times are tough for beasts and their keepers. The mountain peaks that form such majestic vistas throughout much of the state scrub water from the skies and divert it from the parched surface of the grasslands. During the winter, thick blankets of snow mantle the craggy mountains. In July, the whiteness of the mountain summits, twinkling though the summer haze, cools the mind if not the body when viewed from the heat-crazed plains. The melting snow feeds numerous small rivers that carry welcome wetness for a distance across the prairies before losing their identities by discharging into the greedy, swelling Missouri. One of these rivers is the Musselshell, one fork of which arises in the Castle Mountains, the other in the Little Belts, not far from the geographic center of the great state of Montana. The scenic Musselshell, a noted brown trout stream, runs eastward past Two Dot (so named for the cattle brand of its founder), Harlowton, Shawmut, Ryegate, and Lavina, cottonwoods lining its bank, past picturesque bluffs of Late Cretaceous sandstone. After following a course that carries it some 225 km eastward, it makes a right angle bend to the north, where it encounters the Missouri 100 km farther on, only 80 km east of the spot where Cope discovered *Monoclonius* in 1876. Cope would not recognize this part of the river today, because the mighty Missouri is dammed at Fort Peck, forming Fort Peck Reservoir, a large lake that extends westward more than 150 km.

Eddie Cole is a reckless man. If he had asked me for advice, I would never have advised him to visit the Musselshell Valley in south-central Montana to search for Late Cretaceous dinosaur fossils. I would have taken out my map and patiently shown him the spot on the Missouri, 165 km northeast of Shawmut, where Hayden had found *Troodon* and Cope had found *Monoclonius*. I would have explained to Eddie that although the Judith River Formation had been mapped in the vicinity of Harlowton, Shawmut, and Ryegate, no dinosaur fossils had ever been reported from there. Fortunately, Eddie is not one to seek my advice. If I had discouraged his initiative, my scientific life would have been incomparably impoverished.

Eddie went to the Musselshell Valley in 1981 on a vague rumor and sniffed fossils out of the ground where none had been known before. He located an important bonebed on the large Careless Creek Ranch, owned by the Arthur J. Lammers family of Shawmut. With the family's permission, Cole and a small crew, including his wife Ava, worked for a month excavating the bonebed, carefully numbering and mapping each specimen removed. Cole lacked scientific training, and eventually I was asked to visit his fossil shop in Wall, South Dakota, and evaluate the collection. I arrived there in October 1991. I recognized from the specimens laid across the table surfaces in Eddie's shop that this was a more or less typical Late Cretaceous bonebed assemblage, dominated by duck-bills, both crested and noncrested, but also with evidence of meat-eaters large and small, armored ankylosaurs, and even dome-headed pachycephalosaurs. Nondinosaurian vertebrates included abundant turtles, as well as crocodiles and fishes. One find, however, stood out from all the others.

Standing by itself in a corner was a small, fan-shaped solid frill of bone that I recognized at once as the parietal frill of a small ceratopsid. I knew that small ceratopsids are rare, not just in Montana but any-where. The only ones described from Montana had been those of *Brachyceratops*. I was prepared to think that this might be *Brachyceratops*, the fact that the latter was found nearly 400 km away in another part of the state offering no compelling argument against this interpretation. I was very excited and explained to Cole why. As we looked further, we discovered a left squamosal that attached perfectly to the parietal. The squamosal was missing from *Brachyceratops* and thus could not be compared to this specimen. Further rummaging turned up a humerus and femur of approximately the correct size. We consulted the quarry map and determined that all of the bones of the ceratopsid, though disarticulated, had come from a small portion of the quarry, the south-eastern corner. An agreement was reached whereby the bones would come to the Academy of Natural Sciences of Philadelphia. Furthermore, Cole agreed to take me to the site the following summer, introduce me to the Lammers family, and help me search for more bones of the small ceratopsid.

At that time in my life, the years 1980–1983, I was excavating at Dinosaur Provincial Park in Alberta, supported by the National Geo-graphic Society as a guest of Phil Currie and the Tyrrell Museum. In July 1982, I took leave from the park for a few days, loaded the van with my family and several collecting associates, and arrived at the Careless Creek Ranch. Eddie Cole was as good as his word, and the foray was a success in every way. We collected foot bones, ribs, and assorted miss-ing parts of the ceratopsid and secured permission from the Lammers family for further excavations, which began in 1984. Initially, our fund-ing support came in the form of a private grant from the Academy of Natural Sciences; later I received a grant from the National Science Foundation. Anthony Fiorillo became my graduate student and man-aged expeditions to the site and the region for five years, earning a Ph.D. at the University of Pennsylvania in the process. We quickly came to love the ranch for its natural beauty, wildlife, and mountain vistas, and for its inhabitants, the three generations of the Lammers family. They were remarkable hosts who took great interest in the scientific operation on their land. On evening visits to camp, they always brought some delicacy from one of their kitchens and lemonade or iced tea swimming with ice cubes. They never expected compensation for the fossils that went to science—how admirable!

I gave Tony the primary responsibility for taphonomy (the study of the preservation of fossils in their geological contexts), reconstruction of sedimentary environments, and analysis of associated faunas.[13] The ceratopsid was mine! Eventually all the fossils arrived at the Academy of Natural Sciences, and I was able to press on with my analysis. The little ceratopsid was incomplete, but there was nevertheless a substantial amount of material to examine. There were fourteen vertebrae, including fused cervicals, sixteen ribs (plus fragments of others), both scapulae and coracoids, both humeri, a left radius and ulna, and a good representation of the bones of both hands. No pelvic bones had been preserved, but both femora, tibiae, and fibulae were there, as were a good set of foot bones. Much of the skull was present. In addition to the parietal and left squamosal were both upper jaws and one lower jaw, a braincase, a jugal, both quadrates, quadratojugals and pterygoids, and the left premaxilla. All told, there was plenty of material from which to reconstruct the appearance of the little animal, but the bones that were missing (and which remain missing) are frustrating indeed: among them is the top of the skull, including both the nasal and orbital horn cores! Thus, the appearance of these vital structures remains entirely conjectural.

I determined by review of the literature and by careful inspection of specimens at the Smithsonian, the American Museum of Natural History, the Yale Peabody Museum, the Canadian Museum of Nature (Sternberg's old National Museum of Canada), the Royal Ontario Museum, the Royal Tyrrell Museum, and the Museum of the Rockies that the little ceratopsid was *not Brachyceratops*, nor was it any other species that had ever been described. So it was that in 1986 I named it *Avaceratops lammersi*, "Ava's horned face" (Fig. 6.2). The name honors Ava Cole, Eddie's wife, and the Lammers family. It is the most complete dinosaur specimen ever to come from the Judith River Formation of Montana. I published my paper in the *Proceedings of the Academy of Natural Sciences* on December 30, 1986, in the same journal in which Cope had published his report of *Monoclonius crassus*, 110 years and 2 months earlier. (We paleontologists are very sensitive to history.) It was the first new genus of ceratopsid to be described since *Pachyrhinosaurus* in 1950.[14]

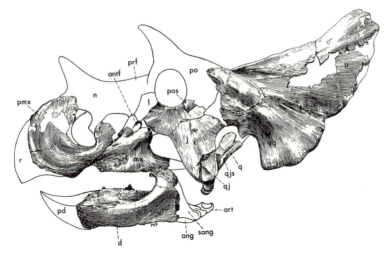

FIG. 6.2. Skull of *Avaceratops lammersi*, Academy of Natural Sciences, Philadelphia. ang, Angular; antf, antorbital foramen; art, articular; d, dentary; j, jugal; l, lacrimal; mx, maxilla; n, nasal; p, parietal; pd, predentary; pmx, premaxilla; prf, prefrontal; po, postorbital; pos, sutural surface on jugal for postorbital; q, quadrate; qj, quadratojugal; qjs, sutural surface on quadrate for quadratojugal; r, rostral; sang, surangular; sq, squamosal. (Courtesy of Paul G. Penkalski.)

The skull was reconstructed and cast by Kenneth Carpenter, and the skeleton was mounted for the Academy of Natural Sciences by Leroy Glenn. The skeleton was unveiled in December 1986 to coincide with the publication, and the mount became a permanent component of the wonderful Discovering Dinosaurs exhibit that opened at the Academy in January 1986. *Avaceratops* may be viewed at the Academy, or a high-fidelity cast may be viewed at the Upper Musselshell Valley Historical Museum in Harlowton, Montana, only 50 km from the Lammers ranch from where it came. The cast was prepared by my student Paul Penkalski and was installed in Harlowton on July 3, 1993, a gift to the people of Montana from my employer, the School of Veterinary Medicine of the University of Pennsylvania, and the Academy of Natural Sciences. The gift was made possible by the generosity of several blessed individuals who have been at my side at important times during my career.

I found several facts about *Avaceratops* very striking, starting with its small size. For an index of comparison, *Brachyceratops* is an obvious choice. Gilmore's skeleton is 2.06 m long. Bones in common between the two skeletons include femur, tibia, metatarsal III, and ulna. These elements in *Avaceratops* average about 12 percent longer than the same

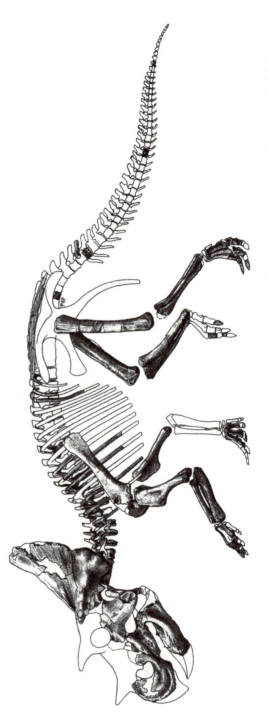

FIG. 6.3. Skeleton of *Avaceratops lammersi*, Academy of Natural Sciences, Philadelphia. (Courtesy of Paul G. Penkalski.)

FIG. 6.4. Reconstruction of *Avaceratops*. (Robert Walters.)

bones in *Brachyceratops*. I therefore estimated the length of *Avaceratops* at 2.3 m (7 ft 7 in.). Key measurements are as follows:

Femur length	41 cm
Tibia length	38.5 cm
Metatarsal III length	11 cm
Height at hips	91 cm

This last figure compares with 71 cm at the hips in *Brachyceratops* (Figs. 6.3 and 6.4).

In the skull, both the parietal and the squamosal are highly distinctive. The indications are that the parietal is solid and without openings. It is thin and delicate. At an estimated 34 cm in length, it is not in absolute terms a small bone (being, for instance, about 70 percent of the length of the parietal in the type of *Centrosaurus apertus*). The rear border is smoothly rounded with little ornamentation, and there is no indentation on the midline, as is universally seen in other centrosaurines, including *Brachyceratops*. The border is 7 mm thick on the midline, thickening to as much as 13 mm at the corners. Along the line of articulation with the squamosal, the thickness is about 16 mm. Where fenestrae might be expected, the parietal thins to as little as 3 mm in thickness, although it is typically 4–6 mm thick. This region is poorly preserved (as is also the case in all specimens of *Brachyceratops*). Eddie Cole, who collected the parietal, was quite adamant that there were no

189

fenestrae, and my strong inclination is to follow his testimony, even though its condition upon discovery in the field cannot be verified. In fact, I have my doubts that *Brachyceratops* had fenestrae either. If fenestrae did exist, they were unusually small.

In contrast to the delicacy of the parietal, the squamosal is robust and relatively large. At first I even doubted that they were associated, but Cole's map shows that they were found in proximity one to the other, and even more importantly, they fit: the articular surfaces are congruent. I found the shape of the squamosal of *Avaceratops* very distinctive. It is ramrod straight. In all chasmosaurines, the dorsal border is strongly curved, corresponding with elevation of the frill. In other centrosaurines, the dorsal border has a stepped profile: the front portion is straight, there is a step-up at the back border of the supratemporal fenestra (which is situated between the squamosal and the parietal), and the rear portion is straight the rest of the way back. This morphology characterizes *Monoclonius*, *Centrosaurus*, and *Styracosaurus*. I presume it is found in *Pachyrhinosaurus* and is unknown in *Brachyceratops*, but it is quite different and unique in *Avaceratops*. The back half of the squamosal in all ceratopsids is wide and robust. It is also longer than the thinner front part of the bone. In chasmosaurines, the rear part of the squamosal may exceed the length of the front part by a factor of 4; in centrosaurines, the factor is 1.5. In *Avaceratops*, the front and back parts of the squamosal are about equal in length. The total length of the squamosal in *Avaceratops* is about 88 percent of the length of the parietal. In specimens of *Styracosaurus* and *Centrosaurus* the squamosal is 81 and 70 percent of the length of the parietal, respectively. The squamosal of *Avaceratops* bears four distinct scallops.

The great question about *Avaceratops* is whether it is a juvenile or an adult specimen. The fact that the skull bones came apart prior to burial is certainly compatible with a juvenile status (although in reptiles generally, and in dinosaurs typically, at least some skull sutures remain open throughout life—with the potential to come apart after death). However, there is an exception. Actually, *all* bones did not come apart. The braincase is the single fused unit in the collection. I was particularly impressed by the spherical occipital condyle, so characteristic of ceratopsids. In *Avaceratops*, this structure measures a puny 38 mm in diameter, yet is completely fused, suggesting that growth was close to complete. (Growth between bone centers was clearly finished, but it is possible that further growth could still have increased the diameter a little bit by addition of bone around the outside, a process of growth called appositional growth, i.e., growth like tree rings.)

For comparison, among specimens of mature *Centrosaurus* skulls that I have measured, condyle diameter ranges from 65 to 78 mm. In the collection of the RTMP are a series of condyles from the remarkable *Centrosaurus* bonebed, including the remains of hundreds of individuals of *Centrosaurus*: juveniles, subadults, and adults.[15] These unfortunates may have perished in a mass crossing of a flooded river. The condyles range in diameter from 50 to 65 mm; all are fused, and they show no indication of the three separate elements (paired exoccipitals, basioccipital) that comprise the condyle. The smallest condyle in the collection measures 47 mm in diameter and is unfused, showing the three separate bones. From this I infer that in *Centrosaurus*, fusion of the cartilage bones of the braincase that compose the condyle occurs at a diameter of about 50 mm. At this point, growth between the bones ceased, and further growth, perhaps as much as 50 percent more, must have occurred by appositional growth. Now in *Avaceratops*, the condyle is already fused at 38 mm. This point seemed sufficiently important to me that I had CAT scans performed, and no trace of sutures could be discerned inside the bone. My conclusion is that although *Avaceratops* may not have ceased growth, it was approaching maturity. By analogy with *Centrosaurus*, diameter of the condyle may have increased by as much as 50 percent, but this would still place *Avaceratops* at the small end of the size range for a mature centrosaurine, with a body length of 3–4 m, rather than the 5 m typical of *Centrosaurus*.

One other line of evidence suggests that *Avaceratops* was not a typical centrosaurine. The unguals of ceratopsids are rounded, even in juveniles. Those of *Avaceratops* are more pointed and blunt, reminiscent of those of *Protoceratops*. This is a character I do not expect to change significantly during growth. In skeletons of *Brachyceratops*, for example, the unguals resemble those of typical ceratopsids, not those of *Avaceratops*.

It is unfortunate that *Avaceratops* remains the only specimen of its kind. The fossil record of ceratopsians is generally a very good one. Unlike that of small meat-eating dinosaurs (coelurosaurs), for which many genera are based only on a single fragmentary specimen, we almost expect that for ceratopsians, there will be at minimum a complete skull, and typically there will be several skulls.

My student Paul Penkalski, who is gifted with superb technical and artistic skills in paleontology as well as with a fine imagination, completed a detailed study of *Avaceratops* in 1994, which I trust will appear in print before long. He pointed out that Hatcher, Marsh, and Lull in 1907 figured two squamosals, collected by Hatcher from the Judith

River beds not far from the type locality of *Monoclonius crassus* at the same time as *Ceratops*, which Marsh described in 1888. In the 1907 monograph, these specimens are labeled as *Ceratops montanus?* (USNM 4802) and *Ceratops* sp. (young) (USNM 2415). Paul points out that these specimens show the same morphological features as the squamosal of *Avaceratops*, most especially the straight dorsal border, but also a line of three prominent bumps on the side. He believes that these old and forgotten specimens constitute proof of the coexistence of *Avaceratops* and *Monoclonius*. He believes that the larger one indicates an animal 4.2 m long.[16]

Furthermore, an intriguing new skull from the same region, known as the Missouri Breaks, has recently come into the collection of the Museum of the Rockies—the rich get richer. It is a bit of a road-kill, as we say, flattened and incomplete, but it shows a most interesting set of features. Penkalski estimates the length at 1.27 m, roughly the size of the ROM skull of *Centrosaurus apertus*, the smallest *Centrosaurus* skull documented. This estimate is very rough because the front of the skull is missing and the frill is not in sutural contact with the rest of the skull. Nonetheless, it is clear that the skull is relatively small, appears to have a short, solid frill and, most interesting of all, has rather tall postorbital horn cores. These measure about 250 mm in height, exceeding by a long shot those of any known centrosaurine. (The skull of *Centrosaurus flexus* in Ottawa has a postorbital horn core about 160 mm in height.) Unfortunately, the squamosal cannot be reliably interpreted from the evidence at hand. It is premature to refer the skull to *Avaceratops*, but this is a possibility to bear in mind.

There are those who take the position that it is unreasonable to define dinosaur genera on the basis of immature material. Considering the history of dinosaur paleontology, I am excruciatingly aware that far worse sins have been committed. My scientific career was founded on studies of growth changes in alligators, lizards, *Protoceratops*, and duck-billed dinosaurs.[17] My experience shows me that although it is important to take growth trajectories into account, it is not true that it is impossible to characterize taxa in their juvenile stages. For example, I was concerned that the details of the squamosal of *Avaceratops* might represent juvenile character states. Perhaps the stepped profile of the squamosal developed during growth? Although most centrosaurine skulls are large, an important small specimen came to light in Dinosaur Provincial Park in 1982. In 1988 Phil Currie and I described this specimen as the smallest ceratopsid skull yet discovered.[18] The tiny parietal frill of this specimen measures 210 mm in length, and the beautifully preserved left squamosal measures 165 mm. The specimen, which we very tentatively referred to *Monoclonius*,[19] shows that the large size of the squamosal relative to the parietal seen in *Avaceratops* is *not* a juvenile character, and also that the stepped profile of typical centrosaurines does not change during growth but is seen in the smallest specimen. The tiny Alberta specimen reinforces the distinctiveness of *Avaceratops* by showing that not all of its important characters are due to its youthfulness.

MORE HORNS IN MONTANA—STRANGER THAN EVER

Jack Horner from the Museum of the Rockies at Montana State University is a duck-bill man. Given his extraordinary skill in locating specimens, it is not surprising that other, non-hadrosaurian, specimens come into his possession, including some of the redoubtable meat-eater, *Tyrannosaurus rex*. It is also not surprising that these specimens fail to grip his imagination in the same way the hadrosaurs do. In 1987, Horner and his parties were prospecting northern exposures of the Two Medicine Formation in the Milk River badlands on the Blackfoot Indian Reservation, near Landslide Butte, where Gilmore had worked over fifty years earlier. As luck would have it, Horner came up with ceratopsian material, including skulls, in several bonebeds. The names Canyon Bone Bed and Dino Ridge Quarry were quickly applied.[20] Graduate student Ray Rogers determined through taphonomic investigation that both bonebeds document drought-related mortality. The herd gathered around a dwindling waterhole until survival was no longer possible.[21] The first impression is that these ceratopsids were styracosaurs of some sort, though not the classic *Styracosaurus albertensis*. Such a discovery would not be altogether surprising, recollecting that this is the very area from which Gilmore had collected *Styracosaurus ovatus* in 1928, not to mention *Brachyceratops montanensis* in 1913. At last, it appeared that some long-standing problems were going to be cleared up. However, it seems that the paleontological equivalent of Murphy's law is at work: the more specimens are found, the less clear the situation becomes.

Horner kindly invited me to examine the specimens, given my interest in horned dinosaurs. I visited the Museum of the Rockies in 1987 and again in 1989. Two morphological features immediately became

clear. One is that the parietal frill of the new "styracosaur" has a relatively simple pattern, with a single prominent spike on each side of the
frill parallel to the midline. *Styracosaurus albertensis* has three spikes and
S. ovatus has two. The new styracosaur, for which there are multiple
specimens from the bonebed, consistently shows the single spike—no
specimen varies in the direction of *S. ovatus*, and we as yet seem to have
no insight into the nature of this animal. The second and extremely
striking feature is a remarkable nasal horn core of the sort seen in no
other ceratopsid, or in any living or fossil horn-bearer to my knowledge. This is a very prominent "floppy" horn core with a 23-cm-long,
solid base that droops forward over the rostral bone. This horn looks
like nothing so much as an old-fashioned bottle opener (an anachronism in the land of Rainier beer pull-tab cans). When one first sees this
bizarre structure, one doubts that this is the horn of a normal, healthy
animal. There are, however, a number of specimens in the museum's
collection that show a similar structure. Thus notions of individual
anomalies or pathology must be dismissed. There are juvenile nasal
horn cores as well that resemble those of *Brachyceratops*. One interpretation is that during growth *Brachyceratops* itself (or an animal very much
like it) saw its nasal horn core transformed from a gracile, delicate,
backward-curving structure into a massive, forward-flopping one. I am
not convinced that this was so, although it is not impossible.

The case is even further complicated: in 1992 Horner, his student
David Varricchio, and Mark Goodwin from the University of California
at Berkeley published a paper in the prestigious British journal *Nature*
in which they intimated the existence of not one but *three* separate
ceratopsids, which they posit form a transition between *Styracosaurus*
and *Pachyrhinosaurus!* Details are very sketchy, but I will convey the
situation as described. The authors note that the Two Medicine Formation is about 650 m thick and covers an estimated six to seven million
years duration. Three fossil-bearing horizons occur in the uppermost
portion of the formation; a radiometrically dated ash not far below the
bonebeds gives a date of 74.3 Ma. The three horizons are separated by
no more than 40 m vertically, which works out to approximately half a
million years or less. This suggests that evolution was extremely rapid
indeed. They hypothesize that habitats were cramped by rapid en-
croachment of the Bearpaw Sea that lay immediately to the east. (The
Bearpaw Sea stretched from the Gulf of Mexico to what is the Arctic
Ocean today and ensured that much of what is now the western United
States and Canada was under water at that time.) The crowding may

have established what geneticists refer to as a population bottleneck, which constitutes strong pressure for rapid evolutionary change.

In any case, all three styracosaurs have one pair of parietal spikes, a character that I would interpret as more basal than the luxuriant three pairs of spikes of *Styracosaurus albertensis*. The lowermost of the three, designated as "transitional taxon A,"[22] with six specimens, is said to be "identical to *Styracosaurus*, except that it possesses only one pair of parietal spikes." I regard this as an important *except*. It is not clear from their illustration that the nasal horn core is identical to *Styracosaurus*, by which I presume is meant *Styracosaurus albertensis*. The succeeding "transitional taxon B" came from Canyon Bone Bed and Dino Ridge Quarry, less than 200,000 years after taxon A. There are twenty specimens of B, which is the "droopy-horned" ceratopsid. "Transitional taxon C," with four specimens, came some 250,000 years later than B, and it shows an extremely rugose nasal horn core, if it can be called that, that seems to prefigure some of the structures seen in *Pachyrhinosaurus*. In addition, the surface of the bone over the eyes is extremely rugose as well. This implies that coalescence of the separate centers of rugosity observed in taxon C formed the definitive nasofrontal boss of *Pachyrhinosaurus canadensis*.[23]

The schema just outlined is an interesting hypothesis, not an undoubted fact. If it is so, it represents a remarkably rapid burst of evolution, probably the best example among the Dinosauria. Other interpretations are possible. I am concerned, given the short interval of time between taxon B and taxon C, and the apparent identity of the parietals of the two forms, that they may be sexual variants of the same taxon. As described the situation seems to me to represent a reversal of trends of elaboration of the parietal and of the nasal horn core that culminated in *Styracosaurus albertensis*. Indeed I suspect the lineage had nothing to do with *Styracosaurus*, but instead represents a separate lineage within the Centrosaurinae. I had been prepared to see *Pachyrhinosaurus canadensis* as a variant on the theme of *Centrosaurus*, but perhaps that is not the case. Or perhaps taxon C and *Pachyrhinosaurus* are merely convergent upon a similar skull shape, although this would not seem to be the most probable solution. Nonetheless, the ecological bottleneck of the St. Mary River Formation—which apparently contains transitional forms of carnosaurs possibly ancestral to *Tyrannosaurus*, a domehead, and a hadrosaur transitional between *Lambeosaurus* and *Hypacrosaurus*—is an extremely interesting evolutionary experiment.

195

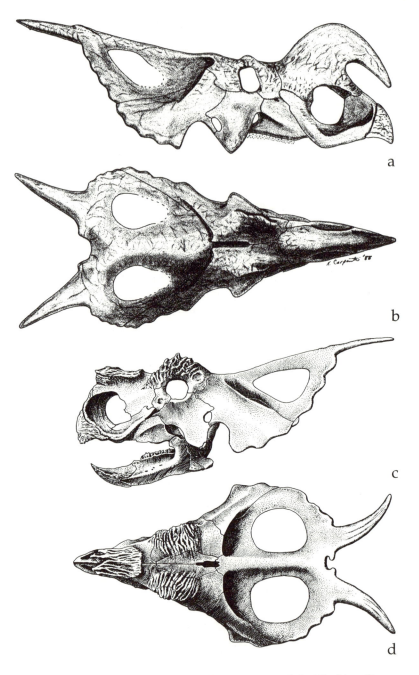

FIG. 6.5. New horned dinosaurs at the Museum of the Rockies, Bozeman, Montana. (a) *Einiosaurus procurvicornis*, lateral view. (b) *E. procurvicornis*, dorsal view. (Kenneth M. Carpenter.) (c) *Achelousaurus horneri*, lateral view. (d) *A. horneri*, dorsal view. (Robert Walters after Kris Ellingsen from Sampson 1995b.)

196

The detailed description of the ceratopsids lies in the competent hands of Scott Sampson, an energetic young man who recently completed his Ph.D. at the University of Toronto and has now taken a position at the New York College of Osteopathic Medicine on Long Island. He has recently named the two new horned dinosaurs. The droopy-horned "taxon B" (Fig. 6.5) is known as *Einiosaurus procurvicornis* ("forward-curved horn buffalo reptile," using a Blackfoot name). The proto-pachyrhinosaur, "taxon C," is named *Achelousaurus horneri*, after Achelous, a form-changer in Greek mythology who broke his horn in battle, and Jack Horner.[24] Clever names, Scott!

WHAT'S THE BIG IDEA?

Now at last we are in a position to review the Centrosaurinae. There is virtual unanimity among paleontologists that the ceratopsids described in this chapter and the preceding one form a natural monophyletic assemblage or clade, which Lambe named the Centrosaurinae in 1915. However, there is a remarkable absence of consensus on what those genera and species actually are. One view, the Barnum Brown/American Museum of Natural History view, is that *Monoclonius* was the most common genus in both Montana and Alberta, and that *Centrosaurus* has no claim to validity; another is that *Monoclonius* is undiagnosable and consists only of subadults of *Centrosaurus*, *Styracosaurus*, and *Pachyrhinosaurus*.[25] Some feel that *Brachyceratops* is a juvenile of *Monoclonius*, *Styracosaurus*, *Centrosaurus*, or *Pachyrhinosaurus*. Some feel that *Avaceratops* is indeterminate because it is too young to show any definitive characters. For a field that for many years was rather sleepy, it is remarkable how much interest there is these days in horned dinosaurs. I may decide to take up butterfly collecting instead!

It is conceivable that each specimen located in Alberta, Montana, or Alaska is potentially a separate genus and species. Clearly the thought of some sixty genera and species has little appeal. Early in this century, this appeared to be the trend, but around 1940, I am happy to say, the practice stopped. What is needed is some rational basis for dealing with the taxonomy of centrosaurines. For me, that rational basis is numbers—biometrics, the study of taxonomy by measurement. Years earlier, I had rationalized the taxonomy of crested duck-billed dinosaurs (lambeosaurines) from Dinosaur Provincial Park by biometric methods. I suggested that thirty-six skulls placed in three genera and twelve

197

species actually consisted of juveniles, subadults, and male and female adults of two genera and three species.[26] I recently followed a similar approach with centrosaurines. At the outset, I will say that the fossil record of centrosaurines, although by objective standards one of the best among dinosaurs, is considerably inferior to the lambeosaurine record. I had only fourteen reasonably complete centrosaurine skulls to work with, and the smallest of these is already a large skull 1.2 m long, more than three-quarters the maximum adult size. By contrast, the smallest lambeosaurine skulls I worked with were only one-third the length of adult skulls. Changes during growth were easy to document with duck-bills; they are much harder to demonstrate with centrosaurine skulls, although the undescribed bonebed material will be extremely valuable for this purpose.

Using two-dimensional (bivariate) plots, I made the following determinations. To me it is as clear as night from day that *Monoclonius* and *Centrosaurus* are separate genera. I found that *Styracosaurus* resembles *Centrosaurus* more closely than either resembles *Monoclonius*. Juvenile *Centrosaurus* may conceivably resemble *Monoclonius* (although this possibility emphatically needs to be documented), but I believe that the growth trajectories of the two animals are completely different. In a sense, *Monoclonius* may be a pedomorphic version of *Centrosaurus*—it retains juvenile characters at large size and presumably sexual maturity. I know of no evidence that *Centrosaurus* lived in what is now Montana, although *Monoclonius* certainly lived in Alberta. Morphology is clearly variable within each genus; in fact, it is difficult to say that any two specimens are similar. For whatever reason, it appears that considerable morphological variability was one of the constructional rules of centrosaurines.

Within *Centrosaurus*, I was impressed by differences in a suite of characters. One morph seemed to be characterized by a longer, thicker parietal frill; a thicker, wider squamosal; a deeper face; a tall, forward-curved nasal horn core; a relatively longer orbital horn core; a longer jugal; and a long caudal process of the premaxilla that separates the nasal from the maxilla. The specimens labeled *Centrosaurus flexus* typify this condition. The second morph is characterized by a shorter frill relative to basal skull length, a lower face, a shorter premaxilla that permits contact between the maxilla and nasal, a nasal horn core lower and variable in orientation, generally weaker orbital horn cores, and a thinner parietal and squamosal. *C. longirostris* typifies this morph. In my judgment, the entire suite of characters is consistent with sexual dimorphism, the male shape being *C. flexus*, the female *C. longirostris* and probably *C. apertus*. This is an inference, of course, because the type specimen is only a parietal frill. Thus I recognize only a single species of *Centrosaurus, Centrosaurus apertus,* the first named species. Buoyed by this result, I looked at *Monoclonius* and hypothesized that *M. sphenocerus,* with its erect horn, was a male and that *M. crassus,* best typified by Sternberg's specimen of *M. lowei,* with its low nasal horn core, is the female type.

Lest these results seem too bland, I followed my numbers to one more conclusion. Brown's *Monoclonius nasicornus,* otherwise clearly a *Centrosaurus* type of animal, clustered with *Styracosaurus* in my analysis. It shares with *Styracosaurus albertensis* numerous quantitative characters, only one of which is the striking similarity of the tall, erect nasal horn core. I suggested that perhaps *Monoclonius nasicornus* is actually the female of *Styracosaurus!* There is one subtle piece of information that hints that this suggestion may not be totally bizarre. That is that *M. nasicornus* shows evidence of the abortive rostral processes on the back of the parietal that appear to be failed homologues of the processes that overhang the parietal fenestrae in *Centrosaurus.*[27]

Am I right about all this? Who is to say? The hypotheses are certainly not right just because I say so. At least they are worthy of consideration. This example illustrates the way in which science is an open-ended process. It involves a dialogue with the data, the specimens. Differing interpretations arise from differing philosophical and analytical approaches. In the past, there was a sense that every morphological variant should be named a new species. Today, there is an attempt to interpret the biological dimensions of variation. There is an a priori expectation that variation is an important part of nature. As an illustration, we can look at taxonomic diversity in modern ecosystems where large-bodied mammals are important. On the African savannah, we do not see multiple species of closely related, large-bodied herbivores coexisting. We see only one species of elephant, one of giraffe, one of hippopotamus, and two of rhinoceros, one a browser on leafy twigs, the other a grazer on grass. This experience conditions our view of how many different species of dinosaurs may have lived together in the same communities. We have a strong preference for a taxonomy that recognizes a low number of biologically variable species, rather than a large number of invariant species. The scheme I present may be overturned when more information from bonebeds is documented. Until this material is described, it is as if the theory does not exist—it can have no effect on the scientific community.

No Horns and No Frills

SMALL BEGINNINGS

HORNED DINOSAURS stand among the mightiest dinosaurs of the Late Cretaceous. Their acute features seem chiseled from stone, their silhouetted heads casting bold shadows upon the tremulous landscape. If any dinosaur could hold a hungry *Tyrannosaurus* at bay, it was a bull *Triceratops*, its lethal horns glinting in the sunlight. A white rhinoceros, transported from the grassy African plains to the fern savannahs of the Late Cretaceous, would have wilted at the sight of a *Torosaurus*, whose physique was superior in every measure. Yet, even as mighty Samson was once an insignificant stripling, so the greatest of horned dinosaurs had humble roots—ancestors whose entire bodies could have perched on the heads of the great ones.

The first fragmentary remains that we can recognize in retrospect as ceratopsian were found and named as early as 1872. The first horned dinosaur to take definitive form was *Triceratops*, in 1889. It was rapidly followed by *Torosaurus* in 1891. Before long, Canadian horned dinosaurs began to come to light, beginning with *Centrosaurus* in 1904; *Styracosaurus* in 1913 was followed quickly by *Anchiceratops* and *Chasmosaurus* in 1914. The Canadian dinosaurs did not show the extreme sizes of their American counterparts, but with skulls 1.5 m long they were still very significant animals. If the United States hosted the greatest of the horned dinosaurs, their ancestors came to light elsewhere: first in Canada, later in Mongolia. In this chapter, we focus on a group of small, generalized horned dinosaurs from North America that must include the ancestors of the great ceratopsids.

LEPTOCERATOPS

In his first season in Alberta, 1910, Barnum Brown found a small partial skeleton high on the side of the canyon, along an old cow trail, about three miles north of Tolman Ferry (now Tolman Bridge). Brown wrote of this discovery four years later, in 1914. Perhaps his interest had been spurred by the discovery of a horned dinosaur in Montana the year before. Charles Whitney Gilmore, the Smithsonian's dinosaur paleontologist, had named the world's first small horned dinosaur, *Brachyceratops montanensis*, from north-central Montana, in 1914. Brown announced his find in a paper entitled "*Leptoceratops,* a New Genus of Ceratopsia from the Edmonton Cretaceous of Alberta." His name, *Leptoceratops gracilis,* means "small horn face, slender," although he felt no need to explain this name to his classically educated readership. He thought it was a "primitive, aberrant type related to *Brachyceratops.*" He described it as an animal "not more than four feet in height" (which is overly generous) and was impressed by its long tail. The specimen was incomplete, having been shattered by weathering during harsh Alberta winters and by the trampling feet of several generations of cattle and mule deer.

Brown did a skillful job of illustrating, describing, and interpreting what he had, but a reconstruction of the graceful small one was not possible. He found "the various skull bones so different from those of related genera that, in view of their fragmentary nature, it seems unwise even to sketch an outline of the skull." He was struck that the nasal bone showed "not the slightest indication of an incipient horn." Here was a horned dinosaur without a horn on its nose! The teeth were relatively simple and did not show the split root that is so characteristic of *Triceratops* and all previously known horned dinosaurs. The bones that formed the front of the skull were not recovered, but the deep lower jaw and the very long predentary were well preserved.

Only a portion of the parietal frill was identifiable. He noted that it had a sharp median ridge and a smooth, nearly straight rear border, remarked that it apparently had no openings in it,[1] and inferred that the squamosal extended back to "the extreme end of the crest." He also inferred that the crest was narrow. The shoulder and forelimb were complete, and both the separate bones and the articulated limb were figured. The little five-fingered hand was beautiful, and Brown noted that it was the only complete forefoot then known among the Ceratopsia. The unguals (or "terminal hoofs" as he called them) that terminated the inner three fingers were narrow and pointed, unlike the large rounded unguals of the big guys like *Triceratops.* The outer two toes ended in blunt nubbins of bone. The hind limb was not well preserved, with half of the femur, half of the tibia and fibula, and little else. A nice segment of tail, consisting of twenty-four vertebrae, was found. Brown was impressed by the height of the neural spines and length of the chevrons, indicating a very deep tail of the sort that he had never seen before in a ceratopsian.[2]

The forelimb elements indicate an animal roughly half a meter high at the shoulder. It is not clear why Brown estimated that *Leptoceratops* was no more than 4 ft high, unless he perhaps thought it might have walked on two legs, although he gave no indication of this. He opined that *Leptoceratops* and *Brachyceratops* were so different from other ceratopsians that the two animals would prove to belong to a new subfamily of the Ceratopsia. Three years later when Gilmore monographed *Brachyceratops,* he repudiated this view. It was clear to Gilmore that his animal was unrelated to *Leptoceratops*. He correctly regarded *Brachyceratops* as a small and possibly young relative of *Monoclonius* (i.e., in our terms a centrosaurine). He wryly observed that "Both, it is true, belong to the Ceratopsia and both are diminutive members of that order, but there, for the most part, their close resemblance ends."[3]

Leptoceratops was a remarkable animal. Here was a tiny dinosaur, 2.7 m long at most, that could have walked under the belly of *Triceratops,* with whom it shared the latest Cretaceous landscape. Though it was a horned dinosaur, it had no horns on its nose—and subsequently it became clear that it had none anywhere else either. It had a simple little frill, which lacked openings and which also lacked the other kinds of lumps, bumps, patterns, or ornaments so typical of large horned dinosaurs.

Brown found no further specimens and *Leptoceratops* remained enigmatic for many years, although by 1933 it was clear that it was related to *Protoceratops,* a dinosaur discovered in Mongolia ten years earlier. Barnum Brown collected another small ceratopsian skeleton, this time in the St. Mary River Formation near Buffalo Lake in Montana, in 1916. In 1942, he and E. M. Schlaikjer described this specimen as a new species of *Leptoceratops, L. cerorhynchus*.[4] Their opinion did not stand the test of time.

In 1947, a spry C. M. Sternberg, sixty-two years young, was working in what was then known as the Upper Edmonton Formation (today the Scollard Formation), the latest Cretaceous badlands located 50–80 km north of Drumheller and 20 km northeast of Elnora—a speck on the map. He had proven that these beds were the equivalent of the latest Cretaceous Lance and Hell Creek formations of Wyoming and Montana, respectively, by demonstrating the presence of *Triceratops* and

Ankylosaurus in the Alberta beds. In 1951 he wrote this account of his discovery:

> toward the end of the season I located a skull and jaws and a considerable part of a splendidly preserved skeleton of *Leptoceratops gracilis* (Catalogue No. 8889). At the same time, student assistant T. P. Chamney located a smaller individual of the same species (No. 8888) which was complete except for most of the head and parts of the left fore foot, which had been eroded away. . . . While working on this specimen, a third and still smaller individual (No. 8887) was located, lying beside it. (It is believed that this is the only absolutely complete ceratopsian skeleton known.) Number 8887 has an overall length of 5 feet, 4 inches; No. 8888 is 6 feet, 7 inches (front of skull estimated); and No. 8889 is estimated to have been 7 feet 8 inches to 8 feet long.[5]

Comparing his extraordinary specimens with Brown's original specimen, Sternberg noted:

> All comparable parts check very closely with preserved parts of the type specimen and with each other, and it is believed that our three and Brown's two specimens[6] all represent the same species. As our specimens represent three sizes, all of which are smaller than the type, and Brown's second specimen is still larger, we have five growth stages. The complete ossification of the bones shows that they were not extremely young, but the open sutures suggests that none represents an old animal. They probably represent half-grown or young adults.[7]

Sternberg noted the irony that *Leptoceratops*, the most primitive known ceratopsian, coexisted with *Triceratops*, the most advanced ceratopsian.[8] He also contemplated the environment that *Leptoceratops* inhabited:

> One might hazard a guess that the upland was preferred by the primitive ceratopsians and that in *Leptoceratops* we have a primitive form that continued on with little change, except for some increase in size. Perhaps the delta country was more conducive to rapid growth and evolution, but the smaller and more primitive forms usually remained on the upland and, therefore, were not so often preserved as fossils.

He carefully elucidated many details of anatomy that were previously poorly known. For Sternberg, "the most surprising feature of the *Leptoceratops* skull is the lack of a parietosquamosal crest which is so characteristic of the Ceratopsia." He clearly recognized how generalized the characters of *Leptoceratops* are, having no evidence of horns

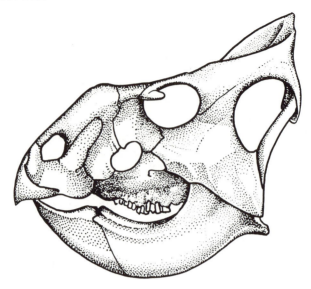

FIG. 7.1. *Leptoceratops gracilis* skull, Canadian Museum of Nature. (From Dodson and Currie 1990. Donna Sloan. Courtesy of the University of California Press.)

over the nose or eyes, and having a squamosal that to him resembled that of duck-billed dinosaurs (hadrosaurs) rather than that of highly derived ceratopsids (Fig. 7.1). He also outlined some very important details about the teeth that had eluded Brown. There are seventeen vertical rows of teeth in each jaw, a low number compared to that in larger horned dinosaurs. This tooth formula is not surprising because of its small size, but because there are only two teeth at each position, the tooth in use and its replacement. The elaborate batteries of teeth of large horned dinosaurs simply were not in evidence. Neither of these characters was perhaps surprising. What was unique to the teeth of *Leptoceratops* is that they were wider side-to-side (technically we say lingual to buccal, that is from tongue to cheek) than long. In addition, the lower or dentary teeth were notched, forming a shelf for crushing instead of simply slicing.

Sternberg's technical description was comprehensive and very good; we need not go into the details. The account ends with a description of the toes of the hind feet. There is no discussion, reconstruction, speculation, or other attempt to sell or to self-promote. The account is illustrated with a number of photographs. Although these photos are useful, they really fail to do justice to the magnificent quality of the specimens. Two skulls are shown, one crushed slightly dorsoventrally in dorsal and palatal views, the other strongly crushed from side to side in profile. It is clear that *Leptoceratops* showed a mix of familiar and unfamiliar details. It had the narrow toothless beak of all ceratopsians, the flaring cheeks with jugal "horns" pointed down and back, and the beginnings of a recognizable frill. The face was short and the skull was short and deep. It looked like a ceratopsian without a frill projecting behind the skull to overhang the neck. Details of the skeleton are shown, and one photo shows two articulated skeletons without skulls on a slab.[9] Nonetheless this stops short of providing an accessible visual image of how the animal actually looked in life. The skeletons were exhibited at the National Museum of Canada—without fanfare or puffery. In consequence of this typically Canadian modesty, *Leptoceratops* has never enjoyed the renown that the quality of its fossils and its evolutionary significance naturally merit (Plate V).

Dale A. Russell eventually succeeded Sternberg as dinosaur paleontologist at the National Museum of Canada, subsequently known as the National Museum of Natural Sciences and currently the Canadian Museum of Nature. Russell began at the National Museum in 1965, a newly minted Ph.D. from Columbia, where he had studied with E. H. Colbert, and fresh from a year at Yale with John Ostrom. As a dinosaur paleontologist, Russell was a bit of a fraud,[10] as his master's studies at the University of California, Berkeley, were on Cenozoic mammals, and his doctoral dissertation was on mosasaurs, reptilian predators of the seas. Then as now, museums unfailingly select curators who have not trained in dinosaur research over those who have. It might be allowed, however, that in the mid-1960s dinosaur research was a very quiet field indeed. Russell is in many ways the antithesis of Sternberg. Whereas Sternberg was short, taciturn, and doggedly descriptive, Russell is tall, loquacious, witty, and manically imaginative. He threw himself into the study of Canadian dinosaurs with passion and abandon, and a steady stream of important dinosaur papers was soon emanating once again from Ottawa.

One of Russell's early papers, published in 1970, was a skeletal reconstruction of *Leptoceratops* (Fig. 7.2). In this paper, Russell restored the skull to compensate for the crushing of the two complete specimens and provided a simple outline reference drawing that portrays for the first time the size of the skull relative to the rest of the body. This graceful little plant-eater is only about 75 cm (30 in.) high at the hips. Its head is 38 cm (14 in.) long.[11]

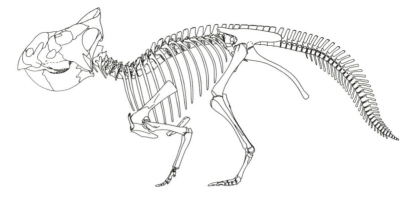

FIG. 7.2. *Leptoceratops gracilis* skeleton, Canadian Museum of Nature. (From Russell 1970. Courtesy of the *Canadian Journal of Earth Sciences.*)

One other report of *Leptoceratops* is noteworthy. This one is of the back half of a skeleton with associated teeth, described by John Ostrom in 1978. The specimen comes from latest Cretaceous rocks in the Bighorn Basin of northern Wyoming. Otherwise, all convincing specimens of *Leptoceratops* are confined to the Red Deer River Valley of Alberta.[12]

PROTOCERATOPS

Henry Fairfield Osborn (1873–1935) was nothing if not ambitious. Born to wealth and privilege, socially connected, highly intelligent, and certainly a man of vision, Osborn became president of the American Museum of Natural History in 1908 and guided it to preeminence among the natural history museums of the world. One initiative that he was pleased to support was an ambitious project to search for human ancestors in the heart of Asia. Indeed, as early as 1900 he had predicted that Central Asia would be revealed to be the cradle of humankind, despite the fact that almost no fossils of any kind had yet been reported from this remote and rugged wilderness. Roy Chapman Andrews was a flamboyant young member of the technical staff of the museum. His wanderlust had led him to visit Mongolia on his own in 1919, and he soon infected Osborn with the idea of mounting an expedition there. With Osborn's social connections and Andrews's energy and charm, a quarter of a million dollars of private money was raised from New York's banking and financial community.

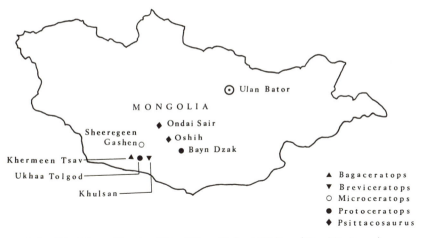

MAP 7.1. Mongolian fossil localities. (Robert Walters/Tess Kissinger.)

In 1922 an expedition of twenty-six adventurers (but only a single experienced paleontologist, Walter Granger) was launched with a maximum of publicity and fanfare. Theirs was a risky and daring enterprise, setting out in search of fossils where none were known to be and where no expedition had ever before ventured. The team was equipped with all the latest technology, even *automobiles*! The expedition roared out of Peking (Beijing) in April 1922, jostling along in three Dodge Brother touring cars and two Fulton trucks. There were no Sinclair stations along the route for fill-ups or "comfort stops," so a caravan of 125 camels carried gasoline and food supplies.

Despite high optimism, the expedition failed miserably in its goal. No human ancestors were ever found. The project did earn at least a footnote in history, however, by discovering some of the richest dinosaur beds in the world, not truly in Central Asia but rather in the Gobi Desert of eastern Asia. A narrative of the Gobi expeditions of the American Museum of Natural History, five in number between 1922 and 1930, would be a pleasant diversion but one slightly peripheral to my goal.[13]

The object of our enquiry came to light late in the 1922 field season, at a scenic locality named the Flaming Cliffs by the Andrews party. The site was known to the locals as Shabarakh Usu, but its name today is Bayn Dzak (Map 7.1). Here the 50-m-high cliffs, composed of soft, bright red sandstone of the Djadochta Formation, extend for 8 km. During a stop to seek directions from locals, expedition photographer J. B. Shackelford wandered off to inspect some rocks and soon found that he was standing on the edge of a vast basin, looking down upon a chaos of ravines and gullies cut deep into red sandstone. He made his way down the steep slope with the thought that he would spend ten minutes searching for fossils and, if none were found, return to the trail. Almost as though led by an invisible hand he walked straight to a small pinnacle of rock on the top of which rested a white fossil bone. Below it the soft sandstone had weathered away, leaving it balanced ready to be plucked off.

Shackelford picked the "fruit" and returned to the cars. . . . Granger examined the specimen with keen interest. It was a skull, obviously reptilian, but unlike any with which he was familiar. All of us were puzzled.[14]

A few more hours of prospecting yielded little else, and the next day the expedition continued onward toward Kalgan and then Peking, not recognizing the significance of their find, or of their site. The fossil was sent back to New York along with other specimens from the successful collecting season. Before the end of the year, Osborn cabled some interesting news back to Andrews in Peking: "You have made a very important discovery. The reptile is the long-sought ancestor of *Triceratops*. It has been named *Protoceratops andrewsi* in your honor. Go back and get more."[15]

Going back to get more was very much on Andrews's agenda. Another large expedition was mounted in 1923. In that year, in addition to Granger the team included three experienced members of the American Museum's paleontology field staff: George Olsen, Peter Kaisen, and Albert Johnson, the latter having worked with Barnum Brown in Alberta. They reached Shabarakh Usu on July 8 and remained there until August 12. Of the site, Andrews wrote in wonderment:

This is one of the most picturesque spots I have ever seen. From our tents, we looked down into a vast pink basin, studded with giant buttes like strange beasts, carved from sandstone. . . . There appear to be medieval castles with spires and turrets, brick-red in the evening light, colossal gateways, walls and ramparts. Caverns run deep into the rock and a labyrinth of ravines and gorges studded with fossil bones make a paradise for the paleontologist. One great sculptured wall we named the "Flaming Cliffs," for when seen in the early morning or late afternoon sunlight it seemed to be a mass of glowing fire.[16]

Of course one needn't travel so far just to admire beautiful sights: it was bones that kept them there—lots of them. In five weeks they collected sixty cases of fossils weighing five tons. Among them were seventy skulls, fourteen skeletons, and twenty-five dinosaur eggs! Besides *Protoceratops* were the meat-eaters *Velociraptor* (Fig. 7.3), recently made famous by Michael Crichton in a piece of fiction whose name I forget, and *Oviraptor*, the toothless but unjustly named "egg thief." Superb specimens were almost ridiculously easy to find. This has continued to be the experience of expeditions up to the present time. One characteristic of bone from this site is that its color is white.[17] Thus the contrast with the red sediment makes fossils positively conspicuous.

The first ceratopsian specimen that was sent back to New York was not spectacular. It was an incomplete skull of small size, about 15 cm long, lacking the front tip but showing all of the upper beak and the entire back of the skull including the frill. The name given it at its scientific baptism on May 4, 1923, *Protoceratops andrewsi*, means "Andrews's first horn face." Its description by Granger and the distinguished anatomist W. K. Gregory was both concise and incisive. They were struck by the flaring triangular shape of the skull; the relatively enormous orbits that, at 50 mm in length, seemed to dominate the skull; and the narrow bar of bone (postorbital-squamosal) at the upper back corner of the skull. Because erosion had destroyed the frill and the infratemporal fenestra, they did not suspect the prominence of these elements. On the basis of the beak-like predentary bone of the lower jaw and teeth, they correctly recognized *Protoceratops* as an ornithischian. They posited that *Hypsilophodon*, a two-legged plant-eater from the Early Cretaceous[18] from England "might well be the starting point for the far more specialized conditions of *Protoceratops*."

To the authors *Protoceratops* was not a ceratopsian, but possibly a link between ceratopsians and the armored dinosaurs or ankylosaurs with which ceratopsians were contemporaneous. Ceratopsians as then known were strictly North American and were dinosaurs of large size (conveniently ignoring *Leptoceratops* and *Brachyceratops*). All had horns, and all had a prominent crest or frill, with a relatively small orbit. Granger and Gregory found that *Protoceratops* presented the opposite of the salient ceratopsian characters. They were so impressed by its generalized nature that they proposed a new family, Protoceratopsidae, "characterized by the lack of horns, the very large size of the orbits, and the narrowness of the postorbital-squamosal bar."[19]

Two years later, a new description of *Protoceratops* appeared, by Gregory and Charles C. Mook, who became a great student of fossil crocodiles. By now a series of skulls had been prepared for study, "starting

FIG. 7.3. *Velociraptor* sizes up two specimens of *Protoceratops*, Mongolia, 76 Ma. Discretion is the better part of valor. (Robert Walters.)

with an extremely young stage not long out of the egg, and ending with a very old stage with a wide frill 511 mm wide. In one of the younger stages, with a total skull length of 283 mm, the frill is not much wider than the skull itself, but is already produced behind the occipital condyle."

Gregory and Mook chose a superb skull, AMNH 6408, to describe. One could only wish that all paleontology were based on such fine specimens. The skull was about two-thirds of maximum size. The sutures among the bones of the skull were very clear, allowing precise anatomical determinations to be made. In addition, skeletons were now beginning to be prepared.

By now, *Protoceratops* had become a ceratopsian. With the addition of a perfectly fenestrated frill, it could be seen as ceratopsian in every way except for the absence of horns. An important new character was recognized: teeth in the premaxilla at the front of the face. This is a normal vertebrate condition, but one not previously known in a ceratopsian. Gregory and Mook confirmed that the teeth had single roots. The description now drew on aspects of skeletal anatomy, including remarks on the vertebrae and the fore- and hind limbs. The femur is shorter than the tibia, unlike the case in large horned dinosaurs, and the hind foot is very long and slender. The tail has very tall neural spines, and they made a tentative suggestion that the foot and tail were indicative of "partly aquatic habits," a judgment that has not found much favor subsequently.

A detailed definition of the Protoceratopsidae was offered, and membership was extended to *Leptoceratops*. An interesting observation was made. "There seem to be two kinds of skulls, a long and a very broad kind, possibly representing males and females." A photo of AMNH 6408 was labeled as "small, young adult skull, possibly a female", and AMNH 6414 was captioned "supposed old male skull." A number of comparisons were made with North American horned dinosaurs. Curiously, the authors regarded *Leptoceratops* as progressive toward *Triceratops* in "secondary closure of parietal fontanelles," rather than considering the possibility that its ancestors never had parietal fenestrae.[20]

The next milestone in the history of *Protoceratops* occurred in 1940, when it was the subject of a lengthy monograph by our old friend Barnum Brown, aided by E. M. Schlaikjer. Brown did not "do" Mongolia. After finishing in Alberta he collected mammalian fossils in India. In the 1930s he excavated Jurassic dinosaurs of the Morrison Formation at the Howe Quarry in the Bighorn Basin of Wyoming and also very important Early Cretaceous dinosaurs from the Cloverly Formation nearby. But fossils from these sites were never described during his lifetime, either by him or by anyone else. By 1940, the sixty-seven-year-old Brown was in the twilight of his distinguished career.

The *Protoceratops* monograph is one of the classics in the literature of dinosaur paleontology, must reading for any serious study of horned dinosaurs. It is not just about *Protoceratops*, but is profusely illustrated with very effective line drawings of series of specimens, and broadly comparative in its scope, with much valuable material on large horned dinosaurs such as *Triceratops* and *Centrosaurus*. Because the preservation of *Protoceratops* is so exquisite and the sutures among cranial bones so clear, a number of problematic identifications in other ceratopsians are definitively resolved, for example, the contribution of the parietal to the frill. Furthermore, the paper is not just descriptive of bony elements but also considers three problems: the origin of horn cores, the composition of the frill, and the origin of the secondary skull roof of horned dinosaurs. A firm conclusion of the study is that, despite the numerous specimens in the collection, only a single species is represented. The discussion of sexual dimorphism is more tentative:

> The determination of skulls as male or female is purely a matter of conjecture, and our reasons for doing so are indeed tenuous. We regard those skulls as males which at any growth stage—as judged purely on a basis of size—are most robust and which give greatest emphasis of the salient growth changes. . . . Another interpretation is that this supposed sexual distinction is nothing more than individual variation. This, however, seems less likely since each of the types presents some rather marked variations. . . . The fact that the two types of skulls are approximately equally represented in numbers also suggests the sex ratio 1:1.

There is an element-by-element description of the skull from the smallest to the largest specimen. For instance, the nasal is described as follows: "As in *Leptoceratops* the nasal is smaller than in any of the other known ceratopsians. In lateral view, it is quite short and fairy deep in the very young individual. In the medium-sized skulls it lengthens and is relatively less deep. In the older skulls it begins to arch up about midway back forming an incipient horn-core, and assumes deeper proportions." The logical conclusion is that the arching of the nasals in *Protoceratops* represents the earliest stage in the development of nasal horn cores. The nostrils of *Protoceratops* are oval slits in a dorsal position, in contrast to the greatly enlarged nostrils of large ceratopsians.

The parietal and squamosal bones of the frill are described in great detail, as might be expected because they form such a spectacular aspect of the skull and seem such an ideal precursor for the condition in later ceratopsians. *Protoceratops* showed without a shadow of doubt that the principal bone of the frill of ceratopsians is the parietal, which contacts the frontal as in all vertebrates. In small specimens, the parietal is rather narrow with large fenestrae and has a horizontal orientation. In large skulls of male form, the parietal is erect and wide. In skulls of the female type, the parietal is not as erect. The squamosal elongates, deepens, and changes orientation as the *Protoceratops* increases in size. The bones of the braincase and palate are beautifully described and illustrated, once again demonstrating what a useful starting point the Brown and Schlaikjer monograph is for any study of horned dinosaurs. The teeth in the upper jaw (maxilla) in *Protoceratops* number from as few as eight in smaller specimens to as many as fifteen in the largest; this number is two fewer than the seventeen in *Leptoceratops*. The description of the skull is among the most complete for any dinosaur. Summary tables of growth changes are provided, as well as a useful table of measurements for ten of the skulls.

An aspect of the elegant preservation of the Mongolian plant-eater is a sclerotic ring. A number of vertebrates, from fishes to birds, have a series of thin bony plates in the sclera, the fibrous skeleton of the eyeball. Dinosaurs were no exception, but these bones do not have a high fossilization potential and are almost always lost. But not in *Protoceratops*, in which several specimens show scleral bones. An incomplete ring of fifteen bones was found in AMNH 6466, a large skull. It indicates a large eyeball, about 50 mm in diameter. A beautiful cast of the brain was made from the same skull, and it was compared with those of *Triceratops* and *Anchiceratops*. Unfortunately, Brown and Schlaikjer did not report the volume of their cast, which would be of interest in comparing the size of the brain with that of other dinosaurs.

There were fewer skeletons than skulls to draw upon. The description again is excellent, but it encompasses less size-related variation. It is frustrating that no perfect specimens are known; none is complete to the end of its tail (Fig. 7.4). The most complete tail has thirty-two vertebrae. Brown and Schlaikjer estimated the total number at forty (there are thirty-eight to forty-eight in *Leptoceratops*), but this count has not been confirmed. The tail is deep, with tall neural spines. Such tails

FIG. 7.4. *Protoceratops andrewsi* skeleton. (Gregory S. Paul.)

are very different from the simpler tails of large horned dinosaurs. It is clear that there are eight vertebrae in the sacrum, versus six for *Leptoceratops*. An interesting bone reported for the first time in horned dinosaurs (*Psittacosaurus* was not then recognized as a horned dinosaur) is the clavicle, a small cylindrical bone located on the front edge of the shoulder bones just in front of the shoulder joint. Subsequently Sternberg recognized the clavicle in *Leptoceratops*. The clavicle seems a legitimate vestigial structure, without function and evidently lost in large horned dinosaurs.[21]

The limb bones of *Protoceratops* are very slender, consistent with the small size of the animal. Brown and Schlaikjer were struck by the shortness of the front limb, barely half the length of the hind limb. The hand is especially small and almost delicate. To them, this was a contrast not just with large horned dinosaurs but even with *Leptoceratops*, and it suggested that *Protoceratops* was less thoroughly quadrupedal than other horned dinosaurs. The unguals of both the hand and the foot are elongated and pointed, but not as long, as sharp, or as curved as those of *Leptoceratops*.

The very generalized nature of *Protoceratops* is clearly seen in the pelvis. The ilium is not at all like those of large ceratopsians, in which there is a prominent horizontal overhang. Here the blade of the ilium is entirely vertical. The pubis shows a very short forward process, which is much more prominent in large ceratopsians. The ischium is long, nearly straight, and rather simple in form. Large ceratopsians have an ischium that is strongly curved downward. Brown and Schlaikjer considered that the pelvis of *Protoceratops* was essentially that of a bipedal animal. The bones of the hind leg are slender and elongated, including those of the foot. In *Protoceratops* and *Leptoceratops* the femur is shorter than the tibia, a character usually taken to indicate swift running ability. The femur has a prominent bony tab, called the fourth trochanter.[22]

Brown and Schlaikjer provided measurements from four skeletons. It is clear that only one of these, AMNH 6424, is anywhere near complete. Unfortunately, they did not provide any estimate of the length of the body or its height. The best skeleton does provide a ready measure of height at the hips: the sum of femur length, tibia length, and length of metatarsal III is 65 cm (25 in.). If we assume that the hind limb of *Protoceratops* bore the same relationship to the length of the body as in *Leptoceratops*, we can use measurements in the more complete Canadian fossil to estimate body length in the Mongolian dinosaur. The smaller skeleton with a femur, AMNH 6471, is estimated at 49 cm (19 in.) at the hips and 1.5 m long. The larger specimen, AMNH 6424, would have been about 2.0 m long. There are pelvic bones associated with a very large skull, AMNH 6466, that suggest an individual 2.5 m or longer. At the other extreme, the humerus of AMNH 6419 is only 55 mm long, barely a third of the length of the humerus in AMNH 6471. This individual would have been a little tyke, 56 cm (22 in.) in length and 16 cm (6.3 in.) at the hips! Years later, E. H. Colbert estimated the weights of a series of dinosaurs, including *Protoceratops*, presumably AMNH 6424. Using the technique of water displacement of scale models, he estimated the weight of *Protoceratops* as 177 kg (389 lb).[23] Taking this figure and understanding that weight increases as the cube of linear dimensions, we can estimate the weights of individuals of different sizes. For instance, an animal that is half the length of AMNH 6424 would weigh only one-eighth as much ($1/2^3$). Thus the *Protoceratops* 1.5 m long and 49 cm at the hips would have weighed 36 kg (79 lb). The tyke 56 cm long would weigh 1.7 kg (3.8 lb). A large animal 2.5 m long and 75 cm at the hips would weigh 260 kg (570 lb). Frankly these figures seem a little high to me, but you get the idea.

EGGS—THE INSIDE STORY

Protoceratops is known from an outstanding series of skulls. For years they formed a stunning display along the wall of the Cretaceous Hall of the American Museum of Natural History, in the shadow of the *Tyrannosaurus*, which stood as a silent sentinel over its diminutive charges.[24] It is also known from an equally remarkable series of eggs and nests— maybe. Fossil eggs were a great find during the 1923 Mongolian field

FIG. 7.5. "*Protoceratops*" eggs at the American Museum of Natural History. Recent discoveries in Mongolia suggest that these eggs belong to the supposed "egg thief," *Oviraptor,* instead of *Protoceratops.* (Photograph by Robert Walters.)

season. On July 13, when George Olsen reported at afternoon tea that he had found fossil eggs, his account was initially greeted with skepticism but then confirmed by paleontologist Walter Granger. The elongated eggs measure 20 cm long and 17 cm in circumference, one end being slightly smaller than the other. The color is reddish brown, the originally white shell having been stained brown by the oxidized iron in the sediments. The shell is 1 mm thick; it has a striated or finely ridged surface externally and is smooth internally. A number of years later Granger added helpful details. He noted that the eggs were buried in circles with the large end up (Fig. 7.5). In the most complete nest, there were five eggs in the lowest circle and eleven eggs in a higher circle, and two eggs in a damaged upper circle, which he estimated to have contained twenty eggs when complete. Thus he estimated more than thirty eggs in a complete nest.[25]

Also found was the skeleton of a toothless meat-eating dinosaur, whose proximity to the nest "immediately put the animal under suspicion of having been overtaken by a sandstorm in the very act of robbing the dinosaur egg nest" in Osborn's interpretation. As the eggs were *assumed* to be those of *Protoceratops*, the theropod was elegantly named *Oviraptor philoceratops*, "egg thief, lover of ceratopsians."[26] Thirteen eggs were found in the first block of sandstone uncovered. A second cluster of five eggs and a third cluster of nine were soon found, and twenty-five eggs were triumphantly carried back to the United States. Never shy about publicity, Andrews found in due course that it could be a double-edged sword. News of the remarkable discoveries preceded him on his return to the United States, and when his ship arrived in Seattle streams of reporters brandishing fists full of dollars swarmed on board, each howling for exclusive rights to pictures. Andrews wisely resisted the lure of quick bucks but in due course orchestrated an auction. He described it thus:

> Purely for the sake of publicity, we decided to sell a single egg to the highest bidder, the proceeds to go to the Expedition. This directed attention to the fact that we were urgently in need of funds with which to continue our work. Offers for the eggs were received from England, France, Australia, New Zealand and many parts of the United States. It was finally purchased for five thousand dollars by the late Colonel Austin Colgate, and presented to Colgate University.

With such a worldwide campaign, word of the sale inevitably reached China and Mongolia, and with it the notion that commercially valuable fossils were being removed without compensation—scientific imperialism in the finest tradition. Understandably, Chinese officials were extremely displeased. Life became more difficult for the explorers, who managed only one more season in the dinosaur beds, in 1925. Andrews referred to the "cupidity" of the authorities and complained disingenuously: "They could never be made to understand that that was a purely fictitious price, based upon carefully prepared publicity; that actually the eggs had no commercial value."[27] He had made his own life more difficult. Nonetheless, the discoveries of eggs were of surpassing interest. The eggs extended through 200 ft of section vertically and probably spanned hundreds of thousands of years. Thus dinosaurs had found the region a favorable breeding ground for a long period of time, returning to the same site year after year—a concept that paleontologist Jack Horner later called "nest site fidelity."[28]

In 1925, Belgian paleontologist Victor Van Straelen was invited to study the microscopic structure of the fossil eggs. He found the pores in the egg shell to be small and sparse, consistent with the structure of the eggs of birds and turtles that lay in dry regions.[29]

Further details were added by Brown and Schlaikjer in 1940. The entire collection consisted of over fifty more or less complete eggs and two nests, plus large quantities of fragments. They estimated one nest to have contained thirty to thirty-five eggs, deposited in three circular layers, as Granger had described. The other had eighteen eggs. The eggs vary in length from about 9 cm to 20 cm. In the smaller eggs the surface texture is less pronounced than in the larger ones. Brown and Schlaikjer did not consider that the variations in size and texture were sufficient to indicate the existence of more than one species of egg-layer. Owing the abundance of *Protoceratops* skulls and skeletons, there seemed no reason for the authors to suspect that *Protoceratops* was not the egg-layer.[30]

In 1979, Australian physiologist Roger S. Seymour performed further study of *Protoceratops* eggs. The egg he studied measured 9.6 cm in length, and thus was small. It had an estimated volume of 335 cm^3, about that of a can of soda. Its weight would have been 360 g (13 oz). The egg's surface was covered with pores that allowed the developing embryo inside to breathe. Look carefully at a hen's egg and you will see similar pores. The pores of the *Protoceratops* eggs were 0.12–0.15 mm in diameter, and some seven thousand of them covered 0.37% of the surface area of the egg. This would be equivalent to an opening 1 cm × 1 cm in an otherwise solid egg. It turns out that *Protoceratops* has eight times more pores than a comparable bird egg. Its shell thickness averages 1.3 mm, nearly twice the 0.75-mm thickness expected in a bird egg of similar size. These measurements all suggest that the rate of water loss from the egg, which is a normal process for a respiring organism, was four times higher than that for a bird egg of that size. What all of this means is that conditions encountered in the nest of a modern bird would have been fatal to a clutch of *Protoceratops* eggs. The eggs would have dried out, and the embryos would have died. Instead, the eggs were either covered by a mound of loose soil or vegetation or buried in the ground. In either case, fatal water loss would have been prevented while access to life-giving oxygen was maintained.[31]

The eggs just described provide an unprecedented snapshot of dinosaur life. However, there is one misfortune. Despite claims to the contrary, no embryos have been found inside *Protoceratops* eggs. If embryos had been recognized, a different story might have emerged. It now appears that the famous *Protoceratops* eggs *do not belong to* Protoceratops *at all!* Several researchers have independently reached this startling conclusion. It was reported by Gradzyna Mierzejewska at a meeting in Warsaw in 1981 that the so-called *Protoceratops* eggs extend vertically through three different formations: the Djadochta, the Barun Goyot, and the Nemegt, from oldest to youngest. However, skeletons of *Protoceratops* are found only in the Djadochta beds, the oldest. More recently, Karol Sabath confirmed in 1991 that the eggs of *Protoceratops* have been misidentified. He described *Protoceratops* eggs as smaller, thinner, and less ornamented. He believes that the larger, more heavily ridged and striated eggs usually attributed to the little horned dinosaurs are actually those of small meat-eaters, which are remarkably common and diverse in the Late Cretaceous rocks of Mongolia. In 1994, Philip Currie reported on his experience in strata from Inner Mongolia, that part of China adjacent to Mongolia. There the Canadian-Chinese parties found a skeleton of *Oviraptor* straddled atop a nest of familiar eggs in an apparent brooding position. Moreover, Mark Norell and his colleagues James Clark, Luis Chiappe, and Mongolian geologist Demberelynin Dashzeveg reported a similar finding in Mongolia in the British journal *Nature* in December 1995. The discovery of multiple specimens of *Oviraptor* apparently engaged in the same nesting behavior makes it unlikely that the first report was a fluke or an incorrect interpretation. What a twist fate has taken—the much-vilified egg-thief may actually have been a good mother after all, not a connoisseur of omelettes on the fly![32]

The coup de grâce came in 1993, when American Museum of Natural History/Mongolian Academy of Science field parties at long last found an embryo inside a *Protoceratops* egg, and out came an *Oviraptor*![33] A historic injustice has now been set right. *Oviraptor* was a good mother after all, protecting her nest to the end, possibly even brooding her eggs like a hen. Could it be that the cylindrical teeth in the beak of *Protoceratops* were actually for puncturing eggs of *Oviraptor,* and that the erstwhile pacific "first horn" was actually a dastardly oviphile? Bambi-like images sometimes have to be revised! It has been claimed that *Protoceratops* eggs have been discovered, but until embryos are found inside them, I reserve my right to skepticism. Clearly the story from dinosaur eggs is more complicated than has previously been allowed.[34]

SEX IN THE CRETACEOUS

If we don't have eggs and we don't have embryos, we still have remarkable information about "first horn." As we learned previously, the idea that male and female skull shapes could be recognized arose very quickly, but not a great deal was done with the concept. In 1972 the

Soviet paleontologist Sergei M. Kurzanov published an important study on sexual dimorphism in *Protoceratops*. His study resulted from collections made by Soviet paleontological expeditions to Mongolia during the years 1946–1949 and also in 1969. He was also influenced by a Soviet evolutionary theorist, L. S. Davitashvili, who wrote an important monograph on sexual selection in 1961.

Kurzanov had seven partial skulls to work with, plus a cast of the American Museum specimen AMNH 6408. He carefully analyzed the specimens he had available. He believed that the nasal horn of *Protoceratops* had neither a defensive nor an aggressive function, because it was too low and blunt to inflict injury. It makes sense, however, as a result of sexual selection. The mere existence of a nasal horn in males serves to distinguish males from females, a distinction essential for the propagation of the species. Second, a nasal horn, even a blunt one, could have been used in combat between males. Such a horn would have produced pain but not injury, an outcome desirable for the well-being of the species. Kurzanov noted that the bony frill of *Protoceratops* was very thin, in some cases only 1 mm thick, and therefore unlikely to have been very useful for protection against sharp teeth. However, he also pointed out that the specimens with a nasal horn had erect, wide frills (and wider faces), whereas those without nasal horns had lower, narrower frills (and narrow faces). For Kurzanov, it was clear that *Protoceratops* was sexually dimorphic and that the frill was used to attract females and to engage other males in aggressive encounters. Skeletons of the proto-horn-bearers are so common, Kurzanov speculated, because they lived in herds, and secondary sexual characters were exaggerated, as they are in polygamous animals today.[35]

Kurzanov's study was very important and stimulating. I also found it frustrating because his analysis was poorly quantified. I had the good fortune to study *Protoceratops* myself during my graduate studies at Yale University. I had access to the marvelous study series of skulls at the American Museum of Natural History. This treasure house by Central Park has more skeletons of dinosaurs in its research collections than any other museum on earth. One of its most stunning displays for more than fifty years was a graded size series of *Protoceratops* skulls running along the wall of the old Cretaceous Hall, which now lives on only in memory (replaced by the glittering Hall of Ornithischia). The largest specimens were bulky heads half a meter long, a fivefold size range from the smallest to the largest. I studied every skull I could get my hands on—a count that, despite claims of hundreds of skulls and many skeletons, came to only twenty-four. These included some of the most exquisite skulls ever collected. I thrilled to hold the smallest skull, 115 mm long, in the palm of my hand. As sunlight streamed through the windows into that hallowed hall, I noticed that the frills, even of the large skulls, were so eggshell-thin that they were actually translucent— even the beam of a flashlight could pass through the frill!

So many superb skulls distributed across such a great size range were available. I chose to follow a rigorous biometric approach as an excellent way to document size-related change of shape. Armed with calipers and measuring tape, I gently measured a lengthy series of variables (forty in all) on each precious skull.[36] The next step was to keypunch my measurements onto cardboard data cards (the clay tablets of the computer age) and then to hand in the bulky stack of cards to the operator of the room-sized IBM mainframe computer. I waited patiently for several hours until my job was processed, and reams of oversized fanfold paper labeled DODSONP disgorged through the window of the air-conditioned operating room. It is indeed marvelous that we possess today on our desks the computing power that twenty years ago filled entire climate-controlled rooms.

What did I learn from these exercises? I detected a strong and rigorous signal of dimorphism. I used a powerful statistical method called principal coordinates analysis that considered all forty variables for each specimen simultaneously and clustered specimens according to their similarity to each other. I discovered that on a plot of the results, there were two general clusters, one larger than the other. When I considered the twenty-four specimens in order of increasing size, I began to see a pattern in the clusters. The first eight specimens, ranging in size from basal lengths of 76 mm to 170 mm, formed part of the larger cluster. The first specimen to separate from the larger cluster and to plot on the other side of the chart was specimen number 9, which, at 191 mm basal length, was just over 50 percent of maximum size—how marvelous! Moreover, this fine specimen, AMNH 6409, showed some very interesting features. On the basis of all four of Kurzanov's hypothesized sexual characters, this specimen was a perfect "male." After specimen number 9, all of the remaining specimens plotted on one side of the chart or the other, with nearly equal numbers in the male field or the female/juvenile field. Not only did I confirm Kurzanov's four dimorphic characters, but I was able to add eight others, including length of the frill, size of the openings in the frill, height of the eye socket, size of the nostrils, and height of the lower jaw.[37]

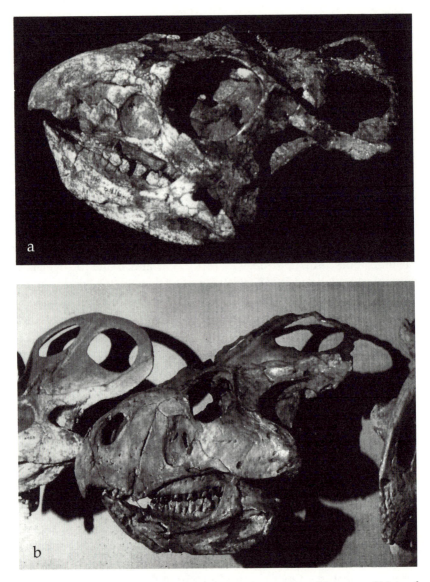

FIG. 7.6. *Protoceratops andrewsi* specimens at the American Museum of Natural History. (a) Juvenile (AMNH 6409); (b) subadult female (AMNH 6408); (c) large adult female (AMNH 6466); (d) large adult male (AMNH 6438). (Photographs by Peter Dodson.)

FIG. 7.6. (*continued*)

These results do not *prove* that the dimorphism is a *sexual* dimorphism, and if it is a sexual dimorphism they do not prove which are the males and which are the females. In falcons and blue whales, for example, the females are larger than the males. In a small shorebird known as the red phalarope, the female is brightly colored and the male's plumage is dull, for it is he who hatches and raises the chicks. Nonetheless, these examples stand out in the biological world because they *are* exceptions. In *Protoceratops*, juvenile specimens are nondescript, but when specimens are about half grown, they differentiate into two groups: individuals characterized by widely flaring faces with arched nasals (the proto-horn) and crests that are showy, being long, wide and erect; and individuals with a lower profile of the nose region, a narrower skull, and a crest that is shorter, narrower, and lower. It seems reasonable to conclude that *Protoceratops* was sexually dimorphic, as had long been suspected, and that the animals with the bulkier, showier skulls were the males. I am willing to bet that males displayed with brightly colored frills, just as peacocks court peahens with their stunning tails (Fig. 7.6).

Protoceratops is one of the most abundant, best-studied dinosaurs on earth. Every expedition that goes to Mongolia collects specimens of this dinosaur. Unfortunately, no embryos have yet been found, but we certainly know a great deal about its posthatching existence. We have no reason to infer that a hatchling had a good mama to look after it. Individuals of all sizes are found singly, with no indication that young ones remained together in the nest. Because so many individuals of all sizes are found, I infer that *Protoceratops* grew throughout life, a reptilian trait. Studies of the microscopic structure of *Protoceratops* bone currently underway by Dr. Anusuya Chinsamy may confirm or refute this speculation. The ratio of males to females seems to have been 1:1, and males and females appear to have attained the same size. The largest skull in my study series measures 357 mm in basal length and is female; the second largest is 5 mm shorter and is male. We have not yet heard the last of old "first horn face."

PSITTACOSAURUS

Not all treasures of the Mongolian expeditions were equally celebrated. In 1922, two small and rather well preserved dinosaur skeletons were found in Early Cretaceous sediments in the western Gobi, up to 200 km

Protoceratops and the Legend of the Griffin?

The correct interpretation of fossils is a difficult human endeavor. To gaze upon a fossil is not necessarily to comprehend its significance as a biological entity. The reason that our European medieval forebears did not have knowledge of the world of fossils was not because they were ignorant, superstitious unfortunates cowed by the authority of the church. It was because they had so little knowledge of the living world around them. Europe is not particularly informative regarding the biological diversity of the modern world, especially that of tropical regions or of marine realms. Imagine that the great eighteenth-century Swedish botanist Karl von Linnée, who in his lifetime classified some ten thousand species of plants and animals, believed that he was classifying *all* of the biological world (variously estimated today to comprise between ten and forty million species)!

Knowing these matters as I do, I have been terminally skeptical about attempts to claim a transmission from fossils to folklore. But a novel idea put forward by Dr. Adrienne Mayor has put my skepticism to the test. She has analyzed the legend of the griffin or *gryps*, which first appears in Greek writing about 675 B.C., around the time that the Greeks first made contact with Scythian nomads of Central Asia. The essential features of griffins are sharp beaks, four legs, and prominent claws. They live in wilderness areas and are said to guard deposits of gold. Aeschylus (460 B.C.) called them "silent hounds with beaks." Ctesias (ca. 400 B.C.) described them as "a race of four-footed birds, almost as large as wolves, with legs and claws like lions." Pliny the Elder (A.D. 77) noted the "terrible hooked beak" and added long ears (i.e., something sticking above and behind the face) and wings. Although they are often pictured with wings, most accounts seem to agree that griffins did not actually fly, although they may have taken to the air with short hops during combat. Digging behavior is frequently described, associated both with nesting and with gold mining.

What could be the basis in fact for such an animal? It is not associated in myth with the exploits of any named heroes. No commentator ever claims to have seen one alive. It is not an obvious hybrid of any known animals. Mayor makes the startling case that the griffin represents an attempt to interpret *Protoceratops* skeletons observed in the ground by ancient traders whose caravan routes crossed the Gobi Desert or by gold miners crossing the Gobi to reach the Altai Mountains (whose very name means gold). It is true that skeletal remains of *Protoceratops* are very abundant. Their bones are white, that is, obviously bone colored to the most casual observer. Furthermore, as the sediments of the Djadochta Formation are bright red, the bones are conspicuous in the ground. The size of the animals is true to legend, as is the combination of beak, four legs, and claws. The vagueness about wings or ears could represent the obvious bafflement arising from attempts to interpret the bony frill at the back of the skull. Even the large eye sockets of *Protoceratops* are consistent with reports of baleful glaring eyes that added a corroborating detail of ferocity to the legend (Mayor, 1991, 1994).

Although I stop short of giving Mayor's interpretation a wholehearted endorsement, I honestly cannot detect a major flaw that allows me to dismiss it out of hand—it is something fascinating to ponder.

west of the Flaming Cliffs of Shabarakh Usu, which had not yet been discovered by the Americans. The specimens were described in preliminary fashion the following year by Henry Fairfield Osborn. The two skeletons came from what seemed to be separate geological formations and from localities about 100 km distant from each other. The first-found specimen, discovered around July 30, 1922, came from the Ondai Sair Formation, near Uskuk, in the Tsagan Nor Basin. The second specimen, collected a month later, came from the Oshih in the Artsa Bogdo Basin. Despite similarities, Osborn named one *Psittacosaurus mongoliensis* ("Mongolian parrot reptile," pronounced "sih-TACK-oh-SORE-us," with a silent *P*),[38] and the other, the first one found, *Protiguanodon mongoliense* ("Mongolian first *Iguanodon*"). The skull of the former he illustrated elegantly in five views. He sketched the skeleton of the latter but showed only the lower jaw and some teeth, the skull being less well preserved.

The following year, preparation of both specimens having been completed, Osborn produced a scientific treatment that was to suffice for nearly sixty years. In this paper he illustrated the two skeletons both as

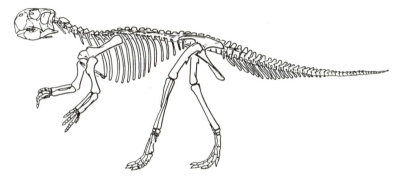

FIG. 7.7. *Psittacosaurus mongoliensis* skeleton, American Museum of Natural History. (From Sereno 1990a. Courtesy of the University of California Press.)

they were found in the field and as reconstructed skeletons walking in bipedal (two-legged) pose. He also presented details of the pelves, both of the two animals in question and of representative ornithopod dinosaurs, including *Hypsilophodon, Iguanodon,* and *Edmontosaurus,* with which he chose to compare his new dinosaurs. Osborn's descriptions were excellent, and he frequently referred to the opinions of W. K. Gregory, with whom he wisely consulted. What is fascinating from our perspective is the failure of Osborn then or subsequently to comprehend the ceratopsian nature of *Psittacosaurus.*

Let us first describe these little herbivores. Both were very small animals. *Psittacosaurus* was 131 cm long and about 60 cm high at the hips; *Protiguanodon* was 135 cm long and about 57 cm high at the hips (Fig. 7.7). The robust skulls are short, wide, and tall, measuring about 15 cm in length and almost as wide. The upper and lower beaks are pointed and toothless and, combined with the height of the snout, give the animals a truly parrot-beaked appearance (Fig. 7.8). The openings for the eyes and the infratemporal fenestra are very prominent. The nostrils are circular and small, situated high on the snout. The squamosal is simple in form, forming the upper back corner of the skull in lateral view. There is not the slightest indication of a frill. A striking feature of the skull is the prominence of the jugals, which form flaring, ventrally directed processes. The teeth are simple in form and few in number and form only a single functional row.

The hind limbs are unremarkable, with the expected phalangeal formula, digits ending in claws, tibia slightly longer than femur, and long foot (about the length of the femur), proportions to be expected in a

227

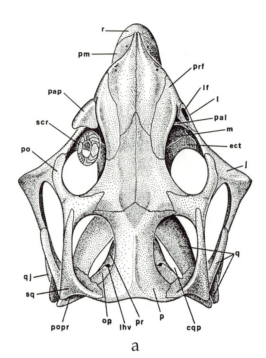

a

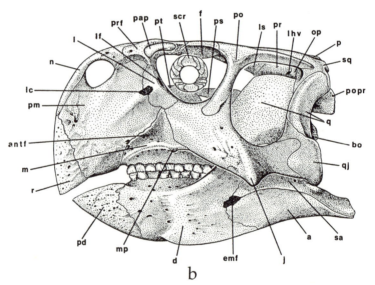

b

Fig. 7.8. Skull of *Psittacosaurus mongoliensis*, American Museum of Natural History. (a) Dorsal view, (b) lateral view. a, angular; antf, antorbital foramen; bo, basioccipital; cqp, cranioquadrate passage; d, dentary; ect, ectopterygoid; emf, external mandibular foramen; f, frontal; j, jugal; l, lacrimal; lc, lacrimal canal; lf,

228

small running animal. The front limbs are quite interesting, however. First of all, a clavicle is recognized for the first time in any dinosaur. Second, the forelimbs are strikingly shorter than the hind limbs: only about 55 percent the length of the hind limbs. Because the humerus is up to 78 percent of the length of the femur, it is evident that the forearm of *Psittacosaurus* is significantly shortened. Furthermore, the hand, which is only half the length of the foot, has a very idiosyncratic form, with only three functional digits; the fourth digit on the outside is strongly reduced, and the fifth digit is lost. Both *Leptoceratops* and *Protoceratops* had all five digits on the hand, as is typical of horned dinosaurs generally. The conclusion is readily reached that *Psittacosaurus* walked habitually on its hind legs. Osborn was so convinced of this fact that he lavished great effort on establishing similarity of the pelvis with that of the two-legged plant-eaters known as ornithopods. The ilium is long and low and shows none of the specializations of either highly derived ornithopods such as duck-bills (hadrosaurs) or horned dinosaurs (ceratopsians). The ischium is long and straight, unlike that in the ornithopods *Iguanodon* or *Camptosaurus*, and lacks the bony tab (obturator process is the technical term) that in all ornithopods points downward to support the pubis. The pubis is small and slender and resembles that of no ornithopod. A mass of bony tendons found along the vertebral column extends from the base of the tail as far cranially as the sixth dorsal vertebra. Elsewhere among dinosaurs abundant tendons supporting the vertebral column are highly characteristic of the two-legged ornithopods, including *Hypsilophodon*, *Iguanodon*, and the hadrosaurs. Tendons are known in four-legged animals as well, including horned dinosaurs.

Osborn frankly was puzzled by the parrot-beaked dinosaurs. He was disposed to see *Psittacosaurus* and *Protiguanodon* as separate from each other. Both the name and the comparisons he made reveal his belief that the latter was related to *Iguanodon*, and he suggested a subfamily Protiguanodontinae for it within a family Iguanodontidae. In contrast, because an interesting series of small, rounded, bony cones was found in

---

lacrimal foramen; lhv, lateral head vein; ls, laterosphenoid; m, maxilla; mp, maxillary process; n, nasal; op, opisthotic; p, parietal; pal, palatine; pap, palpebral; pd, predentary; pm, premaxilla; po, postorbital; popr, paroccipital process; pr, prootic; prf, prefrontal; ps, parasphenoid; pt, pterygoid; q, quadrate; qj, quadratojugal; r, rostral; sa, surangular; scr, sclerotic ring; sq, squamosal. (From Sereno et al. 1988. Reproduced courtesy of the Society of Vertebrate Paleontology.)

the throat region of *Psittacosaurus*, he thought its affinities were to be found with the ankylosaurs or armored dinosaurs. He defined a family Psittacosauridae but did not place it in any specific higher taxon. Yet, recognizing how very similar the two animals were in so many respects (a view that W. K. Gregory urged on his friend), he made clear his preference: "In case the animals prove to be the same, *Psittacosaurus* will have preference."[39]

Other species of *Psittacosaurus* have been described. The great Chinese paleontologist C. C. Young (Yang Zhong Jian as he is now called) published on Chinese dinosaurs for more than 50 years (1930–1982) and is numbered among the greatest students in the history of dinosaur studies. In 1931 Young described two new species from China, *Psittacosaurus osborni* and *P. tingi,* but neither appears to differ significantly from the type, *P. mongoliensis.* Many years later in Shandong Province, he found abundant remains, which he described under the name *P. sinensis* in 1958. Four years later, a young protégé, Chao Shichin (Zhao Xijin), described *P. youngi* from Shandong Province and named it in honor of his mentor.[40] But the problem always remained: what is *Psittacosaurus?*

Alfred Sherwood Romer was a giant in American vertebrate paleontology. He was an expert on late Paleozoic mammal-like reptiles, but his paleontological and anatomical expertise ranged widely. His textbook *Vertebrate Paleontology,* which went through three editions (1933, 1945, 1966) was canonical until the publication of Robert Carroll's text in 1988. Romer organized the great mass of paleontological data from fishes to mammals into a coherent whole, and he often provided penetrating insights into fields rather far from his expertise. He dutifully classified *Psittacosaurus* as an ornithopod; however, he made it very clear that he regarded it as "highly suggestive of ceratopsian ancestry." In particular, he stated in 1956 that the beak was "apparently formed by separate rostral bone."[41] More than any other bone, the rostral is unique to horned dinosaurs. The conclusion seemed to be that psittacosaurs were ornithopods that were ancestral to ceratopsians.

Finally the Gordian knot was cut and the obvious conclusion was articulated by Teresa Maryańska and Halszka Osmólska in 1975. These resourceful women were veterans of some of the most successful of all paleontological expeditions to Mongolia. They enjoyed great success in collecting protoceratopsids in Late Cretaceous sediments, as we shall see shortly. In analyzing the relationships of various protoceratopsids, they found it desirable to reexamine the morphology of *Psittacosaurus*. Their analysis was very clear: *Psittacosaurus* could not itself have been ancestral to known protoceratopsids or ceratopsids. However, *Psittacosaurus* has important features that are found only in horned dinosaurs and nowhere else. Most crucially, they confirmed the existence of the rostral bone, which had generally been misidentified previously as an enlarged premaxilla. They presented compelling reasons for their identification of the bone. In addition, they emphasized the widening of the skull across the jugals, another trait very characteristic of all horned dinosaurs. The narrow, bird- or turtle-like beak and the widely flaring jugals give the ceratopsian skull a triangular appearance when viewed from above. As ancillary characters, they noted similarities of the teeth with those of protoceratopsids and a small nose horn in *P. youngi*. They even interpreted an incipient frill in *Psittacosaurus*, on which no previous author had commented. To them the conclusion was unavoidable: the Psittacosauridae belong within the Ceratopsia and nowhere else. They modestly allowed that they were not the first to make this suggestion, but their case was entirely convincing and has been universally accepted since then.[42]

American paleontologist Walter P. Coombs, Jr., is an expert on armored dinosaurs or ankylosaurs. Because of Osborn's belief that *Psittacosaurus* had something to do with ankylosaurs, Coombs had occasion to examine some fragmentary fossils collected in 1922 at the same site as *Psittacosaurus*. The specimens proved to be of great interest, and he described them in 1980 and 1982. Included were two extremely small skulls that evidently represented hatchling *Psittacosaurus*. As he reconstructed them, the larger skull measures 42 mm in length, and the smaller skull measures 28 mm, making it possibly the smallest dinosaur skull ever found! Sitting on a 25-cent piece, the tiny skull would cover George Washington's face, barely overhanging the margins of the coin. The previous smallest skull was the 30-mm-long skull of a baby prosauropod from Argentina named *Mussaurus*, the "mouse reptile." The smallest *Protoceratops* skull measures about 62 mm in basal length, and that of *Bagaceratops*, another protoceratopsid, measures 47 mm. Although the skulls are poorly preserved, such juvenile characters as enormous orbits and short round faces are very evident.

The larger skull has an associated partial skeleton from which important information may also be gleaned. For instance, there is a right humerus about 30 mm long. You would not enjoy nibbling on a chicken wing as small as that, unless your taste ran to quail. The foot (metatarsal III) was about 20 mm long. Coombs pieced together the information available to arrive at an estimate of the length of the smaller specimen of 230 mm.[43] By comparison, *Compsognathus* is often described as the smallest dinosaur; its length is estimated at 750–800 mm. Moreover, observed Coombs, the ends of fourteen broken tibiae reveal that at least seven individuals of the same small juvenile size were present. Thus a group of small juveniles, which Coombs termed a "sibling group," remained together after hatching. At this time, no evidence has been brought forward that they were cared for in a nest. Coombs suggests that this behavior is within the range of that seen in modern gregarious lizards, such as Galapagos marine lizards (*Amblyrhynchus*).[44]

From a modest beginning, our understanding of *Psittacosaurus* has evolved in an interesting way as its importance has become realized. Paul C. Sereno enjoys one of the highest profiles of any dinosaur pale-ontologist in the United States. A genial, curly-haired, boyishly hand-some phenom, who is now associate professor of anatomy at the University of Chicago, Sereno created quite a stir as a graduate student at Columbia University during the 1980s. Whereas in the 1990s, travel to China and Mongolia seems commonplace, this was not so a decade earlier. In 1984, Sereno spent three months in Beijing studying fossils previously seen by no other westerner, including more specimens of *Psittacosaurus* than anyone dreamed existed. Following his highly pro-ductive sojourn in Beijing, he performed successively less probable feats by visiting Mongolia (not only the museum in Ulan Bator, the capital city, but also traveling in the countryside and actually wrestling a Mongolian wrestler in front of a yurt); and finally entering the former Soviet Union and traveling aboard the Trans-Siberian Railroad across the vast expanse to Moscow. When he arrived in Tübingen in Septem-ber 1984 to attend an international scientific meeting on Mesozoic ter-restrial ecosystems, he had the most impressive accounts to give of his travels, of the specimens he had seen, and of his cogent analysis of the significance of these specimens.

Sereno went on to complete his Ph.D. at Columbia on the anatomy and relationships of *Psittacosaurus*. Two of his early published papers, with the coauthorship of several Chinese colleagues, described two new species of *Psittacosaurus*. *Psittacosaurus xinjiangensis* comes from the Tugulu Group in the Junggar Basin in northwest China, more than 1,000 km west of the classic sites in the Gobi Desert. The specimen is incomplete but shows some significant features, including peculiarities of the teeth, jugal, and vertebral column (in that ossified tendons appear to run halfway down the tail). *Psittacosaurus meileyingensis* comes from the Jiufotang Formation in Liaoning Province in northeast China, nearly 1,500 km east of *Psittacosaurus* sites in the Gobi. This species has an unusually tall skull with a rounded profile, rather than the rectangular profile typical of *Psittacosaurus*.[45]

In 1990, Sereno analyzed the species described to date to assess the diversity of the genus *Psittacosaurus*. As he was clearly the authority on the subject, his conclusions were of great interest. The first topic to consider was whether or not *Protiguanodon* was a valid genus. Many had cause to doubt whether it was distinct from *Psittacosaurus*, but no study had considered the question. One of the supposed differences was in maxillary tooth number. Sereno pointed out that the smallest skull of *Psittacosaurus* that we discussed previously had five teeth each in the upper and lower jaws, that a slightly larger skull had seven teeth, that subadults had eight to ten teeth, and that adult skulls had ten to twelve teeth. It was easy to see that tooth number increases during growth and may be slightly variable among adults; thus this cannot be a taxonomically significant character for *Psittacosaurus*. Sereno could see no characters to suggest that *Protiguanodon* differed from *Psittacosaurus mongoliensis* in any significant way; both genus and species are thus synonyms of the latter. He affirmed the validity of Young's species, *P. sinensis*, but regarded Chao's species, *P. youngi*, as a synonym of *P. sinensis*. All told, Sereno recognizes four valid species of *Psittacosaurus*.[46] In 1992 French paleontologist Eric Buffetaut and his Thai colleague Varavudh Suteethorn described a partial jaw from Early Cretaceous sediments in Thailand as a new species of *Psittacosaurus, P. sattayaraki*.[47] I am a little reluctant to accept this referral on the basis of such an incomplete specimen, but it certainly draws attention to a potentially important and very welcome locality almost 3,000 km south of the classic Mongolian sites.

With an estimated 120 specimens, most of them in China, *Psittacosaurus* appears to be one of the most abundant dinosaurs that we know. A few years ago, I listed dinosaurs with one hundred or more specimens in order of decreasing abundance, beginning with the hadrosaur *Maiasaura, Psittacosaurus*, the ceratosaur *Coelophysis*, and the prosauropod *Plateosaurus*. *Protoceratops* comes in not far behind with an estimated eighty specimens.[48] Although it is difficult to be precise about exact numbers, the sense is that *Psittacosaurus* was abundant, diverse, and geographically widespread. Few dinosaur genera have more than a single species that has stood up to critical scrutiny. The fact that *Psittacosaurus* apparently had four or five valid species is quite interesting and in a general way probably reflects the small size of the animal— as a rule, the number of species in a genus of living animal decreases as body size increases.

One of Sereno's most important accomplishments was to draw the Psittacosauridae into the Ceratopsia, thereby expanding the concept of the Ceratopsia. Thus the word *ceratopsian* has now taken on a new meaning, slightly different from that used by Romer in 1956. When the word is used to support a general statement, it is now understood to include psittacosaurs. Sereno added a very useful new term, *Neoceratopsia* (literally "new Ceratopsia"), that encompasses all animals previously held to constitute the Ceratopsia. Neoceratopsians consist of protoceratopsids (*Leptoceratops, Protoceratops,* and other animals we are soon to describe) and ceratopsids, the typical horned dinosaurs of western North America.[49]

It is accepted that the Early Cretaceous psittacosaurs that we know could not themselves have been ancestral to neoceratopsians. They show many specialized structures. The pattern of the hand, for instance, shows considerable reduction of the outer digits, with the fifth digit absent and the fourth digit strongly reduced. Neoceratopsians have five complete digits, and it may be assumed that their ancestors did as well. Psittacosaurs were evidently successful dinosaurs in their own right. Although they were very small, it appears that they had very powerful jaw-closing muscles. They must have cropped and chewed a rather tough vegetation. In addition, up to fifty small stones have been found in the stomach region of some specimens. It appears that these were unusual ornithischian dinosaurs that made use of stones to help digest plant matter. Psittacosaurs lived at a time from which dinosaur faunas around the world are comparatively poorly known, and those of Asia are no exception. Several large relatives of the familiar European *Iguanodon*, one named *Iguanodon orientalis* and the other *Probactrosaurus gobiensis*, lived in the region at the same time, as did an ankylosaur, *Shamosaurus scutatus*. Meat-eating dinosaurs are very poorly documented, although there are a number of names on record based on various fragmentary and isolated specimens. The most notable of these is probably *Harpymimus okladnikovi*, a toothed relative of the ornithomimids. The quality of the psittacosaur record stands by itself.

NEWER FINDS

Males and females are both well represented in the fossil record of dinosaurs. Quite in contrast is the sexual representation of the scientists who study dinosaurs. Women certainly played a role in the early history of fossil discovery. Mary Anning collected fossil ichthyosaurs, plesiosaurs, and pterosaurs at Lyme Regis, England, during the first half of the nineteenth century, and she is famous in paleontological lore as the inspiration for the tongue twister "She sells seashells by the seashore." Mary Ann Mantell may have found the teeth in 1822 upon which her husband, Gideon Mantell, based his dinosaur, *Iguanodon*, in 1825. For the ensuing 150 years, however, most dinosaurs were described by dead, white males—in fact, by dead, white, English-speaking males (known on a tectonically unstable plate of the North American continent as DWEMs), primarily by British, American, or Canadian scientists.[50] Since 1970, dinosaur paleontology has become an international enterprise. Its leading practitioners, measured by the criterion of most descriptions of new kinds of dinosaurs, are Argentine, Mongolian, Chinese, Russian, and Polish.[51] The foremost Polish dinosaur experts are Teresa Maryańska and Halszka Osmólska. Magdalena Borsuk-Bialynicka and Ewa Roniewicz have also made important contributions. Zofia Kielan-Jaworowska has not studied dinosaurs, though she has collected more than her share; rather, she is one of the world's greatest authorities on Cretaceous mammals. The common link among the aforementioned Polish paleontologists? They are all women.

Poland, like most countries in the world, has been blessed with paleontological treasures other than dinosaurs—trilobites, Baltic amber with insects, but barely a hint of dinosaurs. Polish paleontologists are typically experts on trilobites or other Paleozoic fossils. But like paleontologists everywhere, they are also dreamers. Kielan-Jaworowska had heard of the American expeditions to the Gobi, an account of which had been published in a Polish magazine before World War II. She had also read Roy Chapman Andrews's book *Ends of the Earth*, which had been translated into Polish. Furthermore, Kielan-Jaworowska had visited the Paleontological Institute of the Soviet Academy of Sciences in Moscow in 1955 and had seen for herself the mounted skeletons of the carnosaur *Tarbosaurus* and the hadrosaur *Saurolophus* collected by the Soviet-Mongolian expeditions of 1946, 1948, and 1949. The opportunity to realize her dream arrived in the form of an agreement between the Polish and Mongolian academies of science signed in 1962. Despite her expertise on trilobites, Kielan-Jaworowska found herself in charge of Polish paleontological expeditions to the Gobi Desert. Between 1963 and 1971, eight expeditions were mounted, some large, some small, but all immensely successful, yielding remarkable collections of dinosaurs, mammals, lizards, and other vertebrates. The scientific results of the expeditions have been flowing ever since, principally in a series of magnificent technical volumes in the journal *Palaeontologica Polonica,* published by the Polish Academy of Science, beginning in 1968.[52] The first dinosaur to emerge in print was the enigmatic theropod *Deinocheirus mirificus,* described by the erstwhile trilobite expert, Halszka Osmólska, and Ewa Roniewicz in 1970.

However, for our purposes, we turn our attention to Part 4 of the results, published in 1975. Our eye is drawn to a paper by Teresa Maryańska and Halszka Osmólska, entitled "Protoceratopsidae (Dinosauria) of Asia." Although many specimens of *Protoceratops* were collected (including a famous skeleton whose skull is surrounded by the grasping outstretched hands of a nasty *Velociraptor*), Maryańska and Osmólska described instead a new genus and two new species of protoceratopsids and reported new material of a previously described but very poorly known protoceratopsid as well. *Bagaceratops rozhdestvenskyi* (*baga* meaning small in Mongolian) is named in honor of A. K. Rozhdestvensky, a prominent Soviet dinosaur paleontologist. More than twenty skulls, partial skulls, and scattered skeletal elements were collected from the Khermeen Tsav Formation, which is presumably slightly younger than the Djadochta Formation. *Bagaceratops* is unmistakably a protoceratopsid, but it is a slender, dainty one (Fig. 7.9). The largest skull of *Bagaceratops* is about 17 cm long, and the smallest is 47 mm, making it in 1975 the smallest dinosaur skull on record. It has a small frill, initially reported to be without openings, but Osmólska has reported to me that specimens with openings in the frill are now known. In several respects *Bagaceratops* is more derived than *Protoceratops.* For instance, there is a modest, low, blunt "horn" (actually more of a boss) on the nasal, and there are no teeth on the premaxilla. Other characters are more basal—the shortness and narrowness of the frill and a very low tooth count of only ten teeth in each jaw (the smallest skull has six teeth). Other characters are unique to *Bagaceratops,* such as an extra opening in the skull just behind the nostril. The lower jaw has a straight lower margin, unlike the bowed margin of *Leptoceratops* and *Protoceratops.* In the baby, the skull is rounded; the orbit is enormous, accounting for nearly half the skull in profile; the

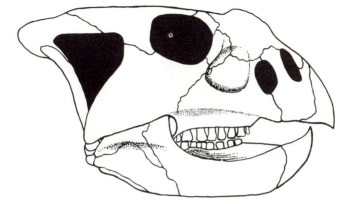

FIG. 7.9. Skull of *Bagaceratops rozhdestvenskyi*, Polish Academy of Sciences. (Robert Walters after Maryańska and Osmólska 1975.)

horn is already in evidence; and the extra opening is considerably larger than the nostril. There is little information from the skeleton, except that six vertebrae in the sacrum have been confirmed, compared to eight for *Protoceratops*.

A second new species in the collection was designated ?*Protoceratops kozlowskii*, in honor of Professor Roman Kozlowski, the dean of Polish paleontologists, the man whose lectures had inspired Kielan-Jaworowska with the dream of going to the Gobi herself. The question mark expressed the authors' doubt that the species really pertained to *Protoceratops*, but the small specimens are poorly preserved and they conservatively (and perhaps admirably) declined to name a new genus.

The problem is that when you show such restraint somebody else will very likely come along and rename your dinosaur. Sure enough, in 1990, Russian paleontologist S. M. Kurzanov obtained new material that convinced him that the animal was unique, and he proposed the new name *Breviceratops kozlowskii*, meaning "short horn-face" (Fig. 7.10).[53] I agree that the name is justified. *Breviceratops* comes from the Barun Goyot Formation (roughly equivalent to the Khermeen Tsav Formation) from the important locality of Khulsan (pronounced with a silent K), more than 100 km east of Khermeen Tsav and more than 200 km southwest of Bayn Dzak. The best skull was only 70 mm long, but a fair partial skeleton added details. In addition, there is a less complete skull that is about 25 percent smaller, as well as other fragmentary remains. When there is so little material and the individuals represented are

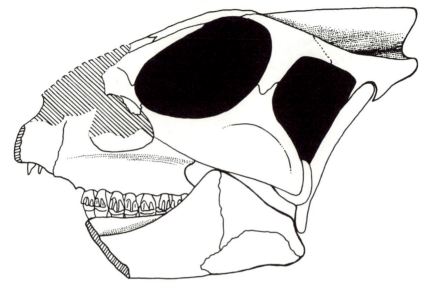

FIG. 7.10. Skull of *Breviceratops kozlowskii*, Polish Academy of Sciences. (Robert Walters after Maryańska and Osmólska 1975.)

young, it is hard to evaluate the genus. The biggest problem is that we have no idea how large the animal grew.

It is clear that the animal is neither *Protoceratops* nor *Bagaceratops*. There is no trace of a nose horn as there is in *Bagaceratops*, and there are two premaxillary teeth, so it cannot be *Bagaceratops*. The frill is very narrow, and the skull does not flare widely at the jugals the way it does in *Protoceratops*.[54] The lower jaw is straight rather than bowed as in *Protoceratops*. There are only seven teeth in the maxilla, but one dentary contains eleven teeth (the smallest contains seven). A highly derived character in the pelvis is the structure of the ilium, which has a dorsal border that hangs over the hip, foreshadowing the complete eversion of the dorsal ilium in the Ceratopsidae. There are eight sacral vertebrae, a high count also seen in *Protoceratops*. The femur is 41 mm long. If the little *Breviceratops* had the same body proportions as in our reference specimen of *Leptoceratops*, it would have been about 11.5 cm (4.5 in.) at the hips and 33 cm (13 in.) long. These were small dinosaurs. We have much more to learn about *Breviceratops*.

The third subject of study by Maryańska and Osmólska was not new but had previously been described by the Swedish paleontologist Anders Bohlin in 1953. *Microceratops gobiensis* was collected in Gansu,

Inner Mongolia (China) by the Sino-Swedish expeditions of 1927–1931, but the material was so fragmentary that description was a decidedly low-priority matter. Bohlin actually named two species, including *M. sulcidens*, from a separate locality. As the name implies, it was clear that the animal was a very tiny protoceratopsian, but little more could be said. In 1971, the Polish-Mongolian expeditions collected a specimen of *Microceratops gobiensis* at Sheeregeen Gashoon, in beds that are believed to be of early Late Cretaceous age.[55] The single specimen is still very incomplete, but the partial skeleton provides significant information that allows the mysterious little animal to be characterized for the first time. The skull is about 20 cm long; although small, it is not tiny or microscopic by any means. In fact, it is the largest of the eleven skulls listed in the monograph. The frill seems well developed but short, with prominent parietal fenestrae. The most important characters come from the skeleton. There are seven unfused sacral vertebrae. The slender forelimb is complete to the wrist. The forelimb was about 70 percent of the length of the hind limb. The pelvis is fragmentary and somewhat uninformative, but the rod-like pubis is unique to *Microceratops*. There is an excellent hind limb and foot. The tibia is 15 percent longer than the femur (versus 10 percent in *Protoceratops*), and the foot (as represented by metatarsal III) is 55 percent of the length of the femur (versus 50 percent in *Protoceratops*). The stretching out of the lower segments of the leg compared to the femur suggests that *Microceratops* was capable of running swiftly. This is not surprising in such a small animal, one that stood about 26 cm at the hips and was about 90 cm long, much tinier than *Protoceratops*. Maryańska and Osmólska hypothesized that *Microceratops* was bipedal during swift locomotion and quadrupedal at rest and during slow movement.[56]

Soviet paleontologists have named several other protoceratopsids recently. In 1992, S. M. Kurzanov named *Udanoceratops tschizhovi*, on the basis of a moderately well preserved skull of large size (approximately 60 cm long) with very little horn and frill (Fig. 7.11). He regards it as occupying a more basal position than *Leptoceratops*, the usual candidate for the most basal protoceratopsid. In 1989, the Soviet paleontologist Lev Nessov from St. Petersburg, with his colleagues, L. F. Kaznyshkina and G. O. Cherepanov, described two very fragmentary taxa, *Asiaceratops salsopaludalis* and *Turanoceratops tardabilis*. These specimens come not from Mongolia but from Uzbekistan, formerly Soviet Central Asia. The former consists of teeth, skull fragments, and a toe bone. The latter consists of several skull fragments, one of which may be a horn core.[57]

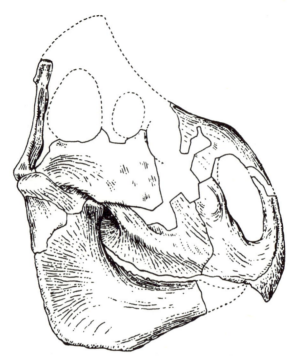

FIG. 7.11. Skull of *Udanoceratops tschizhovi*, Russian Academy of Sciences, Moscow. (Robert Walters after Kurzanov 1992.)

Although these animals are potentially interesting, there is so little material at present as to make the taxa unusable, and they have been provisionally designated as "nomina dubia," a state of taxonomic limbo. The practice of naming genera on such a slim basis is a highly undesirable one, greatly to be discouraged.

MONTANOCERATOPS—BACK HOME AGAIN

We have taken a tour of the world, from Alberta to Mongolia and China, and have barely mentioned the United States. How can that be? At last, we have occasion to come back home. As you recall, when *Protoceratops* was discovered, it seemed terribly primitive. As the years have gone by, it has become more and more clear that it occupies a highly derived position in the phylogeny of protoceratopsids. Only one dinosaur remains to be considered. In 1942 Barnum Brown was at the tail end of a dazzling career that had spanned two centuries, several

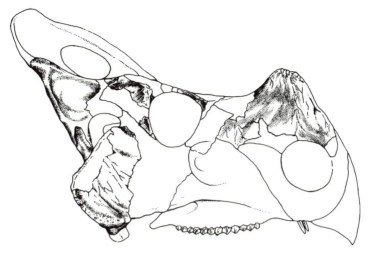

FIG. 7.12. Skull of *Montanoceratops cerorhynchus,* American Museum of Natural History. (Robert Walters after Brown and Schlaikjer 1942.)

continents, scores of specimens collected, and many described. Times had changed. Osborn had retired nine years earlier and died in 1935. The Asiatic expeditions had ended. The American Museum had emptied out during the war as scientists and staff alike had assumed war-related duties. It was a quiet place. Yet Brown still had papers to write, and he had a young assistant, E. M. Schlaikjer, to help him. He returned to a skeleton he had collected nearly three decades earlier, in 1916, at Buffalo Lake, Montana, in the St. Mary River Formation of early Maastrichtian age (equivalent to the Horseshoe Canyon Formation of Alberta). The specimen consisted of a fine skeleton with a fragmentary skull. It was mounted for exhibition in 1935. Brown and Schlaikjer were so impressed with its characters that they designated it a new species of Brown's old protoceratopsid, *Leptoceratops.* They called it *Leptoceratops cerorhynchus,* the species name meaning "horn beak."

The skull bones preserved include a nasal showing a prominent blunt horn and fragments of maxilla, squamosal, jugal, quadratojugal, quadrate, and jaw. In the mount, the fragments were made up into a whole skull, an approach that to me is an exercise more in fiction than in fact and that makes the individual bones inaccessible to study (Fig. 7.12). I do not believe we know at all what the skull looked like, apart from the nasal horn. A postorbital shows that there were no horns over the eyes. The vertebral column was beautifully preserved. It shows the four fused cervical vertebrae that so characteristically support the head in ceratopsians. There are six further cervical vertebrae, twelve dorsal vertebrae, and eight sacral vertebrae. Only fifteen caudal vertebrae were found. Brown and Schlaikjer inferred by comparison with *Protoceratops* and *Leptoceratops* that the complete vertebrae represented segments out to caudal segment 30 and that two distal centra represent segments 38 and 41 of a hypothesized tail of fifty or fifty-one vertebrae. One striking feature of the tail vertebrae is that the spines in the mid-tail series are very tall, up to five times the height of the centrum. This feature, also found in *Protoceratops* but not in *Leptoceratops gracilis*, gives the tail a characteristically deep appearance and exaggerates the taper at the end. Brown and Schlaikjer did note a shortening of the vertebral centra of the tail, a derived feature found in ceratopsids but not noted elsewhere among protoceratopsids. Several of the tall spines show evidence of injury and healing.

Unfortunately, nothing of the forelimbs was preserved. The pelvis is in excellent condition. The ilium is long and robust. The forward blade bends outward and its upper edge rolls slightly laterally. The forward blade of the pubis is slightly more developed than in *Protoceratops,* and the backward rod of the pubis less developed. The ischium is long and quite robust, and it shows a distinct if gentle downward curve. The femur is rather robust, and the tibia is barely longer than the femur, with greater expansion of the proximal and distal ends than is typical of the more gracile protoceratopsids. The metatarsals are not preserved, but several of the digits are. The unguals are sharper and more claw-like than in *Protoceratops,* the only important feature that Brown and Schlaikjer found to be more primitive than in *Protoceratops.* Their conclusion was that *Leptoceratops* "is slightly more progressive than *Protoceratops.*"[58]

Brown and Schlaikjer provided a very useful description of *"Leptoceratops"* cerorhynchus. Their study was flawed, however, in that they mistakenly assumed that their animal could be correctly referred to *Leptoceratops.* Recall that Brown had described *Leptoceratops gracilis* from Alberta in 1914. His specimen was fragmentary, and no further specimens had yet been recovered. In 1947, C. M. Sternberg came upon three skeletons of *Leptoceratops* that really allowed the genus to be characterized. It was clear to Sternberg in his 1951 paper that true *Leptoceratops*—that is to say, *Leptoceratops gracilis*—was a very different animal from its Montana cousin. Sternberg wrote of *L. cerorhynchus,* "this species belongs to a genus quite distinct from *Leptoceratops.* In almost every respect the Montana species is more advanced than *Protoceratops*, whereas *Leptoceratops* is the most primitive of all."

It was clear to Sternberg, as it is to us today, that the Montana specimen needs a generic assignment separate from *Leptoceratops*. Sternberg was an old-fashioned gentleman. He apologetically explained in a footnote, "I wrote to Dr. Schlaikjer, pointing out the great differences in the two species and suggested that he propose a new generic name for *cerorhynchus*, but no action has been taken." By this time, Brown had retired from science and Schlaikjer had embarked on a new career as a mining engineer in Saudi Arabia. Sternberg thus quite properly baptized the dinosaur *Montanoceratops cerorhynchus*, and so we know it today.[59]

Unfortunately, *Montanoceratops* is no better known now than it was in 1951, nor have further protoceratopsids been described from North America. David Weishampel of the Johns Hopkins University has worked recently in the St. Mary River Formation, in cooperation with the Museum of the Rockies. He found a partial skeleton of a small specimen, which may include a skull. However, it has not yet been prepared sufficiently to determine its significance. Similarly, Jack Horner has reportedly found protoceratopsid skeletal material from slightly earlier Late Cretaceous formations in Montana, but these specimens too have not yet seen the light of day and thus tantalize rather than inform.[60] The potential is clearly there for interesting new finds—which is precisely what makes paleontology so exciting.

Sisters, Cousins, and Aunts

DECIPHERING THE FAMILY TREE

FOSSILS do not come out of the ground with labels on them. Defining and naming species and genera is a major intellectual activity. When the first horned dinosaur teeth were examined by Leidy in 1856, and when the first skeletal remains were collected by Cope in 1872, 1874, and even 1876, still no concept was gained as to the nature of these animals. Marsh identified the first remains of a horned dinosaur as bison horns in 1887. When at last he finally got it right, naming *Triceratops horridus* in 1889, Marsh compared the animal to *Stegosaurus*, a favorite of his from the Late Jurassic Morrison Formation that he had named in 1877. Only as more specimens and more taxa were described did it become evident that the Ceratopsia represent a distinctive group by themselves. Large animals tend to demand attention first. In 1913, Gilmore discovered a small ceratopsid skeleton, which he named as *Brachyceratops* in 1914, the same year that Barnum Brown described the diminutive *Leptoceratops* from Alberta. The discovery of *Protoceratops* in Mongolia 1922 proved truly electrifying, and it was recognized immediately by Osborn as "the long-sought ancestor of *Triceratops*." *Psittacosaurus* too was a fruit of the American Museum of Natural History expeditions to Mongolia, but it was a remote cousin, all but forgotten for many years and only admitted into full membership in the Ceratopsia in 1975.

An important synthetic activity of the paleontologist is the determination of relationships. In *H.M.S. Pinafore*, Sir Joseph Porter's admiring flock of sisters, cousins, and aunts was available to remind him of his family tree, but for us paleontologists this activity is more difficult. This process is called phylogenetic reconstruction, and although it causes the eyes of my students to glaze over when I describe it, it represents some of the highest intellectual endeavors of the paleontologist today. A family of dinosaurs cannot simply be read from the rocks, particularly for horned dinosaurs. The fossil record simply doesn't permit it.

Clear ancestor-descendant relationships are not unequivocally preserved. Whereas a nearly continuous record of horse evolution or rhinoceros evolution spanning fifty million years may be discerned in North America, the very finest dinosaur record in the world, that preserved in the Red Deer Valley of Alberta, spans only about eleven million years. It has been estimated that the average duration of a genus of dinosaur is five to seven million years.[1] The time available is simply too brief for the observed diversification of dinosaurs. Ancestors and descendants—or rather sisters, cousins, and aunts—are commingled in the same faunas.

It would be most convenient if a lengthy series of time-successive formations were found in the same sedimentary basin, each with a descendant species or genus of dinosaur, but this is not found in practice. There is a suspicion that reduction of habitable land areas by spreading seas, as described in Chapter 1, may have represented an evolutionary "bottleneck" that changed the genetic composition of dinosaur faunas, leading to the evolution of new kinds.[2] Much of the evolutionary activity thus may have gone on "off stage," perhaps in upland areas closer to the mountain uplifts, which are not well represented in the fossil record. Presumably new kinds of dinosaurs then migrated down into the lowlands. This is an interesting model, but one for which there is as yet little empirical evidence.

In any case, how were relationships among horned dinosaurs perceived by their describers, and how do we view them today? The first attempt to review all of the Ceratopsia was in 1907. At that time, only two stratigraphic levels for the horned dinosaurs were recognized: the Laramie beds (now the Lance Formation) of Wyoming and Denver beds of Colorado, containing *Triceratops,* and the Judith River Formation of Montana and Belly River beds (now the Judith River Group) of Alberta, with *Centrosaurus* and *Monoclonius.* Lull addressed the question of phylogeny explicitly, quoting Osborn: "It is not at all improbable that the horned dinosaurs will prove to be diphyletic, one line, with persistent open fossae, leading from *Monoclonius* to *Torosaurus,* the other leading to *Triceratops* with closed fossae."[3] Lull favored the provisional recognition of two phyla or lines of descent, with *Centrosaurus* and *Monoclonius* preceding *Triceratops,* and *Ceratops* preceding *Torosaurus.* He wrote:

> Of the two most primitive genera *Monoclonius* seems to be the more generalized and represents the earliest known stage in the evolution of

the Ceratopsia. Because of the gap between the Judith River beds and those of the Laramie the series is by no means complete, nor are we yet aware of the characteristics of pre-Judithian ancestors. The earliest known Ceratopsia are endowed with the main distinguishing characters, the horns and parietal crest.[4]

It is hard to be too critical of this "phylogeny," because at this time so little material was known. In only a few years, much progress had been made, especially because of the work in Alberta. Lambe in 1915 recognized three subfamilies, two of which, the Centrosaurinae and the Chasmosaurinae, are quite modern in concept.[5] By 1933, when Lull considered "Progressive Evolutionary Change and Phylogeny" in his monograph, the picture had fleshed out considerably. He recognized, as we do today, a third time interval, represented by the "Edmonton" (Horseshoe Canyon) and Fruitland formations in addition to the "Belly River" and Lance. He observed:

> The recorded evolutionary history of the Ceratopsoidea [sic] opens abruptly, with several highly interesting chapters not yet revealed, for when these grotesque animals appear in our records they are already ceratopsians with the essential features established. Not only so, but their differentiation into at least three phyletic lines has already been brought about.
>
> The Protoceratopsidae end where they begin, for the two isolated genera, *Protoceratops* and *Leptoceratops*, each with its single known species, tell all we know of the group. With them there is no *recorded* phylogeny.
>
> The Ceratopsidae, on the other hand, while established as such at their initial appearance, do nevertheless show a very considerable series of changes, some of which are common to all phyla, others peculiar to one or more.

Lull hypothesized the origin of horned dinosaurs "from some unarmored stock of ornithischian dinosaurs, probably bipedal in gait," even though he knew of no probable ancestor. The trends he inferred include the assumption of four-legged posture, increased size of the head, shortening of the neck, fusion of the first three neck vertebrae, modification of the pelvis, and transformation of the head into a skull with a narrow, wedge-shaped snout and a flaring crest.

Lull recognized two subfamilies, which he did not name, and for which he gave no credit to Lambe or to anybody else. His basis was the "length of the squamosal and its reaction upon the crest" and "the persistence or closure of the parietal fenestrae." His long-crested or

long-squamosal line began with *Chasmosaurus* in the "Belly River" and
continued in the Edmonton with *Anchiceratops*—but he noted that that
genus "seems to depart from the main evolutionary line" because of a
highly vascularized crest and small parietal fenestrae. By contrast, *Pentaceratops*,
though far removed geographically from Alberta, "could be
the lineal descendant of *Chasmosaurus kaiseni* with little alteration, other
than size and the more abundant vascularity of the squamosal." He
found that *Arrhinoceratops* presented certain difficulties, resembling as
it did *Triceratops obtusus* in the nose region. He believed that *Triceratops*
was a member of the short-frilled or short-squamosal phylum, and
clearly *Arrhinoceratops* was a member of the long-frilled group. Appro-
priately, *Torosaurus* was the closing member of the phylum. "In the
length of the nasal horn, but not its position, and in the ornamentation
of the posterior part of the crest, the long-squamosal phylum lags
behind that of the short-crested types." The short-crested forms, according to Lull, are *Centrosaurus, Styracosaurus,* and *Triceratops,* with no
unequivocal Edmonton representative. An innovation of Lull's was a
phyletic chart showing his lineages through time. He cautioned that his
phyletic conclusions were not final. "The figure is to be regarded
merely as a trial sheet which is imperfect, due perhaps to misinterpreta-
tion, but more, I hope, to the incompleteness of the fossil record." Two
features are noteworthy. Lull represents *Triceratops* as descending directly from *Centrosaurus flexus,* and he drew another direct line from
Chasmosaurus kaiseni to *Pentaceratops* to *Torosaurus.*[6]

Fresh from his description of the new skull of *Torosaurus,* E. H. Colbert wrote a major review paper on the Ceratopsia in 1948. Though he
pondered many aspects of ceratopsian biology, it is his consideration of
phylogeny that concerns us for the moment. He adhered to the two-
phylum approach and mapped the distribution of nasal horns, postorbital horns, frill length, and frill fenestration. He recognized long-
frilled Ceratopsidae and short-frilled Ceratopsidae, but again no formal
group names were offered. He presented a phylogenetic diagram under
the title "Adaptive radiation of the Ceratopsia, as shown by evolution
in the skull." It was labeled as "modified from Lull 1933," as indeed was
the case. Whereas Lull had shown seven species of *Triceratops* and two
species each of *Torosaurus* and *Chasmosaurus,* Colbert showed but a
single specimen of each. Whereas Lull left the two lineages separate,
Colbert brought them both together in a line that descended from
Protoceratops. He also added *Leptoceratops* to the diagram but incorrectly
(if conveniently from a graphic design point of view) placed it in the

Belly River rather than in the Lance. Apart from the severing of the questionable link between *Arrhinoceratops* and *Triceratops,* the form and information content of Colbert's diagram closely resemble those of Lull's.[7]

John Ostrom wrote a very important paper on the evolution of jaw mechanics in horned dinosaurs in 1966. In this paper, phylogeny was of decidedly secondary concern. Nonetheless, a very familiar phylogeny appears once again, with *Triceratops* arising out of *Centrosaurus* and *Torosaurus* out of *Chasmosaurus* via *Pentaceratops. Leptoceratops* at least has risen one fauna in position, appearing in the Edmonton, and thus only one faunal zone shy of its correct Lancian position.[8]

The unanimity was not complete, but the Canadian voice was largely ignored. C. M. Sternberg published a key paper in 1949, in which he described *Triceratops albertensis* and drew attention to the distinctive faunal character of the Upper Edmonton member compared to the Lower and Middle members. Several decades later the members were formally recognized as separate formations within the Edmonton group, the faunal break occurring between the Horseshoe Canyon Formation, representing the Lower and Middle members, and the Scollard Formation, representing the Upper Member. Sternberg too considered the phylogeny of the Ceratopsidae, and he also presented a version of Lull's diagram. He began, "Students of the Ceratopsidae have not always been in complete agreement regarding the phylogeny of the family, but all seem to believe that *Anchiceratops* from the Edmonton and *Torosaurus* from the Lance are probably derived from the Belly River *Chasmosaurus* or some closely related form."

Sternberg characterized the group in a familiar manner, as having (among other features) a long, flat, rectangular crest with large parietal fenestrae and a squamosal extending to the back of the crest. But he added an important new detail: "The thin, vertical parts of the premaxillae that divided the external nares (the osseus septum) were pierced by a well-defined fenestra." He then considered the second group, which included *Monoclonius, Brachyceratops, Centrosaurus,* and *Styracosaurus, all* of "Belly River" (i.e., Judithian) time. This group is characterized by a crest that is short and round with fenestrae of moderate size; short, roundish squamosals; nasal horn always longer than the brow horns; and the premaxillae *not* pierced by a fenestra as in the rectangular crested forms. He was quite definite that short-squamosaled forms had not been documented in the Edmonton or Lance (though two years later he himself was to document *Pachyrhinosaurus* from the Edmonton). Finally Sternberg discussed *Triceratops:*

In *Triceratops* the crest is generally round, but the squamosals flank the crest and reach to the back of the crest as in *Chasmosaurus*. There is no evidence of fontanelles in the crest. The brow horncores, which arise from behind the orbits, are very large, and unlike earlier genera, the base is hollow. The nasal horncore was much smaller than the brow horncores, was far forward, and in some species it was very small. . . . The premaxillae were pierced by a fenestra, as in the rectangular crested forms. . . .

The writer cannot believe that *Triceratops* could have evolved directly from *Centrosaurus* or from *Brachyceratops* as shown by Schlaikjer, but sees no objection to deriving it from *Arrhinoceratops*.

Sternberg thought that Schlaikjer's species, *Triceratops eurycephalus*, showed a suite of characters, including breadth of the squamosal and thinness of the parietal, and small size of the nasal horn core, that linked it with *Arrhinoceratops* as a logical connection:

The close resemblance of these two forms seems to justify Lull's alternate suggestion that *Triceratops* was derived from the long squamosal forms through *Arrhinoceratops*. It would also suggest that *T. eurycephalus* was a primitive species of *Triceratops* rather than an advanced form as suggested by Schlaikjer.

He concluded with an ecological speculation on the ancestor of *Triceratops*:

The lack of an adequate ancestor for *Triceratops* in the Oldman formation on the Red Deer, may signify that the delta and flood-plain habitat was not favorable, but that they preferred somewhat higher country that existed a few miles to the west. The complete lack of eggs and the almost complete lack of juveniles, of forms of which adults are so common, suggests that the delta was the habitat of the more or less adult forms only.[9]

In recent years, there has been fresh, independent corroboration of Sternberg's assessment of the phylogenetic position of *Triceratops*. Wann Langston, Jr., in assessing the position of *Pachyrhinosaurus*, made it clear that *Triceratops* did not belong with the short-frilled ceratopsids.[10] Except for the shortness of the solid frill, *Triceratops* is otherwise a good chasmosaurine. Thomas Lehman in 1990 completed the process in rigorous fashion. He formally defined the two valid subfamilies proposed by Lambe in 1915.[11] Lehman proposed a series of characters by which the members of the separate subfamilies may be recognized:

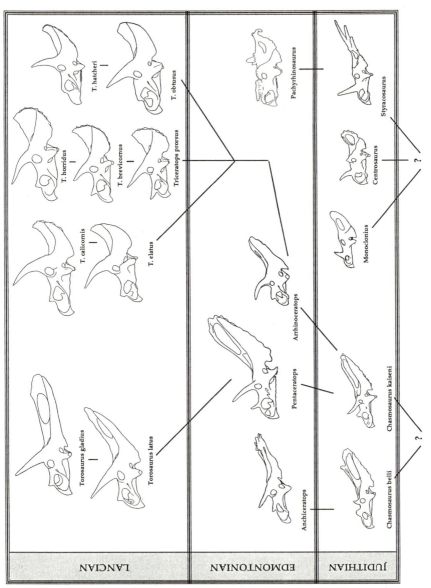

FIG. 8.1. Phylogeny of the Ceratopsidae with emphasis on "species" of *Triceratops*. This figure represents the culmination of a now-outdated process of constructing phylogenetic trees specifying ancestor-descendant relationships. This figure supports Sternberg's recognition of *Triceratops* as a chasmosaurine and disavows Lull's view that *Triceratops* derived from *Centrosaurus*. (Robert Walters after Lull 1933, Sternberg 1949, and

Subfamily Chasmosaurinae Lambe 1915

Large ceratopsian dinosaurs with long, low facial region (preorbital length/height = 1.4 to 3.0); inter-premaxillary fossae present; premaxillary and predentary with horizontal or poorly developed lateral cutting flange; nasal horncore, if present, small and formed in part by a separate ossification;[12] supraorbital horncores usually large. . . .

cranial frill long (0.94 to 1.70 basal length of skull, except *Triceratops*); triangular squamosal bones (length/height = 2.0 to 3.5). . . .

ulna with large olecranon process; strongly curved ischium.

The characters of the other subfamily contrast with the first:

Subfamily Centrosaurinae Lambe 1915

Large ceratopsian dinosaurs with short, deep facial region (preorbital length/height = 1.2 to 1.4); no inter-premaxillary fossae; finger-like processes of the nasal bones projecting into the narial apertures; wide laterally inclined shearing surfaces on the premaxilla and predentary bones; nasal horncore, if present, large and formed primarily by upgrowth of the nasal bones; poorly developed supraorbital horncores. . . .

short cranial frill (0.54 to 1.00 basal length of skull; quadrangular squamosal (length/height about 1.0). . . .

ulna with reduced olecranon process; relatively straight ischium.[13]

Thus, not only do the two subfamilies differ in the pattern of the horn cores and in the frill, they also differ in the structure of the nose region. Centrosaurines have short, deep faces with a curious bony tab projecting into the nostril from behind, whereas chasmosaurines have long, low faces, with either a depression in front of the nostril or a frank opening. The function of this fossa is uncertain. Lehman speculates that it may have held an organ such as the vomeronasal organ (formerly known as Jacobson's organ), but this is far from certain. Whatever its function, it is a very useful taxonomic feature that allows us to discriminate between the two subfamilies. *Triceratops* presents a challenge to Lehman's scheme because the frill is short (0.62–0.89 times basal length, according to him), and this is a centrosaurine attribute. But the triangular form of the squamosal is chasmosaurine, the horns and the nasal region decidedly chasmosaurine. It therefore seems to be a good chasmosaurine with secondary closure of the parietal fenestrae. Figure 8.1 represents a synthesis of the phylogenetic diagrams of various workers, particularly Lull (1933) and Sternberg (1949).

RECONSTRUCTING THE FAMILY TREE
OF THE CERATOPSIA

Early on, workers assumed that phylogeny could be interpreted more or less directly from the record of fossils in the rocks. Phylogenetic reconstruction was thus an exercise in connecting forms that looked most similar. Perhaps this caricature is slightly unfair, but it was an exercise that was not altogether rigorous. With the advent of computers, it is not surprising that greater emphasis should be placed on analytical rigor that produces precise (though not necessarily accurate) results. A method of phylogenetic reconstruction that has had great impact on our understanding of relationships among all organisms is called phylogenetic systematics or cladistics. It was developed by a German entomologist, Willi Hennig, whose book on the subject was translated and published in the United States in 1966. For a number of years there was acrimonious debate about the merits of the cladistic method compared to the more classical methods used by systematists for the previous two centuries. Today there is general agreement, whether or not one subscribes to all aspects of cladistic philosophy, that the methodology of cladistic analysis is a useful one.[14] Briefly, cladistics attempts to reconstruct phylogenetic relationships on the basis of shared characters that represent evolutionary novelties—derived characters or synapomorphies, in cladistic jargon.

The parieto-squamosal crest of horned dinosaurs is an evolutionary novelty: no other dinosaurs have them. If we compare two horned dinosaurs, say *Chasmosaurus* and *Centrosaurus*, with a duck-billed dinosaur, for example, *Corythosaurus*, we would be justified in concluding that *Chasmosaurus* and *Centrosaurus* are more closely related to each other than either is to *Corythosaurus:* horns on the face would tell us this. In cladistics, not all characters may be used. The use of primitive characters is specifically disallowed. Looking at our example once again, we find that *Chasmosaurus* and *Centrosaurus* each have five digits in the hand, whereas *Corythosaurus* has only four digits in the hand. Although the horned dinosaurs resemble each other in this character, we are not permitted in the cladistic system to conclude that this is evidence of relationship between *Chasmosaurus* and *Centrosaurus*, because the possession of five fingers on the hand is a primitive character that was found far earlier in phylogenetic history (in fact, although some of the earliest land vertebrates had more than five fingers, this number has been stabilized at five for more than 300 million years).

Such a character is termed a *plesiomorphy,* and it is not decisive in establishing evolutionary relationships.

The practice of cladistic analysis involves the examination of a large number of characters (or morphological conditions) in the taxa of interest. Each character is scored either as 1 for a derived character or as 0 for a basal or plesiomorphic character. For our purposes, we are interested in all members of the Ceratopsia—*Psittacosaurus,* the Protoceratopsidae, the Centrosaurinae, the Chasmosaurinae—all told about twenty-three genera of horned dinosaurs (not all of which had horns). It is also necessary to look outside the Ceratopsia, at one or several animals that are potentially related to the Ceratopsia but which themselves show no ceratopsian characters. Such taxa are termed outgroups. By definition, the outgroup possesses no derived characters, and therefore scores 0 across the board (As George Olshevsky once put it, "One worker's outgroup is another's life work!"). A score is developed for each taxon studied, and a computer manipulates the 1s and 0s to find the best match among taxa, the most parsimonious way to evolve each species or genus—that is, the way that requires the fewest transformations from one character state to another. The results are presented as a cladogram, a branched diagram that arrays taxa from the most generalized basal taxa to the most highly derived. Cladograms never specify ancestor-descendant relationships. This highly prized goal is regarded as unattainable. A closely linked pair of taxa is said to constitute a sistergroup. The authorized, conservative inference is that members of such a pair of taxa are more closely related to each other than either taxon is to a third one. This cautious statement lends itself nicely to paleontology, where the record is notoriously incomplete.

An unnerving effect of the cladistic method is the abolition of familiar, even cherished, basal groups. Cladistics strives to discover monophyletic groups or clades that consist of an ancestor and *all* of its descendants. Ceratopsia and Ceratopsidae are examples of monophyletic groups, but Reptilia, for example, is not. As historically understood, the class Reptilia gave rise to both the class Mammalia (mammals) and the class Aves (birds). In a cladistic sense, therefore, the class Reptilia is not a natural clade or group, because it does not contain all of its descendants. It is therefore a *paraphyletic* assemblage. The natural clade to which "reptiles," birds, and mammals belong is the Amniota, defined by an the evolutionary novelty of the amniote egg. Reptiles could be defined as amniotes lacking hair or feathers, but defining a group by what it does not possess is not the way to go about it. Cladists at first

253

took delight in declaring that reptiles do not exist. But there is a way that reptiles are legal in the cladistic system. It was realized that the synapsids (previously considered a subclass of the Reptilia) were historically the first group of reptiles to evolve from basal amniotes (i.e., the earliest "reptiles"). All living reptiles—that is, turtles, snakes, lizards, and crocodilians—are more highly derived than the synapsids that led to mammals and thus form part of a perfectly good, natural clade: Reptilia. However, there is an interesting consequence of this. Dinosaurs are nested within the Reptilia, and birds are nested within the Dinosauria. Therefore, birds are reptiles—in fact, they are dinosaurs!

There is just one final point to make before proceeding. It makes no difference at all to a chicken or a pigeon or an ostrich whether we call it a bird or a dinosaur. It is useful for us to be able to make generalizations about the organisms we wish to discuss. But the names we choose to discuss them are purely human artifacts. Again, it has sometimes been urged that pterosaurs (flying reptiles) be included within the dinosaurs. The term *dinosaur* is a human construct, not an objective fact of nature. The name *dinosaur* designates one node on a cladogram. By including pterosaurs as dinosaurs, we would merely be applying the term to a slightly more basal node. We can define our "Club Dinosauria" more broadly or more narrowly, and it makes no difference to the animals inside. It is a matter of taste, not of science. And good taste *never* goes out of style!

Cladistic analysis began to be applied to dinosaurs in 1984. At a meeting in Tübingen, Germany, independent cladistic analyses were presented by Jacques Gauthier, Paul Sereno, David Norman, and Michael Benton. Two very important complementary analyses were published in 1986 by Gauthier on saurischian dinosaurs and by Sereno on ornithischian dinosaurs, including ceratopsians. In 1990 Benton published a cladogram of the entire Dinosauria.[15] For our purposes, I wish to consider Sereno's analysis, particularly the part that relates to ceratopsians.

The origin of the Ceratopsia had long been a problem. In the grand Romerian scheme, ceratopsians were seen in a general way as derived from a two-legged ornithopod resembling *Hypsilophodon*. Sereno came up with a novel arrangement that links the Ceratopsia with the Pachycephalosauria (the dome-headed dinosaurs) as the Marginocephalia ("shelf-heads"). Derived characters linking these two groups include a narrow parietal shelf overhanging the back of the skull; the participation of the squamosal in this shelf; a small contribution of the premaxilla to the hard palate; lack of a pubic symphysis; and a very short postpubis. He postulated that marginocephalians shared a common ancestor with the ornithopods, but this common ancestry might date back to the Early Jurassic (ca. 190 Ma), even though the earliest known representative of either group occurs about 125 Ma in the Early Cretaceous.[16]

Sereno found many characters to unite the Ceratopsia (including the Psittacosauria) as a monophyletic assemblage. Characters unique to the Ceratopsia include triangular head with a very narrow beak; a rostral bone at the front of the snout; jugals that flare laterally; tall snout with a deep premaxilla and maxilla; broad parietal overhanging the back of the skull. A second clade within the Ceratopsia Sereno named the Neoceratopsia ("new ceratopsians"), a term that I find very useful. This designates the Protoceratopsidae plus the Ceratopsidae.[17] Among other characters, neoceratopsians have the following: large heads relative to body size; upper teeth with a prominent ridge; teeth compressed along the tooth row with replacement teeth between functional teeth; tooth roots split to receive crown of replacement tooth; rostral bone with keel terminating in a sharp point; plane of frill confluent with supratemporal fenestra; parietal frill longer and broader; epijugal bone; predentary beak keeled and pointed; fusion of first three cervical vertebrae. Sereno failed to find any evidence for the Protoceratopsidae as a monophyletic group. Rather, he saw them as a paraphyletic assemblage of successively more highly derived forms, beginning with *Leptoceratops* as the most basal member and ending with *Montanoceratops* as the most highly derived, and the probable sistergroup of the Ceratopsidae, which he did not analyze.[18]

The only other cladistic analysis of the Ceratopsia as a whole was published by Philip Currie and me in 1990 (Fig. 8.2). We countered Sereno's suggestion of paraphyly of the Protoceratopsidae by suggesting, albeit tentatively, three potential derived characters uniting protoceratopsids: a shallow, circular fossa in front of the orbit; a sinus in the maxilla; and an inclined process of the palatine bone. The matter is unresolved, and the anatomy of the skull of *Montanoceratops*, putatively the most highly derived member, is inadequately known. There are also some interesting biogeographic aspects of the matter, as the most basal and most derived member of the Protoceratopsidae are from North America, but all other members are from Asia. How closely related are the North American and Asian members?

Phil and I characterized the Ceratopsidae with these derived characters: large heads; prominent parietosquamosal frills; variable but often

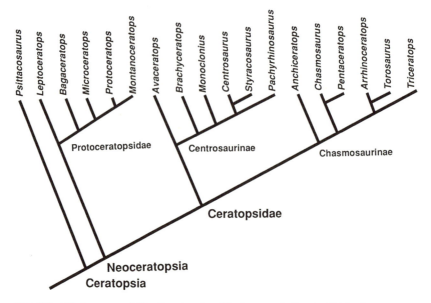

FIG. 8.2. Cladogram of the Ceratopsia. (Dodson and Currie 1990. Courtesy of the University of California Press.)

strongly developed nasal and postorbital horn cores; enlarged external nostrils set in recesses; small infratemporal fenestrae; squamosal strongly expanded caudal to quadrate; free edge of squamosal often ornamented with or without epoccipitals; ten fused sacral vertebrae; robust humerus; blade of ilium horizontal; ischium curved downward; femur considerably longer than tibia; unguals blunt, rounded, and hoof-like. A large body of evidence corroborates the monophyly of the Ceratopsidae. Lehman's analysis, already discussed, supports the division of the Ceratopsidae into the Centrosaurinae and the Chasmosaurinae. The short-frilled Centrosaurinae are smaller than the large chasmosaurines (except for the last representative, the Edmontonian *Pachyrhinosaurus*) and more restricted both in time and in space. The smaller, less ornamented centrosaurines (*Avaceratops, Brachyceratops*) occupy a more basal position within the Centrosaurinae. *Monoclonius* is larger but relatively unornamented, whereas *Centrosaurus* is roughly the size of *Monoclonius* but well ornamented; its sistergroup is undoubtedly *Styracosaurus*, the same size but extremely ornamented. *Pachyrhinosaurus* is clearly the most highly derived of centrosaurines, as well as the largest and the latest. Scott Sampson believes that two transitional forms leading to *Pachyrhinosaurus* are represented in the Two

Medicine Formation of north-central Montana.[19] *Einiosaurus* seems closer to *Styracosaurus,* but with a simpler frill; *Achelosaurus* is alleged to be closer to *Pachyrhinosaurus,* with an incipient nasal boss.

The Chasmosaurinae are more difficult to resolve. Most representatives are Edmontonian or Lancian in age. The earliest one, *Chasmosaurus,* is already highly derived. No known representative is a good candidate for basal status, and there is no simple trend in any morphological character. *Anchiceratops* seems generalized, of moderate size, with ornamented frill and parietal fenestrae of moderate size. *Chasmosaurus* and *Pentaceratops* form a tight sistergroup, emphasizing hypertrophy of the parietal fenestrae and, in *Pentaceratops,* great increase in size. *Arrhinoceratops* and *Torosaurus* are derived in simplification of the squamosal with loss of ornamentation. They may form a sistergroup, with *Torosaurus* the more derived by virtue of great extension of the frill correlating with a remarkable increase in size. *Triceratops* is the most highly derived chasmosaurine, but also occupies an isolated position. Its derived characters represent evolutionary reversals, including relative shortening of the frill (despite massive increase in size of the whole animal), and secondary closure of the parietal fenestrae.[20]

The foregoing discussion provides an overview of ceratopsian relationships using the most powerful and respected scientific tools of the trade. Why stop here? Fools rush in where angels fear to tread. I therefore experimented with another technique. Evangelists for cladistics insist that their technique is the only one that can yield objective, reliable results, but it is also true that cladistics requires prior knowledge and makes certain explicit assumptions. Suppose I tried an analytical technique that simply examines a series of ceratopsian skulls, measures a series of points on each skull, and compares the positions of the points on each pair of skulls? Such a technique is resistant-fit theta-rho analysis, mercifully known as RFTRA (pronounced "RIFF-trah"). One of the developers of this method is my friend Ralph Chapman, a jovial computer expert at the Smithsonian and a boundlessly energetic dinosaur enthusiast.[21] This computer-based procedure is called a landmark-based mapping technique because it maps homologous points from one specimen onto the next and computes a distance function between each pair. Finally, all of the results are clustered to determine which specimens are most like each other. The method is a phenetic one: the results express "similarity" rather than unique sharing of derived characters. How far do the phenetic results take us toward an understanding of phylogenetic relationships? I published my answer in 1993.

I selected fourteen skulls of ten representative genera of ceratopsians: *Psittacosaurus, Leptoceratops, Bagaceratops,* two skulls of *Protoceratops* (a juvenile and an adult), *Styracosaurus,* two skulls of *Centrosaurus, Arrhinoceratops, Pentaceratops,* and two skulls each of *Chasmosaurus* and *Triceratops. Anchiceratops* and *Torosaurus* were not included in the analysis because the skulls do not show the sutures required as landmarks. I digitized a set of twenty-one homologous landmarks on the lateral profiles of each skull. The data were then subjected to RFTRA transformation, and the resulting output data were clustered.

There were two major clusters, one of which corresponds to the Protoceratopsidae plus *Psittacosaurus,* the other of which corresponds to the Ceratopsidae. Not a bad start. Within the Ceratopsidae, there were two more clusters, one of which corresponds to the Centrosaurinae, the other of which is the Chasmosaurinae, with *Triceratops* exactly where it belongs, although it maintains an isolated position within that cluster. Within the protoceratopsid cluster, *Psittacosaurus* occupies an isolated position, allowing the interpretation that it is "more like" protoceratopsids than it is like ceratopsids, but is separate from the former. This too is good. The two specimens of *Protoceratops* did not plot with each other, but the juvenile, in which the crest is weak, plotted with *Bagaceratops.* It seems to make sense. There was another "error" in that the long-horned species of *Chasmosaurus, C. kaiseni,* plotted with *Pentaceratops* rather than with short-horned *C. belli,* but this underscores the close relationship between *Chasmosaurus* and *Pentaceratops.*

I believe that the results of this exercise are outstanding. If I were to maintain a cladistic view of the world, I would say that the changes of shape highlighted by the morphometric analysis captured the essence of the derived characters among taxa. From my less sophisticated point of view, I think a rather simple phenetic technique has ordered the taxa into a very convincing pattern of relationships. It is at the very least a good, "quick and dirty" overview, which may then be examined more carefully by rigorous cladistic analysis.

I went one step further. Is it possible that our understanding of ceratopsian phylogeny is weighted too heavily in favor of horns and crests? I repeated my analysis but removed those measures that reflected the size of the horns and the length of the parietal crest. The results are substantially similar, even improved by virtue of correctly pairing the two species of *Chasmosaurus,* which are then paired with *Pentaceratops.* There is one interesting new error: *Triceratops* now plots with the centrosaurines. Without the horns, which are clearly chasmosaurine in nature, it superfi-

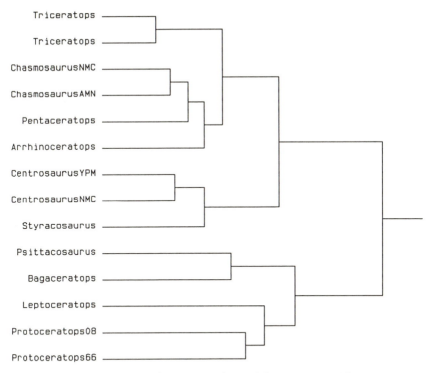

FIG. 8.3. Clustered output of RFTRA analysis of the Ceratopsia. The two major clusters correspond to the Protoceratopsidae plus *Psittacosaurus* (below) and the Ceratopsidae (above). The Ceratopsidae in turn are divided into the Centrosaurinae (below) and Chasmosaurinae (above).

cially resembles the short-frilled ceratopsids. In a third analysis, I removed the front of the skull ahead of the orbits and retained the orbit, cheek, frill, and postorbital horns. With these partial skulls, *Triceratops* is restored to its position among the chasmosaurines, and in fact all four pairs of the same genus plot as each other's nearest neighbors (Fig. 8.3).[22]

These results encourage the interpretation that our understanding of ceratopsian phylogeny is robust. It is derived from all parts of the skull and is not obsessed with unimportant characters. It also encourages the understanding (however desirable "perfect" fossils are) that part of a fossil tells us much more than no fossil at all. In the years to come, new kinds of ceratopsians will come to light in the United States, Canada, Mongolia, China, and perhaps elsewhere as well. The challenge will be to accommodate these new finds within the framework of our current understanding of evolutionary relationships.

259

The Life and Death of Horned Dinosaurs

DATA FROM BONEBEDS

OUR general impression is that ceratopsians were abundant, group-living, gregarious animals. The nature of the skull is one line of evidence that makes us believe this is so, and the other is the fossil record itself. Ceratopsian fossils are abundant. Many specimens are found because they were common both in life and in death. Although *Psittacosaurus* and *Protoceratops* tend to be found as single specimens, there are many of each of them. Ceratopsids have a strong tendency to fossilize in monotypic (single-species) bonebeds, typically with juveniles, subadults, and adults of only a single species.

If bonebeds document death—death on a large scale—they also provide extremely important snapshots of life in prehistoric times, and the Mesozoic is no exception. There are various kinds of bonebeds in the fossil record. Some are formed more or less slowly and represent the accumulation of carcasses of various kinds of animals, for instance on a river bar, as the animals die over a period of several years. The great quarry face, 50 m long, exposed in the visitor's center at Dinosaur National Monument in northeastern Utah is just such a deposit.[1] This monumental quarry preserves Jurassic giants such as *Apatosaurus, Diplodocus, Camarasaurus*, and *Barosaurus;* other herbivores such as *Stegosaurus, Dryosaurus*, and *Camptosaurus;* and predators such as *Allosaurus* and *Ceratosaurus*. The dinosaurian community has been sampled broadly. In addition, certain types of bonebeds may also document very welcome nondinosaurian components that shared the Mesozoic world. These help to flesh out our understanding of the living communities of which dinosaurs formed merely a part, albeit a spectacular one. In contrast, some bonebeds contain only a single species of animal, possibly overwhelmed by a natural catastrophe such as a drought, flood, volcanic episode, or unidentified cause. Herding ungulates today perish in mass mortalities, flood being a common agent of disaster for wildebeest in Africa and caribou in Quebec. Our Ice Age forebears hunted mammoth and bison, large mammals whose herding behavior made them susceptible to mass death. We assume that a monotypic bonebed informs us about the biology of a species, whereas a polytypic bonebed tells us something about a community of species, often averaged over a period of time. The study of bonebeds is a relatively new one. In the early days of our science, collectors concentrated on "trophies" (complete skulls and skeletons) and rarely even noted the existence of bonebeds.[2]

In 1984 Philip Currie and I cataloged some of the monotypic bonebeds in the province of Alberta. At Dinosaur Provincial Park are found monotypic bonebeds of *Centrosaurus, Chasmosaurus,* and *Styracosaurus.* Elsewhere in the province are Edmontonian bonebeds that contain *Pachyrhinosaurus* and *Anchiceratops.* One bonebed in particular at Dinosaur Provincial Park is noteworthy. Bonebed 143 contains the remains of scores or even hundreds of individuals of *Centrosaurus.* I worked there with crews from the Royal Tyrrell Museum in the early 1980s, but as work proceeded thereafter, the bonebed continued to increase in size. When I visited it with Phil in the summer of 1994, what he showed me was staggering. It now appears the bonebed stretches for nearly 10 km, almost horizon to horizon within the confines of Deadlodge Canyon, which is chiseled out of soft Judith River sediments on either side of the Red Deer River. In 1984, Phil and I described our interpretation that this deposit probably documented the drowning of a herd of animals in a river in flood. The deposit averages twenty bones per square meter excavated, although in some quadrants the density reaches sixty bones per square meter. We estimated in 1984 that the deposit contained the remains of at least three hundred individuals of *Centrosaurus* from juvenile, subadult, and adult size classes. Today, it is clear the catastrophe, whatever it was, brought down a herd of ten thousand or more individuals. There is no volcanic ash associated with the deposit to suggest that that sort of cataclysm played a role.[3] Not to belabor a cliché, this is a *mystery*!

It is easy to get the impression that *Centrosaurus* was an abundant, gregarious animal. If the deposit sampled animals that had died randomly on the landscape over a period of time, we would expect to find abundant evidence of hadrosaurs, ankylosaurs, meat-eaters, and other animals that shared the ancient terrain. This is emphatically not the case for the *Centrosaurus* bonebed. Besides *Centrosaurus,* a number of other ceratopsids occur in bonebeds, but by no means is this true for all species. No ceratopsian bonebeds of any kind have yet been documented in China or Mongolia, although I do not rule out their existence. It may be that bonebeds have simply not been exploited there. *Triceratops* appears to have been very abundant. Recollect that Barnum Brown claimed to have seen five hundred skulls in the field in Montana. However, no *Triceratops* bonebeds have ever been described. Some horned dinosaurs appear to have been rather rare, at least as best we can determine from the perspective of a century-long search. Among these are *Torosaurus* and *Arrhinoceratops,* which contrast with *Triceratops* and *Anchiceratops* from their respective faunas.

Feeding Large Tummies

Horned dinosaurs formed important parts of the Cretaceous communities in which they lived. In the Judith River of Alberta, the most abundant dinosaurs were the duck-bills, and horned dinosaurs were next in abundance.[4] In the Hell Creek Formation of Montana and equivalents in Saskatchewan and Wyoming, *Triceratops* is by far the most commonly found dinosaur.[5] In Mongolia, *Protoceratops* is the most abundant dinosaur where it is found.[6] From the dinosaur-bearing formations of western North America and from Eastern Asia we have learned about other components of the dinosaur faunas as well as about the nondinosaurian vertebrates with which the dinosaurs lived. Aquatic and semi-aquatic components are well represented by fishes, some amphibians (although numerically these are usually a minor component), turtles, crocodilians, and an interesting archaic, superficially crocodile-like reptile, *Champsosaurus.* Less common and very interesting components are lizards and mammals. Mongolia is rich in both of these, and it is famous for the supply of complete skulls of shrew-sized creatures. In North America, Cretaceous mammals are generally rare, although they are known from deposits from Alberta to New Mexico.[7] At a few sites, particularly at Lancian sites in Montana (Bug Creek Anthills and adjacent sites) and Wyoming (Lance Creek), mammals are very abundant and account for 25 percent or more of small vertebrate fossils recovered. Three groups of mammals are important in the Late Cretaceous: multituberculates, marsupials, and a small number of placental mammals, which increased in abundance at the very end of the Cretaceous.

Plants are enormously important to dinosaur communities because they lie at the base of the terrestrial food web. Unfortunately, paleobotanical evidence is not always easy to come by from the beds that produce dinosaur fossils. Flowering plants (angiosperms) appeared in the Early Cretaceous and underwent a significant evolutionary diversification in the Late Cretaceous. Nevertheless, they remained subordinate plants in the overall plant community structure, even though they probably formed an important weedy ground cover, particularly near watercourses.[8] Although good-sized trees formed an important part of the landscape, these were predominantly conifers, particularly dawn redwood (*Metasequoia*) and bald cypress (*Taxodium*), as well as ginkgoes. There were small angiosperm hardwood trees, but these for the most part did not exceed 10 cm in diameter and 5 m in height. Their leaves superficially resemble those of modern oaks, poplars, maples, and sycamores, but most are not members of modern genera. As grasses had not yet evolved, ferns—lacy plants of venerable, ancient pedigree—still formed an important part of the ground cover and undoubtedly accounted for a significant portion of dinosaur nibbles. Mosses, fungi, and scouring rushes were there as well, though they probably were not highly palatable.[9]

What did ceratopsians eat? Large animals are today rarely selective feeders and presumably were not so in the Cretaceous either. Ceratopsians chopped and munched large quantities of greenery—whatever was available, including all of the plants just mentioned. The teeth of ceratopsids are unusual, superficially resembling those of duck-bills. The duck-bills or hadrosaurs had marvelous dental batteries that contained seven hundred to a thousand teeth, or sometimes even more. They represent one of the peaks in the evolution of vertebrate teeth and chewing mechanisms. Quite independently of hadrosaurs, ceratopsids too developed dental batteries, but they had their own peculiarities. John Ostrom, who is an authority on the teeth and dental batteries of hadrosaurs, also has done the definitive study on ceratopsid tooth batteries, and he notes significant differences. The maximum number of tooth positions in a ceratopsid dental battery is thirty-five. The maximum number of teeth stacked one upon another in a single tooth position is five, and this number decreases at either end of the jaw (Fig. 9.1). Thus, the maximum number of teeth, even in *Triceratops*, is fewer than six hundred. As described by Ostrom, ceratopsids had a very unusual method of chewing. When the jaws slid past each other, the plane of the opposing upper and lower tooth surfaces was vertical. As a consequence of the angulation of the teeth, only a single tooth in each tooth position was in wear. This almost seems to defeat the pur-

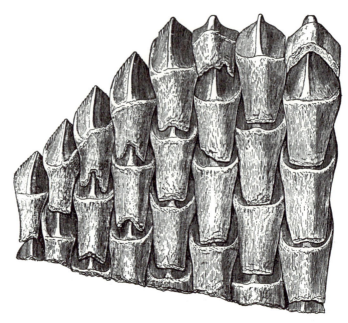

FIG. 9.1. *Triceratops* dental battery. Lower jaw, inner view. Each functional tooth in the battery (top) has three or more replacement teeth waiting to replace it, one by one. (From Hatcher et al. 1907.)

pose of having a dental battery. In hadrosaurs, the angle of the chewing plane between upper and lower teeth is a more reasonable 40–55°, and the chewing plane may cut across as many as three teeth in different stages of wear at a single tooth position (Fig. 9.2).[10]

Ceratopsid jaws thus present an apparent contradiction. At first blush, they resemble those of hadrosaurs. Each has a jaw with large numbers of teeth, and the individual teeth of each are compressed for close packing in the battery. The split roots of ceratopsid teeth seem more highly derived for providing stability during tooth replacement. But the resemblance quickly ends when the vertical cutting plane is recognized. Whereas hadrosaurs have highly developed grinding tooth planes adapted for chewing quantities of resistant plant material over a long functional life,[11] ceratopsid dentitions are dominated by a scissor-like shearing action. Because the opposing upper and lower teeth meet on their narrow edges rather than over a broad surface, forces are concentrated on the tooth edge, and wear must have been relatively rapid compared to that in hadrosaur teeth.[12]

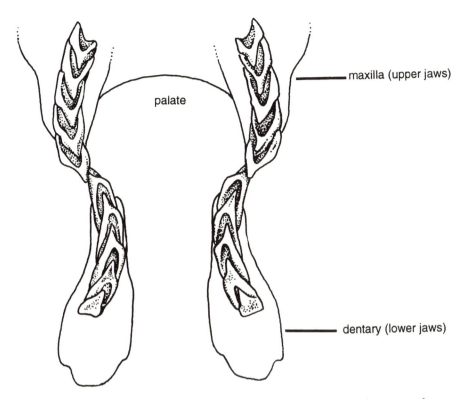

FIG. 9.2. *Triceratops* tooth occlusion, as shown by a transverse section across the upper and lower jaws. The lower jaws are inside the uppers, and the plane of contact between upper and lower teeth is nearly vertical. (Robert Walters after Hatcher et al. 1907.)

It is hard to visualize ceratopsids as sophisticated herbivores after all. It is as if they ate their salad with scissors. Obviously they *could* do it—they were very successful animals. But it hard to comprehend *why* they did it. Ostrom was fully aware of the problem that this view of ceratopsid dentition presented. Although both fossil and modern herbivores may show some degree of shear in their dentitions, in no other case has shear completely dominated over crushing and grinding. One possible interpretation is that the ceratopsids consumed fleshy fruits that did not require much chewing effort. Ostrom rejected this interpretation and opted instead for the alternative: that they ate food that required prodigious effort to chew, in effect a food that was inaccessible to other animals that lacked such teeth and jaws. The suggestion was that they ate fibrous palm and cycad fronds.

I object categorically to both of these food choices. My biases about ceratopsians are firmly rooted in the rocky soil of Alberta, where my paleontological career sprouted. I have seen many ceratopsian fossils in the field, have collected many hundreds of ceratopsian teeth, and have seen plant fossils in abundance. But I have *never* seen a fossil cycad or palm in these beds.[13] I hasten to add that I am no authority on fossil plants in the field. However, cycads and palms are a pair of plant groups that form an evolutionary contrast. Both have important fossil records, and both survive today. Cycads, however, were vacuum-tube-dependent, room-size ENIACs in the Late Cretaceous, whereas palms were Pentium-based, two-gigabyte, fax/modem-equipped, 166-megahertz laptop computers. Cycads were plants of the past that had nourished the sauropod giants of the Jurassic and were now on the wane; palms were plants of the future, part of the great radiation of flowering plants that was then underway. Palms spread from south to north. *Pentaceratops* from New Mexico certainly could have munched on palms, but in the north, if a herd of *Triceratops* had depended on palms for survival, they never would have made it.

Another, more general, point also bothers me. Ceratopsians were very successful animals. Why should they be confined to a marginal food source that nobody else could exploit? This does not make tremendous sense to me from an ecological point of view. Do I have a compelling alternate food source? Not really. But one new food source that might have required some biomechanical prowess to exploit were the angiosperm trees mentioned previously. Perhaps ceratopsians used their horns and heads to knock over the small trees, about as long as their bodies, and then sheared off leaves and twigs, swallowing masses of chopped (but not chewed) leaves and sticks. Yummy! I will say that the small size of basal ceratopsians, the comparative simplicity of their teeth (which do not form dental batteries), and the delicacy of their beaks suggests that *Psittacosaurus* and the protoceratopsids may have fed more selectively on more lush, tender edibles. Unfortunately, we know almost nothing about the plantstuffs available to them.

Another aspect of Ostrom's argument was that ceratopsians had incredibly powerful jaw muscles. The parietosquamosal frills served as frameworks for greatly enlarged jaw muscles. This idea has a long pedigree, extending from Lull in 1908 on *Triceratops* to L. S. Russell in 1935 on *Chasmosaurus* to Georg Haas in 1955 on *Protoceratops*.[14] The idea is not without appeal, but there are difficulties. One of these is the problem of locating the jaw muscles on the external surface of the frill, where they would be exposed to the punctures of rival horns and the toothsome jaws of predators. Putting extensive soft tissue on the frill negates the role of the frill as a defensive structure (although arguably so does the hypertrophy of the parietal fenestrae, as in *Chasmosaurus* and *Pentaceratops*). In a reptile without a frill, the jaw-closing muscles, technically known as the muscles of mastication, are enclosed within the box of the bony skull in the cheek region behind the eye. The two pairs of temporal fenestrae (infratemporal below and behind the eye, supratemporal above and behind the eye) provide some space for the jaw muscles to bulge as they contract.

The frill of ceratopsians is continuous with the supratemporal fenestrae, and it is plausible that the jaw muscles spread from the supratemporal fenestrae out onto the *base* of the frill. In *Leptoceratops*, with its very brief frill, it is likely that the entire frill carried jaw muscles. I have my first difficulty with *Protoceratops*. The frill is very smooth and could conceivably have been covered with jaw muscle, but the bone is also astonishingly thin, genuinely translucent, apart from the median parietal bar and the lateral squamosal bars. But most important to me is the obvious sexual dimorphism of the frill. This would imply that presumed males had more muscles, stronger muscles, and muscles with a different mechanical arrangement than females. It almost implies that males and females behaved like different species. Although this is not impossible, it seems a bit much to me.

My difficulties increase when I consider ceratopsids. The extreme case, of course, is presented by the long-frilled giants, *Torosaurus* and *Pentaceratops* (Fig. 9.3). The problem is, what is the value of jaw muscles 1.5 m long? There are lots of successful mammalian herbivores, including those of large size (e.g., elephants, rhinos, hippos, giraffes). Not one of them needs jaw muscles that long. Few if any have muscles even a third that long. Length of muscles does *not* impart strength. A long muscle will produce the same force on contraction as a muscle half or a quarter of its length with the same cross-sectional area. Length does two things for a muscle: it provides the ability to stretch longer and to resist fatigue. Stretching is good for leg muscles, but it is irrelevant for jaw muscles inserting close to the center of the opening and closing motion, the jaw joint. An animal can open its mouth only so wide, and then there is no further advantage. In general, herbivores do not open their mouths very wide; cheeks prevent it, and ceratopsians probably had cheeks, which are muscular pouches outside the toothrow to catch falling food particles. Strength is enhanced by increasing the *number* of

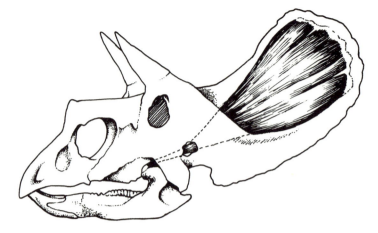

FIG. 9.3. *Torosaurus* jaw muscles. In this hypothetical (and in my judgment improbable) reconstruction, the jaw muscles occupy the entire frill. (Robert Walters.)

muscle fibers. A large number of short fibers inserting on a central tendon makes for a strong arrangement. Short muscle fibers do not contract very far, but this short distance is ideal for jaw muscles. So the frill could have been packed with short, powerful fibers. But why would it have been? Why did ceratopsians need such extraordinary strength of jaw-closing muscle, particularly when it accompanied teeth that wore down rather quickly? The supposed parts of the ceratopsian jaw system do not add up to the whole.

The other piece of the puzzle requires a detailed examination of the frill. We know from comparative anatomy that in a region where a muscle originates or where it passes by, the bone surface is smooth, even slippery, in a living animal. By contrast, where the bone surface is textured with grooves for blood vessels or pitted, tough skin adheres to the bone, and there is no muscle underneath the skin. Alligator skulls illustrate this beautifully. A bleached skull shows the interesting texture of bone sculpture covering much of the skull, except around the temporal fenestrae as described. A fresh alligator skull is covered with black skin, which is surpassingly difficult to remove because it sticks so tightly to the textured bone except around the temporal fenestrae, where the underlying muscle makes the skin more compliant. Ceratopsian frills, beginning with *Leptoceratops*, are always smooth adjacent to the supratemporal fenestrae, and for a certain distance back onto the base of the frill. In certain species of ceratopsids (e.g., *Arrhinoceratops*, *Triceratops*), the frill is strikingly textured, a feature that strongly suggests

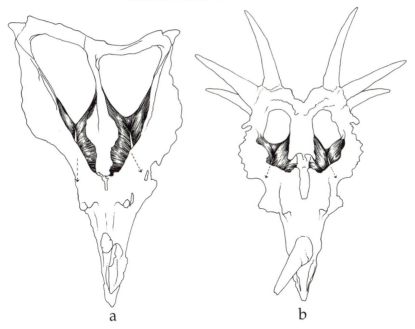

FIG. 9.4. Jaw muscles in (a) *Chasmosaurus* and (b) *Styracosaurus*. In these reconstructions (which I believe to be reasonable), the jaw muscles occupy only the base of the frill. (Robert Walters.)

to me that the surface of the parietal was *not* covered with jaw-closing muscles. I do not believe the frills of these dinosaurs in particular were covered with jaw muscles. Ostrom was aware of this problem and postulated that "a longer pars medialis of the M. adductor mandibulae externus may have attached by a thin sheet of fascia to the frill margins and left little or no indication of its attachment."[15] Although Ostrom may have been correct, I reserve my right to be skeptical.

Furthermore, I want to consider the skull of *Chasmosaurus*, which was examined by Ostrom in particular. Ostrom illustrated the Yale skull of *Chasmosaurus belli*, YPM 2016, in dorsal view.[16] Annoyingly, Lull in 1933 figured this skull in ventral view, but not in dorsal view. My comments are based on two other specimens of *C. belli*, NMC 2280 from Ottawa and AMNH 5402 from New York. Lull's photos of the skull of *Styracosaurus albertensis* are also germane.[17] In dorsal view, it may be seen that there is, beginning at each supratemporal fenestra and expanding backward, a long, triangular track or channel that focuses on the supratemporal fenestra, through which the adductor muscles pass to attach on the jaw (Fig. 9.4). This I allow. The channel is bounded laterally by the smooth upper edge of the squamosal and medially by the elevated median bar of the parietal. In AMNH 5402, there is an expansion of the median parietal bar between the supratemporal fenestra and the parietal fenestra. This expansion, only hinted at in Ostrom's sketch of YPM 2016, has the effect of restricting the width of the muscle channel and would have redirected muscle fibers coming from the back of the frill along the median bar. Furthermore, the median prominence overhangs the muscle channel. There is a sharp and irregular, almost ragged, lip that demonstrates that muscle did not ride over its edge; muscle did not arise from the median bar. I believe that paleontologists have ignored the ragged bone edge that would preclude muscle passage. The lip suggests that the jaw muscles were thickest over the supratemporal fenestra and thinned backward in front of the parietal fenestra. In *Styracosaurus* and *Centrosaurus,* it certainly seems to me that the jaw muscles ended in front of the parietal fenestra. I suspect the same was true in *Chasmosaurus,* and quite likely in ceratopsids generally. Ceratopsians did just fine with the muscles they had. I do not think they went to extraordinary lengths to modify the muscles they inherited.

SLOW AND PLODDING OR LIFE IN THE FAST LANE?

Ceratopsids were rhinoceros-like animals. But there is a danger in that analogy. They were not rhinos. (For me, they are more interesting than rhinos, but allow me my personal quirks.) It is an easy subliminal error to fall into, succumbing to the pull of the present. Although analogies are very valuable, their downside is that we may fall into the trap of losing our objectivity. The image of the dinosaur has been refurbished, and the sleek, hyperkinetic saurians of today's movies are aesthetically appealing, irrespective of whether or not they are accurate. Museum mounts have been remade, bent tails straightened and shortened, soaring heads brought back toward earth. The new *Tyrannosaurus* posture, with gaping maw just out of reach of the visitor to the Philadelphia Academy of Natural Sciences, is ever so much more threatening than the remote *Tyrannosaurus* of the old American Museum of Natural History, with its head 15 ft in the sky, unmindful of the insignificant human flesh beneath it.[18]

What about *Triceratops* and its ceratopsid friends? Artists have striven to rescue ceratopsids from the ignominy of the past. Ceratopsids prance and dance on canvas, sleek as greyhounds, lithe as ballerinas (Fig. 9.5).

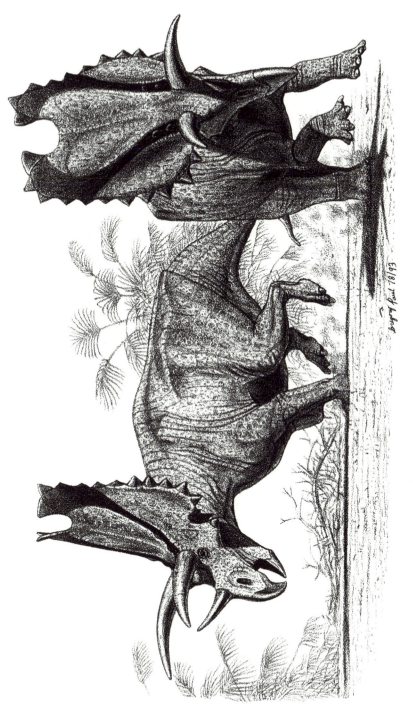

FIG. 9.5. *Pentaceratops* dances. (Gregory S. Paul.)

FIG. 9.6. *Chasmosaurus* sprawls. (Photograph courtesy of the Royal Ontario Museum.)

In the museums they are not doing so well. Their hind legs are fine, underneath the body as they should be. But there is a problem with ceratopsid posture, and it is centered on the front legs. They sprawl. When Marsh first reconstructed *Triceratops*, on paper, it came out with erect, columnar limbs, front and back. But when skeletal mounts began to be erected, first in Washington, then in New York and Ottawa, the front limbs just didn't work out the way they were expected. To be blunt, it is *impossible* to mount the forelimbs of ceratopsids with the joints articulated, the limbs erect, and the elbows rotated underneath the body. They simply don't go together that way (Fig. 9.6). Gilmore and Sternberg and Osborn did not have any a priori interest in reconstructing ceratopsians as lumbering, antediluvian throwbacks. They simply followed the rules of the exercise, making the best sense of the fossils they had, and their own very good knowledge of anatomy.

In our attempt to make ceratopsids as sleek as possible, we draw the shoulders under the body as much as we can and rotate the elbows backward. We had some degree of success doing this with our small skeleton of *Avaceratops*, which we mounted in 1986 at the Academy of Natural Sciences of Philadelphia, but decidedly less success with our skeletal cast of *Chasmosaurus belli*, which the gifted technician and scientist Kenneth Carpenter mounted. (Ken also designed the Academy's *Tyrannosaurus* mount.) The limits are imposed by the skeleton itself, especially the humerus. The bulbous head of the ceratopsid humerus is off center, projecting onto the back surface of the shaft. This structure makes it virtually impossible to mount the humerus vertically. In addition, the upper end of the humerus is rather rectangular, with the bulbous head in the middle; the farther the humerus is drawn upright, the greater the tendency of the inside corner (lesser tubercle) to puncture the ribs.

Rolf Johnson of the Milwaukee Public Museum had the good fortune to discover the first partial skeleton of the great *Torosaurus* in 1981. As he prepared the specimen for mounting and exhibition, he discovered anew what everyone who has ever tried to mount a ceratopsid has learned. The forelimb must sprawl. Yet Robert Bakker and Gregory Paul adamantly dispute this pose. Paul argues that mounting the hindlimbs too far apart and misarticulating the first few ribs will hopelessly misalign the front limbs.[19] I do not believe such alterations would provide the desired solution.

A missing piece of the puzzle for reconstructing forelimb posture in ceratopsids is the testimony of trackways—the footprint record of horned dinosaurs. Although trackways for certain kinds of dinosaurs—especially theropods, sauropods, and large and small ornithopods—have proven to be remarkably common, until very recently there were no reliable reports of ceratopsid trackways. Happily, the situation has now changed with the report by Martin Lockley and Adrian Hunt of numerous possible *Triceratops* trackways from the foothills of the Rockies, near Denver, Colorado. Unhappily, I cannot agree with the authors' optimism that the posture problem of ceratopsids is now solved.

The footprints in the Laramie Formation are those of a large four-footed animal, with handprints about two-thirds the size of the footprints. Exceptional prints measure 75 cm in length or 80 cm in breadth, which is very large indeed. Foot impressions are longer than wide, and hand impressions show the opposite relation. Both hand and foot impressions show four digits ending bluntly (the fifth digit on the hand evidently being too small to register its imprint). I agree with Lockley and Hunt on the ceratopsian identity of the tracks. They applied a name, *Ceratopsipes goldenensis*, to the prints, meaning "ceratopsian foot from Golden, Colorado."

For our purposes, it is the footfall pattern recorded in the trackway that is of interest. Two steps are measured, the distances between a right footfall and a left footfall or the reverse; they are 120 cm and 135 cm, respectively. Using other information provided, we can calculate a stride length of 2.2 m. Remarkably, using a magic formula developed by the delightful British zoologist R. McNeill Alexander, it is possible to estimate speed of locomotion based on trackways.[20] The input data are the length of the stride and the height of the animal at the hips. Assuming the hip height of *Triceratops* to be 2.2 m, the running speed comes out to about 4.2 km per hour or 2.6 miles per hour, a slow speed.

A more pressing matter concerns the position of the handprints. Do they show a narrow track or a wide track (Fig. 9.7)? The authors proclaim that the tracks confirm the narrow-track posture and falsify the wide-track posture argued so passionately by Johnson and Ostrom, but do they really? Curiously, in the text, Lockley and Hunt state "trackway wide, about 1.25 m, with a pace angulation of about 110°." It is hard to construe a trackway fully four feet across as indicating a narrow forelimb stance. Moreover, their illustrations indicate that the handprints fall somewhat outside the footprints, with the axis of the hands rotated outward from the direction of travel.[21] This contrasts, for example, with the beautiful sauropod footprints from Texas described by my friend Jim Farlow and his colleagues Jeffrey Pittman and Michael Hawthorne.

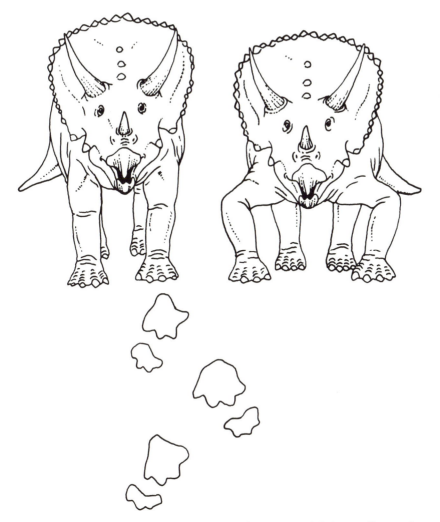

FIG. 9.7 The problem of posture in horned dinosaurs—did they walk upright, as does the *Triceratops* on the left, or did the forelimbs sprawl, as on the right? Which view does fossil footprint evidence support? (Robert Walters after Lockley and Hunt 1995.)

The famous footprints, named *Brontopodus birdi*, show the handprints resolutely in front of the footprints.[22] We learned early on in the book, in Chapter 2, that the front of the rib cage of ceratopsids is straight sided and narrow, whereas the back ribs are rounded, giving a barrel shape to the belly in front of the hips. Thus the shoulders are narrower than the hips. If this inconvenient anatomical fact be so, should the handprints of vertical limbs not at least fall in line with the footprints, if not actually inside them? Should we not expect the hands to point front-ward and not splay outward at some angle? It is well known that in running mammals—whether horses, gazelles, cheetahs, dogs, or rab-bits—the hind legs always fall outside the front legs.

How then can we combine all these observations (narrow shoulders, wide hips, handprints outside footprints, narrow stance) into a coher-ent picture? Answer: we cannot. Something doesn't add up. Again I am forced to conclude that there is a strong element of wishful thinking in the picture of ceratopsids with graceful limbs fully collected under-neath the body. The front limbs did not and could not operate that way. This seems to be a striking example of a case in which the data (wide trackways) are forced to fit the conclusion (upright posture). Science is not supposed to work that way.

Such issues of posture affect the question of locomotion, that is, walking, trotting, or running. Clearly, our assessment of the running ability of ceratopsids depends on our understanding of their posture. Again, our aesthetic preference is clear—we would like to visualize ceratopsians as fleet-footed herbivores, cruising about the savannah, seeking out the luckless leaf they would devour. But our analysis does not necessarily square with our preferences.

I have never been to Africa, but I do enjoy natural history documen-taries on television. Vast herds of placid, grazing game form a stunning panorama, and it is easy in the mind's eye to transform the scene from the troubled present back into the Mesozoic. I draw attention to several points about our savannah scene. The first is that most of the time the animals are not running or trotting: they are standing still or walking slowly, grazing as they go. The second is that when running scenes are portrayed, it is usually smallish animals that are running—zebras, wil-debeests, gazelles. These are animals with weights one tenth to one thirtieth the weight of ceratopsids. Yes, elephants and rhinos and gi-raffes can move quickly, but they do not commonly do so. The high-speed gait is not a gallop but a fast trot or pace. Did herds of ceratopsids stampede like bison? I do not know. I do not believe that ceratopsids defended themselves by trying to outrun predators. Most of all, I do not believe that *Triceratops* or *Torosaurus* or *Pentaceratops* ran away from anyone! Like rhinoceros today, they were far too large and too danger-ous to run out of fear of attack.

What nevertheless are the possibilities for running? Unfortunately for us, ceratopsid trackways are too rare to help us answer this question. Not surprisingly, the trackways of ceratopsids, like those of most dinosaurs, usually document walking behavior anyway. Fortunately other techniques have been developed to deal with this issue. Walter Coombs analyzed limb proportions in ceratopsians and found that they compared with those of weight-bearing (mediportal) mammals such as rhinoceros and hippopotamus rather than those of specialized running mammals.[23] In 1982 Australian paleontologist Tony Thulborn published an extensive survey of potential running speeds in dinosaurs based on a biomechanical analysis of limb proportions, using Alexander's model as a guide. Lacking trackway data, Thulborn used for the necessary datum stride length, "a selected hypothetical figure," which may be worrisome. Thulborn does not believe that ceratopsids generally were capable of attaining a true gallop with all four feet off the ground, but his analysis suggests the ability to attain respectable speeds at a fast trot: 22 km per hour for *Styracosaurus* and 26 km per hour for *Triceratops*. His speed estimates for *Protoceratops* and *Leptoceratops* are 14 km per hour as quadrupeds, but 28 km per hour as dynamic bipeds.[24] I have some doubt about these figures, but I suppose they are like the top speeds of automobiles—more potential than actual.

Much of our understanding about running is based on the anatomy of running mammals and large ratite birds (flightless birds such as ostriches, rheas, and emus). For this discussion, I will ignore two-legged runners, thereby eliminating birds. The first consideration is that small animals get away with things that large animals cannot. The anatomist Milton Hildebrand has observed that foxes are very generalized in their limb anatomy whereas horses are highly specialized runners, yet the fox hunt is a fair contest because the fox can run as fast as the horses.[25] Really large animals rarely show specializations for running. The larger the animal, the straighter and more columnar the limbs, and the smaller their arcs of rotation. Sauropod limbs and elephant limbs are similar in this regard. In running mammals, especially perissodactyls (horses) and artiodactyls (deer and horned stock such as antelopes and gazelles), the feet are elongated, the proximal limb segments (humerus, femur) are short, the chest is narrow, the shoulder moves freely, and the upper end of the humerus is narrow. Running mammals have eliminated the clavicle, a bone that is found in basal placental mammals, including rodents, bats, and primates. (We large primates know the clavicle as the collarbone.) The clavicle is a link between the breastbone (sternum) and the shoulder. It supports and separates the shoulder girdles and limits their mobility.

In none of these respects do ceratopsids emulate running mammals. In ceratopsids, the feet are short, the humerus and femur are strikingly long compared to the ulna and tibia, and the upper end of the humerus has a very broad crest for the attachment of breast muscles. A broad pectoral crest is generally found in animals that sprawl but not in those that run. Mobility of the shoulder is especially suspect.[26] My reservation about running in ceratopsids focuses particularly on the shoulder region. The mammalian scapula is a flat plate of bone that lies against the wall of the chest and rotates during the locomotor cycle to enhance the swing of the limb both forward and aft. The scapula of a horse measures about 30 cm in length. The scapula of little *Avaceratops*, which is less than half the size of a horse, measures 36 cm in length, whereas the scapula of the mighty *Torosaurus* measures a meter in length. Mere size alone may lead us to question the mobility of the very long scapula of large ceratopsids. Furthermore, there is a large coracoid ventral to the scapula in all ceratopsians that curves ventrally and approaches the ventral midline of the animal. The two coracoids presumably did not contact each other across the midline but probably articulated with a cartilaginous extension of the kidney-shaped sternal plates to which the ribs attached. Thus the shoulder girdle was biplanar—partly lateral, partly ventral. The girdle of ceratopsids had to have been capable of some limited movement, but it was not the free movement expected in a running animal. Even if the scapula did swing in a wide arc, another problem is encountered by those who have the privilege of mounting ceratopsid skeletons. It is very difficult to find a position of the scapula that permits the humerus to swing forward of the vertical. Forward movement of the humerus is blocked by the coracoid. There was very little forward reach. This anatomical constraint more than any other seems to preclude genuine running and to have limited ceratopsids to a fast amble or possibly a trot.

Protoceratopsids have the same fundamental anatomy as ceratopsids. However, their limb proportions are much more like those of running animals, with elongation of distal segments, than those of ceratopsids. Walter Coombs has suggested and Tony Thulborn has endorsed the idea that these small dinosaurs may possibly have run bipedally, thus overcoming the forelimb constraint. A bighorn ram may momentarily lunge forward on its hind legs—protoceratopsids could do no less than this. An interesting counterexample is provided by *Avaceratops*, like *Brachyceratops* a protoceratopsid-sized ceratopsid. It shows very strikingly the limb proportions typical of ceratopsids, not those of its forebears. It was not a swift runner.

These are the facts of ceratopsian locomotion. The story does not come out as I might prefer it. But nature is far too rich and interesting to be constrained by my narrow wishes or vision.

AND THEN THERE WERE NONE

Ceratopsians were among the last dinosaurs on earth. If the reign of the dinosaurs ended in a blazing catastrophe, *Triceratops* may have witnessed it. Let us examine this statement a little more closely. The span of the dinosaurs exceeded 163 million years. The oldest dinosaur currently known dates from rocks in Argentina dated at 228 Ma. The last dinosaur skeletons come from Late Cretaceous rocks dated at about 65 Ma, although dinosaur *debris* is found in rocks formed early in the age of mammals, most plausibly derived as sedimentary particles by erosion of soft sediments of Cretaceous age. The last geological division of Cretaceous time is called the Maastrichtian stage. As it lasted six million years or so, it represents too much time to document dinosaur extinction directly. The Edmontonian fauna of Alberta recorded in the Horseshoe Canyon, Wapiti, and St. Mary River formations are early Maastrichtian in age, roughly 70 Ma. The last dinosaurs on earth are those of the late Maastrichtian, particularly well documented in the Hell Creek Formation of Montana and the Dakotas, but also seen in the Lance Formation of Wyoming, the Scollard of Alberta, the Frenchman in Saskatchewan, and the Laramie of Colorado. The Lancian dinosaurs, dating from 68–65 Ma, are the dinosaurs that witnessed the extinction. There were two common Lancian dinosaurs. *Triceratops* is far and away the most abundant, and *Edmontosaurus* is a distant second. *Leptoceratops* and *Torosaurus* are rather rare ceratopsian members of the fauna; *Ankylosaurus, Pachycephalosaurus,* and *Thescelosaurus* are also noteworthy.

The great question of dinosaur extinction is whether they disappeared with a whimper or with a bang. This question is much debated. The "bang" scenario is high profile and has been very popular ever since it was articulated by Nobel Prize–winning physicist Luis Alvarez, his geologist son Walter, and their colleagues Frank Asaro and Helen Michel in 1980. Based on the physical evidence of enrichment of Cretaceous/Tertiary (K/T) boundary sediments at many locations around the world in the heavy metal element iridium, as well as in grains of shocked quartz (in which the crystal structure is disrupted by fracture planes that only occur at extremely high pressures), their scenario involves the impact of a large asteroid, some 10 km or more in diameter, at the very end of the Cretaceous. The impact would have raised great clouds of dust that circled the globe, causing complete darkness for weeks or months, blocking out sunlight, and causing plants to die and animals to starve to death. Additionally there were freezing temperatures, storms, acid rain, tsunamis (tidal waves), earthquakes, volcanism, poison gases, acid rain, *global* wildfires—the four horsemen of the apocalypse were abroad, and dinosaurs could not withstand their wrath! Lest there be any doubt about all this, a 200-km-wide crater named Chicxulub is reported from Yucatan, where it is buried deep beneath the surface of the jungle.[27]

With such a gruesome scenario, who could doubt that dinosaurs became extinct like this? Well, I doubt it! I am a whimperer, a lonely prophet, but I have not yet been mocked to my face. The problem I have is that this grim scenario explains too much, not too little. Although many species on land, in the air, and in the seas became extinct during the Maastrichtian, it is not apparent to me that this extinction was all that sudden. Many extinctions in the seas (of ammonites, inoceramid and rudistid clams) demonstrably began at least six million years before the final extinction at the boundary, and on land there was a pattern of massive survival: of mammals, crocodiles, turtles, lizards, frogs, salamanders, fishes. Plant communities suffered disturbances but seemingly not massive extinctions.

And what of the dinosaurs? There too I see a pattern of dwindling. Ten million years before the end, there were two subfamilies of ceratopsids, the centrosaurines and the chasmosaurines. At the end, only the chasmosaurines were left. Ten million years before the end, there were two subfamilies of hadrosaurs, the crested lambeosaurines and the flat-headed hadrosaurines. At the end, only the hadrosaurines were left. Ten million years before the end, there were two families of armored dinosaurs, the ankylosaurids and the nodosaurids. At the end, only the ankylosaurids were left. By comparison with the fauna of the Judith River Formation of Alberta, the dinosaur fauna of Hell Creek Formation is boring, completely dominated by a single kind of dinosaur: *Triceratops*.

I do not know the cause of dinosaur extinction. I suspect it was due to a symphony of causes, no one of which was unique or totally explanatory. Among them were changing sea levels, increasingly seasonal climate, volcanism, mountain building—the usual physical culprits. Indeed, dinosaurs may have witnessed an asteroid impact that gave them the final nudge into oblivion. But my position is very clear: they were already failing. The impact was at best only the last straw.

Ceratopsian dinosaurs were magnificent animals by any standard. The small ones are found in North America and Asia; the large ones are unique North American treasures. They were not paragons of intelligence, but little stood in the way of their success. No animals on earth have ever had larger skulls, nor more interestingly appointed ones. They were colorful, noisy, gregarious, belligerent, feisty, loveable creatures. They were the last major group of dinosaurs to appear on earth, and one of the last to witness the calamity that erased the last dinosaurs from the face of the earth. How keenly I regret their passing. Why did it have to end this way?

Notes

1. The arrangement I present is a very traditional one, in keeping with my own prejudices. Although many paleontologists would agree with me, others would not accept the taxonomic categories I use. The form that I use is strongly hierarchical, following the practices established by the eighteenth-century Swedish botanist Karl von Linnée (or Carolus Linnaeus, as he referred to himself in his writings, which were in Latin). Dinosaurs have traditionally been referred to the subclass Archosauria of the class Reptilia. Horned dinosaurs have been referred to the suborder Ceratopsia within the dinosaurian order Ornithischia, the bird-hipped dinosaurs. There is no agreement about the taxonomic level or category; some paleontologists believe that we should recognize six or seven orders of dinosaurs, not the traditional two. Because the choice of hierarchical level (order, suborder, infraorder, and so on) is somewhat arbitrary and subjective, reflecting good taste and general consensus rather than strictly objective facts of nature, there is a tendency among some paleontologists to list taxonomic names hierarchically but to leave the taxonomic level unspecified.

2. There are several ways to dispute the interpretation that dinosaurs are reptiles. One is by elevating the Dinosauria to the level of a class, equal to the Reptilia themselves. This approach was proposed by Robert T. Bakker and Peter M. Galton in 1974 and was popularized by Bakker in his 1986 book. Scientifically this idea has met with a cool reception. The other tactic is to say that the Reptilia themselves do not exist. This serious but alarming proposal follows from a very important modern philosophy and method of assessing relationships called phylogenetic systematics or cladistics. The argument is that natural groups of related organisms must include an ancestor plus all descendants. Because birds and mammals are descended from "reptiles" but excluded from the class Reptilia, the term *reptile* does not designate a natural group and must be eliminated. The replacement term is *amniote*, which applies equally to reptiles, birds, and mammals. This matter is discussed in more detail in Chapter 8.

3. Rudwick 1985.

4. Delair and Sargeant 1975.

5. One version of the story is that Mantell's wife Mary Ann found the teeth beside the road while he was attending a patient. However, Mantell told several other versions of the story besides this one.

6. Owen addressed the annual meeting of the British Association for the Advancement of Science at Plymouth in 1841 on the subject of British Mesozoic reptiles. It has long been assumed that he first used the name *dinosaur* in that lecture, which was published in 1842. However, recent research by Hugh Torrens (1992) indicates that Owen in fact did *not* use the term in his talk, but only in the published version. Thus 1842 and not 1841 is the correct date for the introduction of the name *dinosaur*.

7. Desmond 1979.

8. Hatcher et al. 1907.

9. In 1942 Brown and Schlaikjer named a new species, *Leptoceratops cerorhynchus*, but in 1951 C. M. Sternberg renamed the animal as a new genus, *Montanoceratops cerorhynchus*.

10. By *kind*, I arbitrarily mean genus. New species of chasmosaurines have been described as recently as 1989, when Thomas Lehman described *Chasmosaurus mariscalensis* from Texas.

11. A recent useful and authoritative summary of the fossil record of horned dinosaurs is found in Sereno 1990a on psittacosaurs and Dodson and Currie 1990 on the remaining ceratopsians. See especially tables 28.1 (p. 589) and 29.1 (pp. 611–612).

12. Dodson 1990b summarizes the fossil record of dinosaurs. See also Dodson and Dawson 1991.

13. McIntosh 1990.

14. Recollect that James Hutton, in his "Theory of the Earth" of 1788, inferred deep time in earth history (i.e., that the earth had great antiquity, not the few thousand years allowed by a literal reading of Genesis) by his detection of the sedimentary cycle, by which preexisting mountains are eroded, their remains carried to the sea as sediment, and the sediment compressed into rock and uplifted as mountains. Hutton visualized such cycles continuing with "no vestige of a beginning—no prospect of an end" (quoted from Gould 1987, p. 63).

15. The recent report of *Psittacosaurus* in Thailand by Buffetaut and Suteethorn stands as an object lesson that organisms did not necessarily exist only where their fossils are found. On the other hand, the fossil record has a coherence to it that makes it appropriate to be skeptical of inferences drawn from isolated bones found in unexpected places. For example, a dinosaurian jaw lacking teeth from Patagonia was described in 1918 as *Notoceratops*, a presumed protoceratopsian. There is nothing else in the entire fossil record of South America that suggests that any ceratopsian was present there, and there are many good biogeographic reasons to doubt this interpretation. As the fossil record of South America becomes better known, thanks to the research of such inspired workers as José Bonaparte, Rudolfo Coria, Jaime Powell, and Fernando Novas, we understand more clearly that this continent was largely isolated from northern continents in the Cretaceous. Its wonderful dinosaurs, including the magnificent flesh-eater *Carnotaurus*, the "flesh-bull," developed

in what G. G. Simpson called "splendid isolation." It is incorrect to attempt to force southern endemics into the straight-jacket of a taxonomy based on northern dinosaurs. To do so is to deny the creative power of evolution.

16. The marine life of the Cretaceous is beautifully described and illustrated in Dale Russell's marvelous book *An Odyssey in Time*, published in 1989.

17. Lehman 1987.

18. Horner 1984; Rogers 1990.

19. Eberth 1993.

20. MacDonald 1984. There are, of course, two African rhinos, the square-lipped, grass-grazing white rhino, *Ceratotherium simum*, and the pointy-lipped (prehensile-lipped) black rhino, *Diceros bicornis*, which browses on leaves and twigs. Its pointy mouth is a better model for that of the horned dinosaurs, but at 1.3 metric tons maximum weight it is decidedly smaller than the white rhinoceros.

21. Colbert 1962; Anderson et al. 1985.

22. Ostrom 1966.

23. Hopson 1977.

24. Lull 1933. Quote from p. 72.

25. Tait and Brown 1928. Quote from p. 23.

26. A parenthetical note ought to be added about the jugals of ceratopsians. As will be described in the next chapter, the jugal is an important bone of the cheek, situated under the eye socket and extending to the lower part of the back of the skull. From *Psittacosaurus* onward, ceratopsians have prominent, flaring jugals. In ceratopsids, an epijugal, which is really a bony scute, is added to accentuate the lower tip of the jugal bone. In *Pentaceratops*, this trend reaches a peak, with the thickness of the combined jugal-epijugal reaching nearly 15 cm, giving rise to the name "five-horned face." Granted that the jugal plays an important role in stabilizing the cheek region and firmly uniting it with the facial region, no analysis has succeeded in determining a specific role for the jugal in the chewing mechanism or in any other vital physiological function. This is especially striking for a structure that supports nearly 15 cm of solid bone. The conspicuous flaring ventral projection of the jugal suggests to me that it functioned as a "knocker," reminiscent of the knobs and thickened bumps on the skulls of giraffes. I suspect that ceratopsians swung their heads sideways and bumped each other's flanks as a prelude to more serious contention.

27. Colbert 1948, 1961. Quote from p. 176 in 1961 reference.

28. Lull 1933, p. 23.

29. Dodson 1975.

30. Dodson 1976.

31. Lehman 1990.

32. Geist 1966. The term *horn-like organ* is used advisedly. Antlers differ anatomically from horns. Antlers are naked, bony structures that are deciduous; that is, they are grown and shed every year. Horns are organs that have a

permanent bony horn core that is covered with a keratinous protein sheath that is the true horn. Only deer (family Cervidae) have antlers, as far as we know. We presume that the horns of ceratopsians were like those of the family Bovidae, even though we have direct evidence only of the horn cores. Still another pattern is shown by rhinoceros, which have neither a horn core nor a keratinous sheath. Instead, the "horn" of rhinoceros is formed by hair-like tubular keratin. The "horn" of rhinoceros is not preserved in the fossil record. We have no evidence for such a structure in horned dinosaurs, although there is some speculation as to whether *Pachyrhinosaurus* might have carried such structures above the platform of their nasal bosses.

The term *horn-like organ* permits useful generalization among all three types of structures, which presumably correlate with similar types of behavior across taxonomic boundaries.

33. Farlow and Dodson 1975.

34. Tanke 1989.

35. Lull 1933, p. 22; Molnar 1977 (see p. 178).

CHAPTER TWO
SKIN AND BONES

1. Yes, I know there are some lungless salamanders—isn't there always an exception to every grand, sweeping generalization?

2. The concept of the *extant phylogenetic bracket* is a very useful one that has been articulated by Lawrence M. Witmer (1995). The method delimits the degree of confidence we may feel in inferring an attribute that is not preserved in a fossil, be it a soft structure, a physiological trait, or a behavior. Basically, if a given character can be observed in a living representative of a more primitive group (e.g., an alligator) and in a living descendant (e.g., a bird), there is a high probability that the trait existed in the fossil group contained within the bracket (in our case, dinosaurs). If the trait is present in one living representative and absent in the other, then its putative existence in dinosaurs is less certain. If it is absent in both the ancestor and the descendant, it may still have existed in dinosaurs—which certainly had many unique characters without counterparts today—but the degree of certainty is lower. For physicists and mathematicians, uncertainty may be an intolerable state, but for historical scientists (as we paleontologists are), it is an everyday reality. We have long since ceased to expect perfection from the fossil record. We are grateful for what we have!

3. Despite stories to the contrary in *Weekly World News* (often an excellent source for entertaining if dubious dinosaur stories—the cover story of November 23, 1993 ["Earth's Water Supply Came from Dinosaur Wee-Wee"] is my personal favorite), do not look for dinosaurs to be preserved in ice from the hothouse times of the Mesozoic—the time scale is all wrong. The oldest surviving ice currently known is only a few million years old. As far as we know, no

dinosaur ever saw a glacier. Yes, dinosaurs lived in Alaska and Antarctica; no, these regions were not as frigid then.

4. Sternberg 1925.

5. Deer are a bit of an exception to the general rule. Antlers are deciduous structures that are shed and regrown every year. When the antlers are growing, they are alive and covered in velvet, within which the blood and nerve supply is located. When growth is completed, the blood supply is cut off, the velvet is shed in tatters, and bare but now dead bone is exposed in the antlers. There is no reason to suspect that typical horned dinosaurs showed anything of this sort.

6. Paul (1987) has written very articulately about the practice of restoring whole animals from skeletons. His essay synthesizes biological and fossil data and is recommended.

7. The specimens that form the basis for this description include a pair of skeletons of *Chasmosaurus belli* that form the type and co-type (see note 5, Chapter 3), supplemented by a skull of *Chasmosaurus mariscalensis* from Texas. Both specimens of *Chasmosaurus belli* were collected from beds of the Judith River Formation along the Red Deer River by the legendary fossil collector C. H. Sternberg. The first specimen was collected in 1913 and the second in 1914. Both were mounted at the National Museum of Canada (now the Canadian Museum of Nature), under the supervision of C. M. Sternberg, who in 1927 described the mounts, including the measurements cited here. The measurements were repeated by Lull (1933, p. 70). The principal specimen used for description of the skeleton is the one labeled NMC 2245. However, the skull of the second specimen, NMC 2280, is better preserved than that of NMC 2245, and so this specimen is used for the description of the skull.

A disarticulated skull of *Chasmosaurus mariscalensis* from Texas described by Lehman in 1989 serves our purposes well, so the illustrations of the skull pertain to that species. Fortunately, apart from the skull, the anatomy of ceratopsians is generally similar from one species to another. Consequently, the figures reproduced here are cobbled together from the best illustrations in the literature. The purpose of this chapter as a primer on anatomy is thus served.

8. Weishampel et al. 1990.

9. Dodson and Currie 1990.

10. As a good, Yankee-Doodle American, I was raised with feet and inches. However, as a scientist I really find the metric system, with its convenient multiples of ten, infinitely preferable. Really it is not hard to deal with. You just need a few simple conversions. An inch is 2.54 centimeters or 25.4 millimeters. Roughly, then, 10 cm is 4 inches. A foot is roughly 30 cm. A yard is 90 centimeters, and a meter is a yard plus 10 percent (more precisely 39.37 inches, but we often don't need quite that much precision). For longer distances, consider a meter to be roughly 40 inches. Thus 1.5 meters becomes 1.5 times 40 = 60 inches or 5 feet, and 50 centimeters is half a meter or ½ × 40 = 20 inches. See, that isn't so hard, is it?

11. It is useful to have a set of terms to describe the various surfaces of the bones based on their typical posture in the living animal. Paleontologists have tended to use a set of terms developed by human anatomists. However, human posture is unique among all the animals—in fact, rather bizarre from an evolutionary perspective. Imagine: standing on one's hind legs without a tail out back for balance. No wonder we suffer from back problems—horned dinosaurs didn't! Thus I prefer a modified system developed by veterinary anatomists for a four-legged animal. We designate the upper surface of a standing animal as *dorsal* (toward the back) and the lower surface as *ventral* (toward the belly). We designate the surface of a bone closest to the center or midline of the body as *medial* and the opposite surface as *lateral*. We call the surface toward the head *cranial* (paleontologists generally have used the term *anterior* for this direction, but this term for humans implies orientation toward the belly not toward the head), and the surface toward the tail *caudal*. On the limbs themselves, surfaces closer to the body may be designated *proximal* and those farther away may be designated *distal*. On the head, we cannot use the word *cranial* because we are already on the cranium, so instead we use the term *rostral*, meaning toward the very tip of the skull. These convenient adjectives can be combined to designate orientations that are not strictly in one direction or another. Examples would include *dorsolateral* and *cranioventral*. I will, however, try to be sparing in my use of this daunting, specialized vocabulary.

12. For dimensions of the foot I use Lull 1933 (p. 63) on the similar-sized *Centrosaurus nasicornus*, the feet being missing on both specimens of *Chasmosaurus belli*.

13. MC I is convenient anatomical shorthand for the prolix expression "the first metacarpal bone." Similarly, MT III is a convenient substitute for "the third metatarsal."

14. Most vertebrates are not terribly imaginative on this score. Each finger, toe, or digit is composed of a series of small bones called the *phalanges* (singular *phalanx*—*not* phalange, please!). These bones are very similar one to another and are arrayed in orderly rows, like soldiers in battle formation (or so the Romans imagined—remember the Roman phalanx from your history lessons?). The joints between phalanges are what permit the digits to bend—a single bone of the same length of your finger would be stiff. If you look carefully at your hand (provided you haven't stuck it somewhere it didn't belong), you can recognize your knuckles as the joint between your metacarpals and the first or proximal row of phalanges. If you flex your fingers gently, you can notice that your thumb has two phalanges, but the other four fingers (you do have four others, don't you?) have three phalanges each. Thus we can represent the phalangeal formula for our hands as 2-3-3-3-3, which means two phalanges for the thumb and three for each of the other four digits. Most mammals with five digits have the same formula for hands and feet. By contrast, most reptiles and birds have a phalangeal formula of 2-3-4-5-3 for hands and feet, although digits

may be lost or modified, particularly digits 4 and 5 of the hand and 1 and 5 of the foot. Dinosaurs are not immune from these tendencies, either of the fundamental pattern or of the modification.

The last phalanx on each digit is specialized to carry a horny proteinaceous covering, familiar today as nail, claw, horn, or hoof. The Latin word for claw is *unguis,* hence the name *ungual phalanx* or simply *ungual,* in paleontological usage. A claw is compressed from side to side and has a sharp point. The unguals of horned dinosaurs are broad and blunt, more like a hoof than a claw.

15. The size of the astragalus is not given, so it is estimated using the proportion of the size of the astragalus to the distal tibia in *Brachyceratops montanensis* (Gilmore 1917) and the width of the distal tibia of *Centrosaurus flexus* (Lull 1933).

16. Stricty speaking, the skeleton includes both the skull and the postcranial (i.e., behind the skull) skeleton. As a bit of shorthand, I use the term *skeleton* to refer to all the bones that do not belong to the skull.

17. The skull I am describing is an adult skull that was found as a single piece. It would be almost impossible to take it apart into the separate bones that I enumerate here. Nonetheless, for rhetorical purposes I am describing it as if it were a younger skull that had come apart. The elements listed can be recognized and measured in articulated skulls. All of these elements are known from ceratopsian bonebeds. An excellent example of such material, a *Chasmosaurus* bonebed from Big Bend National Park in Texas, was described by Thomas Lehman. In fact, the skull illustrated is that of *Chasmosaurus mariscalensis.*

18. The word *foramen* is a very general anatomical term. In Latin it simply means *hole.* Typically a nerve or vessel passes through a foramen. Something that is not as extensive as a foramen, for instance a pit in a bone surface, is called a *fossa,* which means *ditch* or *excavation. Fenestra* means *window* in Latin (*fenêtre* in French). Anatomically it is an opening often closed by a membrane, meaning that a major structure does not pass through it, in contrast to a foramen. Having two such bony windows or openings in the temporal region is characteristic of the diapsid reptiles, which include snakes, lizards, crocodiles, and dinosaurs. The lower temporal opening between the jugal and the squamosal is designated the infratemporal fenestra; the upper temporal opening between the squamosal and the parietal is the supratemporal fenestra.

19. One of the most amazing stories in the evolutionary history of vertebrates is that when mammals evolved from reptiles some 200 Ma a new jaw joint evolved, and the old reptilian quadrate was put out of work. However, it benefited from a job training program and received a new assignment as a bone of the middle ear that we now know as the *incus* or anvil bone. Mammals have three bones, the auditory ossicles, in the middle ear, whereas reptiles and birds have but one.

20. Well, mistakes can be made. The parietal bone is so extravagant that E. D. Cope, who had not yet seen a complete ceratopsian skull, labeled this bone a breastbone! It somewhat resembles the minute breastbone of certain lizards.

21. In mammals the squamosal bone is often known by another name—the temporal bone. It has a very different function in mammals: it forms the joint with the lower jaw itself, thus contributing to the unemployment of the reptilian quadrate. The temporal region of the skull is the region either caudal or caudodorsal to the orbits where the jaw-closing muscles attach to the skull.

22. The anatomical term *occiput* refers to the back surface of the skull that serves as an attachment surface for the neck muscles. The adjective *occipital* refers to structures on this surface.

23. All vertebrates are remarkably similar in the fundamental structure and organization of the brain. It is true that mammals in general have rather more brain than most other vertebrates, but what mammals have by themselves has been added on to the fundamental structures shared by reptiles and all vertebrates. All vertebrates have the same ten pairs of nerves, called cranial nerves, coming from the brain. The land-living, egg-laying vertebrates we call amniotes (that is, reptiles, birds, and mammals) have added two more nerves and so have twelve pairs of nerves. Although consideration of the evolution of the brain and identification of all the major nerves is not without interest, it is perhaps more than is called for here. The important point is that the similarity of the brain and its nerves among all living vertebrates encourages confidence in identifying and interpreting the bony structures and openings that we see in fossil vertebrates.

24. Several other bones of the braincase are hard to see as separate bones, either because they are so thoroughly fused to adjacent bones, because they are hard to characterize since they lack landmarks such as foramina, or because they are hidden from view and rarely observed. These include the orbitosphenoid, the opisthotic, the supraoccipital, and the presphenoid. I mention these for the sake of completeness, but we generally reserve descriptions of these for advanced graduate students.

25. In mammals there is only a single bone in the jaws, the dentary. In all other vertebrates with bony jaws—that is, excluding fishes without jaws (the lampreys and hagfishes) and cartilage fishes (sharks, skates, and rays)—the jaw is a compound structure formed of a number of separate bones. But the dentary is always the largest, and it is always the bone that bears teeth, if teeth are present.

26. The articular bone was also put on the unemployment rolls by the hostile takeover of the jaw by the dentary as mammals evolved. The quadrate and the articular still contact each other in mammals as they did in reptiles, but now they both serve as tiny sound-transmitting bones of the middle ear. In mammals we know the reptilian articular as the malleus or hammer bone.

CHAPTER THREE
THREE-HORNED FACE

1. In June 1994 the children of Wyoming voted *Triceratops* the state fossil of Wyoming. Good judgment, kids!

2. Brown 1917. Given all of his marvelous contributions to paleontology, let us charitably overlook his grammatical lapse!

3. Marsh 1888.

4. Ibid., p. 477.

5. Hatcher in Hatcher et al. 1907, p. 102. Designating a type specimen is an essential component of describing a new species. When describing a new plant or animal, a scientist describes the characters by which the new organism may be distinguished from all others. It is necessary to refer the species to a designated genus. The specific names may be duplicated—hundreds of species may carry the name *familiaris, domesticus, americanum, canadensis, smithi,* or whatever. Thus a specific name by itself has no unique meaning, just as the name John or Mary does not uniquely designate a single person. The specific name takes on unique meaning when it is paired with a genus name (more properly, a generic name), as in *Canis familiaris* or *Tyrannosaurus rex*. Each genus name may be used only once, so the two-name or binomial system does provide a unique way to designate species. The final step is designation of a type specimen, usually a specimen deposited in a museum with a catalog number. This step is very important because if the original description is inadequate (which is often the case), later workers can restudy the type specimen and achieve better understanding of the author's intention. Sometimes, rather than a single specimen, a series of specimens is designated as a type series. In this case, one specimen is designated the type specimen, and the others are called paratypes. This is not a common practice in dinosaur paleontology, but if the type specimen lacks a feature that is evident in a second specimen, the latter may be designated a paratype. In older literature, the second specimen is sometimes referred to as a co-type. Although the meaning of the designation is clear enough, this practice is formally discouraged by the International Commission on Zoological Nomenclature, which adjudicates such matters. It is often essential for scientific progress to reevaluate old specimens in the light of new discoveries. That process of reevaluation is a major theme of this book.

6. Marsh 1887. Quote from p. 324. Marsh himself had collected Pliocene horses near here earlier, when he still did fieldwork himself. This may have predisposed him toward accepting a younger age.

7. Hatcher in Hatcher et al. 1907, p. 6. In a similar vein he wrote (p. 8) concerning the errors of Cope and Marsh:

Nor should these errors of identification be taken as a reflection upon the sagacity of either of these authors, but rather as additional evidence of the remarkable nature of these newly discovered animals, so different in so many osteological characters from anything hitherto discovered or suspected among representatives of that group. They are, however, striking examples of the many pitfalls that beset the path of the paleontologist when attempting to describe from insufficient or fragmentary material new genera and species—

and especially new families—of extinct animals. They are, moreover, striking examples of that axiom so often disregarded in vertebrate paleontology, namely, *that one observed fact is worth any amount of expert opinion.*

Plus ça change, plus ça reste la même.

8. Quoted in Schuchert and LeVene 1940, p. 215. Chapter 8, entitled "John Bell Hatcher, King of Collectors," is a valuable account.

9. Hatcher, in Hatcher et al. 1907, recounts this discovery in a dry manner (pp. 7–8).

10. Marsh 1889a. Quote from pp. 334–335. This brief paper, which contains descriptions of several new species of existing genera of Jurassic and Cretaceous dinosaurs from Connecticut and various sites in the American West, hardly meets modern standards of scientific publication, at least in regard to *Ceratops*. The specimen is neither illustrated nor designated through a museum catalog number. The entire account of *Ceratops horridus* occupies less than one small-format page.

11. Like Cope before him, Marsh did not feel the need to explain the derivation of his names.

12. Marsh 1889b, pp. 173–174.

13. Marsh 1891b, p. 340. The name *Sterrholophus* means "rigid crest."

14. Hatcher in Hatcher et al. 1907, p. 143.

15. Today we use the names Hell Creek Formation in Montana, North Dakota, and South Dakota; Lance Formation in Wyoming; and Denver or Laramie Formation in Colorado to describe the *Triceratops*-bearing beds. Marsh failed to recognize that the beds in which *Ceratops montanus* was found were older than the *Triceratops* beds and belonged to a different formation, the Judith River. Hatcher did not make this distinction himself in his 1893 paper, entitled "The *Ceratops* beds of Converse County, Wyoming," but he did make it clearly in 1907. In the earlier paper, his first, he did make a significant attempt to reconstruct the ancient environments of the Lance Formation:

> The *Ceratops* beds are thought to afford evidence in themselves of having been deposited not in a great open lake, but in a vast swamp, with occasional stretches of open waters, the whole presenting an appearance similar to that which now exists in the interior of the Everglades of Florida. . . .
>
> The conditions that prevailed over this region during the period in which the *Ceratops* beds were deposited were probably those of a great swamp with numerous small bodies of water connected by a network of water courses constantly changing their channels. The intervening spaces were but little elevated above water level or at times submerged. The entire region where the waters were not too deep was covered by an abundant vegetation, and inhabited by huge Dinosaurs, . . . as well as by the smaller crocodiles and turtles, and the diminutive mammals, all of whose remains are now found embedded in the deposits. . . .

In the sandstones of the *Ceratops* beds hardly a fossil bone of any considerable size is to be found that does not bear evidence of having been dropped in
shallow waters. . . . So shallow were the waters, the bone itself became an
obstacle sufficient to produce an eddy on its lower side, in which the leaves
and other vegetable materials accumulated, and sank to the bottom. (pp. 142–
143)

Hatcher does a lovely job of evoking the ancient landscape.

16. Marsh 1889c, pp. 502 and 506.

17. Hatcher in Hatcher et al. 1907, p. 132.

18. Marsh 1890a. Quotes from p. 81. In the same paper, the first ostrich
dinosaur, *Ornithomimus velox*, from Colorado, was named, as well as the Late
Jurassic sauropod from South Dakota, *Barosaurus lentus*. There is a single plate,
but all of the figures pertain to the leg of *Ornithomimus*.

19. Hatcher in Hatcher et al. 1907, p. 134.

20. The four suborders of the order Ornithischia, the bird-hipped dinosaurs,
are the Ornithopoda, two-legged plant-eaters including the hadrosaurs or
duck-bills; the Stegosauria, plated plant-eaters; Ankylosauria, armored plant-
eaters, some of which had tail clubs; and the Ceratopsia.

21. Marsh 1890b.

22. Marsh 1891a. Quotes from pp. 175–176. Surely the final statement was for
British consumption. Whereas hostile natives were a concern in the 1870s, by
this time the threat was an anachronism.

23. Marsh 1891b.

24. Marsh 1895.

25. Quoted in Schuchert and LeVene 1940, p. 295.

26. Ibid., p. 385.

27. Marsh 1891c.

28. Marsh 1898. Quote from p. 92.

29. Hatcher et al. 1907.

30. Hatcher 1905. Quote from p. 419. However appealing this practice may
be, it is regarded today as a breach of scientific ethics to publish the same report
twice.

31. Lull 1905, p. 419.

32. Lull 1933. Quotes from pp. 127–128.

33. Schlaikjer 1935. Quotes from p. 60.

For the sake of completeness, two other species of *Triceratops* may be mentioned, but I do not wish to clutter up the body of the text by including them
there. In 1915, Lull referred in print to *T.* "*ingens*," but the skull to which the
name applied was not specified. In 1933, he makes this explicit statement in a
discussion of *Triceratops horridus*: "To this species, I would refer a huge skull
No. 1928 Y.P.M. [Yale Peabody Museum], as far as can be ascertained in its
present condition. This skull carries the manuscript name of '*ingens*,' in Profes-

sor Marsh's handwriting, but, while the name has been published by Lull, it was without description and was, therefore, disallowed by Hay. It is the largest of all *Triceratops* skulls, measuring over 8 feet in length, an aged individual." Because the skull was never formally described, the name has never been treated seriously. It is generally ignored. I find the reported size of the skull intriguing—larger by 30 cm than that of nearly any other specimen if accurately determined.

The other name to report is *Triceratops maximus*. For some reason, Barnum Brown was in 1933 moved to report on a specimen consisting of eight vertebrae and two ribs that had been collected from the Hell Creek beds of Montana in 1909. He was convinced of its enormous size, although he regarded it as "wild speculation" to estimate the size of the skeleton from such meager remains (Brown 1933a). Nonetheless, however probable that the remains in question pertain to *Triceratops* (although they may not), the fossil is so poorly comparable to any other species of *Triceratops* that *Triceratops maximus* too is a forgotten species.

34. Sternberg 1949.

35. Tokaryk 1986.

36. Lull in Hatcher et al. 1907, p. 172.

37. Lull 1933.

38. Ostrom and Wellnhofer 1986. The plain fact of the matter is that big, heavy, three-dimensional objects such as the skulls of large dinosaurs are hard to preserve absolutely intact. They may very well be damaged before burial, and when exposed by erosion at the earth's surface they suffer again from the elements. In addition, they may suffer at the hands of the collectors, in transport to the laboratory, or even in storage on a museum's shelves. Few indeed are the skulls that lack the smallest blemishes—often skillfully disguised with plaster and paint.

39. Lehman 1990.

40. Forster 1990, 1996.

41. There is a contradiction in the literature on this point. Hatcher et al. (1907) report the length of the skull as 1.990 m, whereas Lull (1933) reports the skull as 6 ft 1 in. (1.85 m) long and therefore slightly below average. I do not know which datum is correct. As far as I know no subsequent author has listed this measurement.

42. Osborn 1933; Erickson 1966; Schuchert 1905. The Smithsonian skeleton of *Triceratops prorsus* is 6.0 m long and about 2.5 m high above the hips. The skull is 1.8 m long. Few other details are provided. Gilmore, in his notes of 1920, does state for the record that the specimen is a composite, though "the greater part of the skeleton pertains to one individual" (p. 273). In 1920, the Smithsonian mount was still the only mounted specimen of *Triceratops*. In the same 1920 paper, reviewing the other Smithsonian reptilian skeletal mounts (including "*Trachodon*," *Thescelosaurus*, *Ceratosaurus*, *Stegosaurus* and the nondinosaur

Dimetrodon), I came across a statement I had sought for many years. Gilmore states, erroneously, that "*Trachodon*," now named *Edmontosaurus*, had "over 2,000 separate teeth in the mouth of one individual" (p. 272). This factoid has a life of its own, and it may still be found in mediocre books on dinosaurs of the sort that our children so often fall prey to. The datum repeats in Romer's usually reliable text of 1966. The error arose when Gilmore correctly stated that the skull in question had up to sixty tooth positions in each jaw, but incorrectly estimated ten to fourteen teeth in each vertical row. I am not aware that the number ever exceeds six teeth per row, yielding a more reasonable, and certainly still impressive, tooth count of 6 teeth per position × 60 tooth positions × 4 jaws = 1440 teeth. In fact, I believe even sixty-six tooth positions may be reasonable, giving 1584 teeth in an exceptional specimen. A more typical count for lambeosaurines or crested duck-bills is forty or forty-four positions with three or four teeth per position, yielding generous counts of 704 teeth. But I digress.

43. Osborn 1933. Quotes from p. 1. Osborn provides the interesting data that Lang's time amounted to 161½ days and that his assistant, Paul Bultman, contributed 102 days. This is only the time spent mounting, not in freeing the specimen from the rock.

44. Erickson 1966.

CHAPTER FOUR
FIVE-HORNED FACE AND FRIENDS

1. Marsh 1891c; Creisler 1993.

2. Marsh 1892. In the same paper, Marsh described two new prosauropod dinosaurs from the Early Jurassic of the Connecticut Valley, *Ammosaurus major* and *Anchisaurus colurus*—all this in four pages with no figures!

3. Hatcher in Hatcher et al. 1907.

4. Colbert and Bump 1947. Measuring large, three-dimensional, highly complex objects such as the skulls of horned dinosaurs is no easy job. Colbert and Bump gave two measurements of length: one length was measured parallel to the floor with the tooth row horizontal; the other was measured from the rostrum to the end of the frill along the midline. The former is 1.625 m, the other 1.8 m. Neither measurement is "right." If the crest were flat and horizontal, as in *T. gladius*, the two would be the same. The more curved and elevated the frill, the greater the difference between the two. Unfortunately, Marsh and Hatcher did not specify how they measured skull length. One reason that I like measurements such as the width of the occipital condyle is that these measurements are relatively simple to perform and unambiguous.

5. Gilmore 1946.

6. Lawson 1976.

7. Tokaryk 1986.

8. Johnson and Ostrom 1995.

9. Lambe 1902.

10. Lambe 1914a, p. 131. The genus name literally means "before *Toro-saurus*." The species name honors Dr. Robert Bell, then the director of the Geological Survey of Canada.

11. Lambe 1902, p. 66.

12. Lambe 1914a. Quote from p. 131.

13. Lambe 1914b. Quotes from pp. 150 and 155.

14. I am very skeptical that this is a genuine anatomical feature. I see it as an anomaly that has been overinterpreted. As there is no left nasal, it cannot even be demonstrated on the opposite side of the same individual, let alone in any other specimen. Furthermore, there seem to me to be good reasons for *not* constructing a horn out of several separate bones. Animals simply don't do this. The formation of the nasal horn core from separate left and right nasal bones is well established in centrosaurines, and this specimen constitutes good evidence for the same phenomenon in at least some chasmosaurines.

15. Lambe 1915. Quotes from pp. 10, 18, and 19.

16. Lehman 1990.

17. Gilmore 1923.

18. Lull 1933.

19. Sternberg 1925.

20. Sternberg 1927. Quotes from p. 69.

21. Ibid., p. 70.

22. Brown 1933b. The dedication reads: "It gives me great pleasure to name this species after Mr. Peter C. Kaisen, my friend and able assistant during many expeditions, skilled preparator, and a member of the American Museum staff for more than a third of a century."

23. Lull 1933, p. 94.

24. Sternberg 1940.

25. Lehman 1989. Quote from p. 139. This is a beautifully executed paper, full of primary data and excellent illustrations. It contains a minimum of arm-waving and a maximum of fact. I am particularly fond of it because it describes not merely a single specimen but a population of specimens. It is a contender in my view for one of the top papers in dinosaur paleontology in recent years.

26. Forster et al. 1993.

27. Godfrey and Holmes 1995.

28. Brown 1914a.

29. Sternberg 1929. Quotes from p. 34.

30. Lull 1933, p. 103.

31. Lehman 1990.

32. Sternberg 1929, p. 35.

33. Langston 1975.

34. Langston 1959. Quote from p. 9.

35. Osborn 1923a.

36. A chilling note added by Lull is that the skeleton of the type was "discarded." I do not know under what circumstances.

37. Wiman 1929. It is gratuitous to criticize paleontologists of another generation for what they did or did not do, other than offering them as instructive examples to understand the history of our science. However, I believe it is inexcusable in this latter age of enlightenment to neglect to include basic measurements in a scientific description. When authors fail in this regard, it is incumbent upon editors to insist that such data be included. The failure to consider the possibility of growth as a biological factor led in another era to the multiplication of species names. Today we see innumeracy expressing itself in naive statements to the effect that juveniles cannot in principle be used to define taxa of dinosaurs—when simple bivariate plots can easily be used to sort out growth trajectories of related species.

38. Lehman 1993.

39. The comparative narrowness may have been a factor in the relatively common preservation of the *Pentaceratops* skulls in lateral aspect, resulting in the crushing and destruction of the parietal. Ceratopsids with typically broad frills are often preserved upside down, as Sternberg noted, preserving the details of the frill very satisfactorily.

40. Rowe et al. 1981.

41. Wiman 1929.

42. Lehman 1993, p. 287. In an important paper, Lilligraven and McKenna (1986) extend the concept of North American land mammal ages from the Cenozoic, the Age of Mammals, back into the Late Cretaceous. The ages, defined by fossil mammal assemblages, correspond to the Judith River Formation (Judithian), the Horseshoe Canyon Formation (Edmontonian), and the Lance Formation, including all of its equivalents (Hell Creek Formation, Scollard Formation, Frenchman Formation, and so on) (Lancian). The three ages are based on the evolutionary stage of rapidly evolving mammals.

43. Lull (1933, p. 107) does not buy this character. "There is, if anything, more of an indication of a nasal horn than in the type of *Triceratops obtusus* at the United States National Museum. And the position of the horn-like area lies over the anterior margin of the narial opening, not above the posterior margin as in all Belly River forms."

44. My own measurements of the specimen supplement those of Parks.

45. Typically, whatever the texture or ornamentation of the upper surface of the frill, the undersurface is smooth.

46. Parks 1925. Quotes from p. 7.

47. This would be a most unusual, not to say improbable, arrangement, bypassing the premaxilla, as Tyson pointed out.

48. Tyson 1981. Quote from p. 1242.

CHAPTER FIVE
A BIG ONE ON THE NOSE

1. A few years ago, my friend Stephen Farrington pointed out to me that I am an intellectual descendant of Leidy's. The intellectual pedigree passes from Leidy to Cope to Osborn in New York, to W. K. Gregory, to E. H. Colbert, to John Ostrom at Yale, to me—who returned to Philadelphia in 1974, eighty-three years after Leidy's candle was extinguished. Except that I am not worthy to tie Leidy's shoe, I am enormously flattered by the thought.

2. To make a long story as brief as I am able, dinosaur teeth were replaced throughout the life of the animal. Mammal teeth are complex organs that must last an entire lifetime. Reptile teeth in general, and dinosaur teeth in particular, are less complex because they are replaced rather frequently. Typically, dinosaur teeth are distinctive at the level of family but not at finer taxonomic levels. There is a long history of attempts to define dinosaur genera and species on the basis of teeth, and usually these efforts are unsuccessful. The paleontological literature is strewn with the wreckage of names that are not useful because of the difficulty in correctly associating skull or skeletal fossils with tooth fossils. One might wish that this practice were an old one that has quite disappeared from the globe today, but I am sorry to say that it is in fact alive and well. I object bitterly to the editorial practice of permitting poorly grounded names to be published, but I am whistling in the wind. As editors of *The Dinosauria*, Dave Weishampel, Halszka Osmólska, and I insisted that authors examine names critically. In consequence, of some 540 names of dinosaur genera proposed since 1824, we recognized only 285 as valid!

3. Cope 1872. Quotes from pp. 481 and 483. This was five years before the great sauropod dinosaurs were first discovered in Colorado and Wyoming.

4. Cope 1874.

5. Hatcher in Hatcher et al. 1907, p. 113.

6. Sternberg 1909, pp. 32–33.

7. Ibid., p. 75. Sternberg provides an articulate and entertaining firsthand account of the expedition to Montana, summarizing it (p. 97) thus: "This chapter has largely been taken up with adventures and a study of the man Cope; but as a matter of fact, there was little else to tell about, as we were in such haste that we secured few specimens, and the most important result of the expedition was our discovery of many new specimens of dinosaurs, represented chiefly by teeth."

8. Cope 1876.

9. One exception is the freshwater ray, *Myledaphus bipartitus*, whose distinctive hexagonal tooth crown and split roots are very diagnostic, and which are abundant and widespread in Late Cretaceous deposits of western North America. These teeth are familiar to every worker in Late Cretaceous deposits of western North America, and to many amateurs as well.

10. Osborn 1931.

11. Creisler 1992, 1993.

12. Cope 1876.

13. Cope 1877. Quote from p. 593.

14. Cope 1889. Quotes from p. 715.

15. Hatcher et al. 1907.

16. Cope 1889. Quotes from pp. 716 and 717.

17. Details of the ascendancy of the American Museum in vertebrate paleontology under the guidance of Henry Fairfield Osborn are given in a delightful book by Ronald Rainger (1991). Cope was a major influence in Osborn's formation as a vertebrate paleontologist. As a young man at the College of New Jersey (now Princeton University), Osborn was involved in an ill-advised cabal against Marsh. Osborn founded the Department of Vertebrate Paleontology in 1891, and the first paleontology expedition to the West, in search of fossil mammals, took place that same year. The hall of mammalian paleontology opened in 1895. The first expedition in search of dinosaurs took place in 1897, at Como Bluff, Wyoming.

18. Hatcher et al. 1907.

19. Hatcher referred to this element as the frontal and postfrontal. We now know that the horn so formed is composed of the postorbital bone, to which the postfrontal is fused, never being seen as a separate element.

20. Hatcher in Hatcher et al. 1907, p. 76.

21. I live by calipers and measuring tapes, and I have seen more ceratopsian horns than Lawrence Lambe ever did. My measurements tell me that the bone over the orbit is about 60 mm thick (nearly 2 ½ inches). Although I completely agree that the horn is not tall and blunt, I am hesitant to say that there was none at all.

22. Lambe 1902.

23. Lambe 1904a, p. 3.

24. Lambe 1910, p. 149.

25. Lambe's earlier paper on *Centrosaurus* was "read" June 22, 1904, as the title page proclaimed. Hatcher died two weeks later. It is safe to assume that he did not read Lambe's paper.

26. Hatcher in Hatcher et al. 1907, p. 92; Lull in Hatcher et al. 1907, p. 93.

27. Lambe 1904b, p. 82.

28. Lambe 1915.

29. The parietal crest of horned dinosaurs was so luxuriant and unusual that a debate was waged for a number of years as to the true homology of this bone and several others of the skull roof. Let us say that no one was quite so flamboyantly wrong as Cope, who at first thought it was part of the breastbone. The correct view was articulated early on by Marsh in 1889, and contrary opinions by O. P. Hay, Friedrich von Huene, and Barnum Brown were proved false by discoveries of *Brachyceratops* in 1913 and *Protoceratops* in 1922. As Alexander Pope proclaimed long ago, a little learning is a dangerous thing.

30. Brown 1914b. Quotes from pp. 549–551, 553, and 556.

31. Lambe 1915. Quotes from pp. 14 and 21. This is a very valuable paper, and I treasure my original copy. It has fine illustrations of the skulls of *Chasmosaurus belli, Centrosaurus apertus, Styracosaurus albertensis,* and *Eoceratops canadensis,* with anatomical details of skull roofs and nasal horns.

32. If I have made any contribution to the study of dinosaurs, most probably it is because of what I have done with a measuring tape. Size is a fundamental biological attribute that we often take for granted. However, size determines ever so many aspects of the biology of every living thing, starting with metabolic rate. It is sometimes stated, for instance, that every vertebrate has the same number of heartbeats allotted per lifetime. A cold-blooded vertebrate has a slower heartbeat, and so lives longer than a warm-blooded one. A smaller vertebrate has a much higher heart rate than a large one, and so has a shorter span. This is a simplification, of course, but it provides food for thought. It is true that dinosaurs were, for the most part, large animals, but not all dinosaurs were enormous. Sauropods stood out, even among dinosaurs, and I always maintain that sauropods were *not* typical dinosaurs. Similarly, typical centrosaurines were respectable in size, but rhinoceros-sized, not elephant-sized.

I am very sensitive about providing, wherever possible, information on the size of a given dinosaur. Children's books, for example, tend to vary alarmingly in their assessments of dinosaur sizes. For the most part, this is because the authors do not know and do not have access to the primary literature. A body length, a measurement along the vertebral column from the snout to the tip of the tail, is a pretty objective sort of measurement when available. Another complaint I have pertains to the way in which the height of an animal may be measured. The height of an animal's head off the ground is rarely an objective datum. It depends entirely on the inferred posture or behavior of the animal. For example, the rearing *Barosaurus* at the American Museum of Natural History has its head nearly 17 m (55 ft) off the ground, but I am willing to bet that in life *Barosaurus* spent 90 percent of the time with its head browsing 3 m or less above the ground. To me, the objective measurement of height is height at the hips, being approximately the sum of the lengths of the femur, tibia, and metatarsal III. Enough said!

33. The name no doubt honors William Cutler, an Alberta homesteader turned fossil collector, who obtained several specimens of note. However, Brown left this interpretation to our imaginations.

34. Brown 1917. Quotes from pp. 285–286, 300, 302, and 305.

35. Parks 1921. Quote from p. 55.

36. Lull 1933, p. 4.

37. Lull is not remembered as the paleontological genius of the twentieth century, but he did make his contributions. His monograph in 1942 on the hadrosaurs or duck-billed dinosaurs, published with Nelda Wright, is a far more useful document. It is a legitimate classic.

38. Lull 1933. Quotes from pp. 81, 82, and 87.

39. Gilmore 1914. Quote from p. 1.

40. Gilmore 1917.

41. Instead of the name *parietal*, Gilmore used the term *dermosupraoccipital*, favored by Hay and von Huene. Ironically, both his specimen and *Protoceratops*, discovered ten years later, demonstrated that Marsh was correct in determining the major element of the frill as the parietal.

42. Gilmore 1917, p. 37.

43. I am hesistant to call *Monoclonius lowei* a juvenile because it has a very large skull, but it does show a split nasal horn and some wide-open sutures that suggest it is at least subadult.

44. Gilmore 1939. Quote from p. 12.

45. Sternberg 1949, p. 45.

46. Dodson 1990a.

47. Sternberg 1938.

48. Sternberg 1940. Quotes from pp. 468 and 469.

49. Sternberg 1940, p. 473.

50. Does that term have you stumped, too? When my panel of consultants failed me, I had to consult my trusty collegiate dictionary to learn that a *cheval-de-frise* is "a portable obstacle, usually a sawhorse, covered with spikes or barbed wire, for military use," presumably favored by the Friesian military. Ugly-sounding, isn't it? My French is a little too pacific to get me through that one. Nonetheless, the term does seem apropos for *Styracosaurus*.

51. Lambe 1913.

52. Brown and Schlaikjer 1937. Quotes from pp. 3, 4, and 7.

53. Gilmore 1930.

CHAPTER SIX
NEWER DEVELOPMENTS AND MODERN STUDIES

1. Sternberg 1950. Quotes from pp. 109 and 110.

2. Langston 1967. Quotes from pp. 172, 173, 176, 183, and 184.

3. Langston 1968.

4. The term *lower vertebrates* is a quaint expression today, used cautiously if at all, and then only in select company. It denotes all vertebrates except birds and mammals. It is definitely *not* politically correct, because it carries implications both of the ladder of progress and of the "superiority" of warm-blooded vertebrates.

5. Langston 1975. Quotes from pp. 1582, 1583, and 1594.

6. A similar statement may be made concerning dinosaur paleontology in Alberta's neighbor to the south across the forty-ninth parallel. Jack Horner returned to his native state of Montana and established one of the most important dinosaur research programs in the world based in the Museum of the Rockies at Montana State University in Bozeman.

7. Tanke 1988.

8. Tanke 1989.

9. Clemens and Nelms 1993.

10. Spotila et al. 1991 discuss the energetics of this migration.

11. Bentonite is an altered volcanic ash that is common in Late Cretaceous sediments of western North America. One of its properties is that it expands when wet. This property can turn an otherwise passable dirt road in Montana or Alberta into a slick nightmare. Gray weathered surfaces on the outcrop often have what is described as a "popcorn surface," another expression of the expansion of bentonite. Bentonite has commercial value owing to its use by oil drillers as a drill lubricant.

12. Notice how specific this location is. Most of Montana east of the continental divide, which is to say most of Montana, is drained directly by the Missouri River, or by its major tributary, the Yellowstone River, into the Mississippi River and thence to the Gulf of Mexico. However, the relatively small area of Glacier National Park in northern Montana next to the Canadian border drains into three separate basins: the Gulf of Mexico as described, Hudson Bay via the St. Mary River and many intermediate steps, and the Pacific Ocean via the Flathead River and many others. A relatively small portion of Glacier Park is actually in the Mississippi drainage.

13. Fiorillo 1991.

14. Dodson 1986. I have been involved in a bit of a dispute with my friend George Olshevsky over the propriety of the trivial name, as it is called in taxonomic circles. George maintains that I erred in calling it *Avaceratops lammersi*, giving it a singular ending. He maintains that since there is more than one Lammers, the plural ending is required. For example, when Jack Horner and Robert Makela named the marvelous "good mother reptile," *Maiasaura peeblesorum*, in 1979, they correctly used the plural ending *-orum* because they were recognizing "the James and John Peebles families, owners of the land where the specimens were collected." Thus there were *two* families specifically named. I intentionally, and on the direct recommendation of my *chef de protocol*, Mrs. Ann Dillon, chose to honor *all* of the Lammers—and with three generations there are a good many of them. Had I chosen to specify which of the Lammers I specifically wanted to honor and which to slight, I could have specified this Lammers family and that one, but not this other one. Then the plural ending would have been the correct one. But instead I wished to honor all of them, and so deliberately chose the collective singular noun, representing the entire family. It may seem a small matter, and if I had named fifty dinosaurs in my career perhaps I would not be so testy or defensive. But *Avaceratops lammersi* is my one scientific offspring, and I don't take kindly to having its name tampered with on questionable grounds! I might also add that the International Commission on Zoological Nomenclature has entirely supported my position (Dr. P. K. Tubbs, letter, August 5, 1994).

15. Currie and Dodson 1984.

16. Penkalski 1994. In no sense is imagination a dirty word in science. Dull people do dull science. The best science is absolutely a creative enterprise on the same footing as great art—one might even say that science is the great art, or human enterprise, of the late twentieth century. Machines will no more replace human insight in science than they will in art or poetry.

17. Dodson 1975, 1976.

18. Dodson and Currie 1988.

19. Sampson would argue that this specimen is too small to show the characters of adult *Centrosaurus*, and that there is no reason not to refer it to that common genus. I will concede this point.

20. Rogers and Sampson 1989.

21. Rogers 1990.

22. This proper designation cautiously avoids specifying whether the new animals are new species of *Styracosaurus* or whether they are instead new genera.

23. Horner et al. 1992. Quote from p. 60.

24. Sampson 1994, 1995a,b.

25. The views are those of Brown (1914b, 1917) and Sampson (1993), respectively.

26. Dodson 1975.

27. Dodson 1990a.

CHAPTER SEVEN
NO HORNS AND NO FRILLS

1. The lack of openings in the crest may not have surprised Brown. Few horned dinosaurs were yet known, and his most frequent comparison was with solid-frilled *Triceratops*. We have long since come to understand that the rule in horned dinosaurs is to have openings in the frill, the parietal fenestrae. Thus a solid frill is a character that may be expected in an ancestral type.

2. Brown 1914c. Quotes from pp. 568, 569, and 571.

3. Gilmore 1917, p. 37.

4. Brown and Schlaikjer 1942.

5. With our newly minted metric sensibilities we render these measurements as 1.6 m, 2 m, and 2.3–2.4 m, respectively. We can even estimate the size of Brown's skeleton. A bone shared by all three skeletons is the humerus. If we assume that the proportion of humerus length to total length of the skeleton is the same in all specimens (and this is an assumption), we can make a simple calculation. The humerus lengths in Sternberg's three specimens are 185 mm, 242 mm, and 255 mm, respectively, whereas the humerus in Brown's specimen is 290 mm long. Using the ratios of humerus length to body length for Sternberg's skeletons results in estimates for the largest specimen of 2.4–2.7 m in length.

6. Brown alluded to two individuals found at the site, but the only reference to the second specimen is to the tail, said to be the smaller specimen. The type specimen is designated as AMNH 5205. No other museum number has entered the scientific literature pertaining to the second specimen.

7. We should note that one way in which reptiles differ from mammals is that in the latter, when growth ceases, the sutures of the skull become obliterated. That is, the soft tissues that separate bones and permit growth are resorbed, and the separate bones of the skull fuse together. However, in reptiles there is a potential for growth throughout the life of the individual, although actual growth rates may be very low later in life. Thus the presence of open sutures between skull bones is an *expectation* in a reptile. For instance, many years ago I studied alligator skulls. The largest skull I studied was enormous, more than 60 cm (2 ft) long. Yet it also had the widest open sutures of any skull I studied. The skull bones had dried out and warped under the harsh Florida sun, and they had pulled apart. For this reason, it is unwise to conclude from open sutures that an animal is not an adult.

8. I have modernist quibbles with this statement. As we will see later in this chapter, *Psittacosaurus* has now been admitted to the ceratopsian club; thus it holds pride of place as the most primitive ceratopsian. Formerly it was thought too generalized to be considered a ceratopsian. There is also a problem with the vocabulary of *primitive* and *advanced,* as discussed in Chapter 1.

9. Sternberg 1951. Quotes from pp. 225, 226, and 229.

10. I make this comment not to libel Dale Russell but merely to apply a gentle tweak to the proboscis of my oldest and dearest friend in vertebrate paleontology. After thirty years of distinguished service there, Dale has recently left Ottawa and has begun afresh at North Carolina State University and the new North Carolina State Museum in Raleigh.

11. Russell 1970.

12. Ostrom 1978.

13. An excellent, richly detailed account of the expeditions was written by Andrews in 1932.

14. Ibid., p. 162.

15. Andrews 1929, quoted in Wilford 1985, p. 144.

16. Andrews 1932, p. 163.

17. White is the color of bone in life and in the experience of all us, whether we grill T-bones in the backyard or can't resist coming back from a walk in the woods with a deer bone in hand. However, when bone fossilizes, it tends to take up minerals from the surrounding sediment and from percolating ground water. Usually fossil bone is brown or black or beige, but this is not the case with bone from Bayn Dzak or other sites in Mongolia. Presumably this bone has taken up a different suite of minerals with different color characteristics, but I cannot specify.

18. In 1923 scientists did not have the benefit of the powerful radioactive dating methods that we use today. Granger and Gregory thought that *Hypsilophodon* was

of Late Jurassic age. In 1925 the view was that *Protoceratops* was of Early Cretaceous age. Each of these dates is now known to be too old by half a period.

19. Granger and Gregory 1923. Quotes from pp. 3 and 4. A name becomes official when a specimen is described in a publication, a genus and species are designated, and the disposition of the find is indicated, usually as a cataloged specimen in a recognized museum. A name is not official until these conditions are met. Having a name appear in a newspaper before a scientific account is published is bad form.

20. Gregory and Mook 1925. Quotes from pp. 2, 3, 5, and 6.

21. The clavicle appears sporadically among dinosaurs and seems to have had little functional significance. It was thought by Gerhard Heilmann in 1927 to be altogether missing in dinosaurs. However, in the group of dinosaur descendents we know as birds, the clavicles are very important. They have fused together to form the familiar wishbone or furcula.

22. Brown and Schlaikjer 1940. Quotes from pp. 139–140 and 149.

23. Colbert 1962.

24. The new dinosaur halls at the American Museum of Natural History were unveiled in June 1995, ending a three-year dinosaur drought for New Yorkers. *Protoceratops* kept its place, but *Tyrannosaurus* has migrated out of the old Cretaceous Hall and into the new Hall of Saurischia (and birds), consistent with the cladistic fervor of that institution.

25. Granger 1936.

26. Osborn 1924a.

27. Andrews 1932. Quote from p. 230. Was the urbane Andrews really so ignorant of Adam Smith? Of course the eggs had commercial value! As scientists our lives would be far simpler if our fossils were indeed commercially worthless, but if that were the case they would also probably hold little public interest. The value of certain fossils, however inconvenient it may be for us paleontologists, is as fundamental a reality as the roundness of the earth or the force of gravity. It behooves us to be in contact with *all* of reality, not just selected facets of it.

28. Horner 1982.

29. Van Straelen 1925.

30. Brown and Schlaikjer 1940.

31. Seymour 1979.

32. Sabath 1991; Currie 1994; Norell et al. 1995.

33. Norell et al. 1994.

34. Carpenter et al. 1994; Dodson 1995. Dong and Currie (1993) report a *Protoceratops* embryo but no egg!

35. Davitashvili 1961; Kurzanov 1972. Dimorphism is the condition of having "two shapes." Sexual dimorphism is a biological phenomenon by which males and females of a species have differing shapes, a pattern understood and appreciated by our own species.

36. Although it is interesting to know the greatest length of each skull, I selected as a base measurement for each a standard measurement called basal skull length, measured from the tip of the rostral bone to the occipital condyle at the back of the skull. There are several reasons for doing this. One is the practical consideration that the delicate frill is often damaged. In addition, if the standard base measurement is not available, it is difficult to use a specimen for the type of study I like to do. Basal skull length is more frequently available than total skull length. A third reason is that basal skull length is more directly comparable to the total skull length of every other vertebrate. The frill of horned dinosaurs really is an add-on gewgaw behind the true skull. See Figure 2.10a to appreciate the contrast between basal length and total length. A fourth reason for choosing basal skull length is that this measurement makes it easy to quantify growth of the frill, for the technical reason that length of the frill is not included in the standard measurement. This may seem obvious—and if it does I congratulate you, because autocorrelation (as the inclusion of the same term in both the numerator and the denominator is called) continues to plague scientific publications.

In my smallest *Protoceratops* skull, basal skull length was 76 mm and total skull length 116 mm; the largest skull measured 357 mm in basal length and 495 mm in total length. Thus total skull length was 40 to 50 percent longer than basal skull length.

37. Dodson 1976.

38. As noted in the sidebar on p. 164, I object to the practice of linguistic imperialism, that is, of specifying that pronunciation must be this way or that way—the "toe-MAY-toes" versus "toe-MAH-toes" problem. I prefer the gentler "paleontologists usually say this or that" approach to names. I particularly object to the paternalistic tone adopted in some children's books when the author clearly does not know phonics or the usual pronunciation of the dinosaur in question. In this case, however, I do make an exception that I hope will prove helpful to the reader, words beginning with *Ps* not being a staple of the English language.

39. Osborn 1923b, 1924. Quote from 1923b, p. 10.

40. Young 1931, 1958; Chao 1962.

41. Romer 1956, 1966. Quote from 1956, p. 631.

42. Maryańska and Osmólska 1975. Priority is no assurance of anything. In retrospect some fortune-teller somewhere can always claim to have predicted any given event, even though the same would-be soothsayer has failed a hundred other times. Charles Darwin was not the first person to suggest that organic evolution had occurred. He merely made a convincing case that it had, and his reasoning became widely accepted.

43. Sereno (1990b) revised Coombs's estimate to 110–130 mm, but he did not make clear his basis for doing so.

44. Coombs 1980, 1982.

45. Sereno and Chao 1988; Sereno et al. 1988.

46. Sereno 1990a, 1990b.

47. Buffetaut and Suteethorn 1992.

48. Dodson 1990b.

49. Sereno 1986.

50. Dodson and Dawson 1991.

51. I am pleased to report (with due modesty of course) that I have played some small role in the training of women in paleontology, including my students—Catherine A. Forster, an authority on *Triceratops*, Ann R. Bleefeld, an expert on the postcranial skeleton of Mesozoic mammals, and Susan D. Dawson, a superb analyst of the structure and function of whale flippers and the use of computer methods to analyze shape—and my postdoctoral fellow, Anusuya Chinsamy, now back in Cape Town in her native South Africa, who is a world authority on the microscopic structure of dinosaur bone. So noteworthy is my commitment to educating women in paleontology that my wife has occasionally questioned whether I truly am an equal opportunity educator. But I do admit token men to my program now and then.

52. Kielan-Jaworowska 1968–1984. Fortunately for our parochial interests, the papers are entirely in English. See also Kielan-Jaworowska 1969 and Kielan-Jaworowska and Barsbold 1972.

53. The specific name remains the same, but the species is transferred to the new genus under the authorship of Kurzanov (1990).

54. In *Breviceratops*, the width of the skull at the jugals is only about 75 percent of basal skull length, whereas in small skulls of *Protoceratops* the width at the jugals is about 100 percent of basal skull length, and in *Bagaceratops* the width is about 120 percent of skull length.

55. Jerzykiewicz and Russell 1991.

56. Maryańska and Osmólska 1975.

57. Nessov et al. 1989; Kurzanov 1992.

58. Brown and Schlaikjer 1942. Quote from p. 15.

59. Sternberg 1951. Quotes from pp. 226 and 227.

60. J. R. Horner, personal communication (1987) and D. B. Weishampel, personal communications (1987, 1995).

CHAPTER EIGHT
SISTERS, COUSINS, AND AUNTS

1. Dodson 1990b.

2. Weishampel and Horner 1987; Horner et al. 1992.

3. Osborn 1902 quoted in Hatcher et al. 1907, p. 161.

4. Lull in Hatcher et al. 1907, p. 162.

5. Lambe 1915.

6. Lull 1933. Quotes from pp. 24 and 27. A pithy analysis, but how I chafe at the use of the word *grotesque*! How can one call the creatures one loves *grotesque*?

7. Colbert 1948.

8. Ostrom 1966.

9. Sternberg 1949. Quotes from pp. 41, 41–44, and 45.

10. Langston 1975.

11. Recollect that Lambe's third subfamily was the Eoceratopsinae. As we now regard *Eoceratops canadensis* as a species of *Chasmosaurus*, the subfamily Eoceratopsinae is pointless. Even if the case were to be made that it is a valid genus, it would still be regarded as a good member of the Chasmosaurinae.

12. I have expressed my doubts about this one (Dodson and Currie 1990).

13. Lehman 1990. Quotes from p. 212.

14. Hennig 1966. Many aspects of "pure" phylogenetic systematics as propounded by Hennig and early advocates in the United States and Britain are superfluous to the practice of cladistics. There was, for instance, an insistence on the naming of each node on a taxonomic diagram (cladogram) and the specification of a level in the taxonomic hierarchy. The number of nodes in a chart is potentially one less than the number of species themselves, and so this practice quickly became an exercise in *reductio ad absurdum*. The solution has been to abandon *all* taxonomic groups. Cladists have little use for the familiar and comfortable Linnaean hierarchy that we all learned at one point or other (kingdom, phylum, class, order, family). This seems strong medicine indeed. If the Linnaean hierarchy still is useful, as I believe it is, I do not think the baby should be thrown out with the bath water, as Chesterton put it.

Sereno (1990c) makes a nice, reasoned case for the value of cladistics. He is a second-generation cladist: benefiting from the battles fought a generation earlier, free to use the method without polemics, comfortable at modifying it to make it as reasonable as possible. He makes several candid statements that I admire:

> The taxonomic subdivision of living and fossil organisms will never be resolved completely by objective criteria due to the nature of morphology and its particularization by the systematist. Subjectivity will always play a substantial role in distinguishing units of morphology used in taxonomic analysis. No methodology . . . has succeeded in specifying a procedure that would direct systematists, working independently, to subdivide a given organism into the same morphologic units—characters and character states—all of which carry equivalent information. (p. 10)

The lad shows wisdom beyond his years. He explains clearly:

> If unique synapomorphies accompanied the appearance of each new lineage during the phylogenetic history of a clade and were conserved in all descen-

dants, the selection of an appropriate hierarchic arrangement of taxa would involve little controversy. However, an analysis of the majority of actual taxa, such as the Ornithischia, demonstrates that no single arrangement of terminal taxa is completely congruent with all character data. Some character incongruence, or *homoplasy*, is present in every possible arrangement of taxa. Cladists generally invoke *parsimony* as a criterion to decide among alternative arrangements. The arrangement that proposes the least number of homoplasies (reversals, parallelisms, convergences) is preferred—not because evolution necessarily proceeds along the simplest path, but because simplicity is the general criterion among the sciences for preference of one explanation over endless alternatives. (p. 11)

15. Benton 1990.

16. The Marginocephalia represents a very interesting group. It is not necessarily real; rather it, like every phylogenetic conclusion, is a hypothesis, always subject to review in the light of fresh evidence. For the time being, the hypothesis is standing up well.

17. Before the recognition by Halszka Osmólska and Teresa Maryańska in 1975 that *Psittacosaurus* possesses a rostral bone and thus belongs in the Ceratopsia, the contents of the Ceratopsia were identical to the contents of Sereno's Neoceratopsia.

18. Sereno 1986.

19. Sampson 1993, 1995.

20. Dodson and Currie 1990.

21. Chapman 1990.

22. Dodson 1993.

CHAPTER NINE
THE LIFE AND DEATH OF HORNED DINOSAURS

1. Lawton 1977; Dodson et al. 1980. This is both a high-diversity bonebed and a source of articulated skeletal material. Often in bonebeds the bones are largely or completely disarticulated.

2. I think it is fair to say that the first systematic, large-scale exploration of bonebeds as a source of Cretaceous dinosaur fossils has been by the crews of the Tyrrell Museum of Palaeontology, first at Dinosaur Provincial Park and then elsewhere around the province of Alberta. C. M. Sternberg and Wann Langston, Jr., were well aware of Alberta bonebeds, and Langston's study of the *Pachyrhinosaurus* bonebed at Scabby Butte is a very important one.

3. Currie and Dodson 1984. See also Dodson and Currie 1990. It must be stated for the record that most dinosaurian bonebeds are of limited extent, a few tens of square meters perhaps being typical. The *Centrosaurus* bonebed stands apart from almost any other dinosaur accumulation I can think of.

4. Béland and Russell 1978; Dodson 1983.

5. Lehman 1987.

6. Osmólska 1980.

7. Lilligraven et al. 1979.

8. Coe et al. 1987; Wing and Tiffney 1987.

9. Dale Russell (1977) in his book on the dinosaurs of western Canada, does a superb job of recreating the plant cover of the Cretaceous West, with heavy emphasis on living representatives of Cretaceous plants.

10. Ostrom 1964, 1966.

11. Weishampel 1984; Norman and Weishampel 1985; Weishampel and Norman 1989.

12. Dodson 1993.

13. Jarzen 1982. Jarzen reported on the palynology of Dinosaur Provincial Park. He demonstrated that cycads were present as a minor component. He did not report any evidence of palms.

14. Lull 1908; Russell 1935; Haas 1955.

15. Ostrom 1964, p. 13.

16. Ostrom 1966, Fig. 5.

17. Lull 1933: Plate VI, Fig. B (AMNH 5402, photo of skull in dorsal view) and Plate VIII, Figs. A and B (*S. albertensis,* photos of skull in lateral and dorsal views).

18. Yes, I know. As of June 1995, the American Museum has caught up with the rest of the world, and their *Tyrannosaurus* mount is the fully modern "roadrunner from hell."

19. Bakker 1987; Paul 1991; Johnson and Ostrom 1995.

20. Alexander 1989. This is a wonderful book that examines a variety of questions concerning large animals using a nonthreatening biomechanical approach. See also Alexander (1991), an accessible *Scientific American* article.

21. Lockley and Hunt 1995. Quote from p. 602.

22. Farlow et al. 1989.

23. Coombs 1978.

24. Thulborn 1982.

25. Did I really say "fair?" Foxes are good runners but do not have the stamina that horses do.

26. Adams 1991.

27. Alvarez et al. 1980; Raup 1986; Hildebrand et al. 1991; Archibald 1996. The book by Raup is a fine popular account by a superb paleontologist, who with colleague Jack Sepkoski, Jr., added the extra wrinkle that catastrophes such as the K/T boundary impact occur periodically on earth every twenty-six million years. The implication of this is that some astronomic event occurs with this periodicity, caused by an orbiting twin star or by our solar system crossing the plane of our galaxy, the Milky Way, every twenty-six million years, unleashing a storm of comets or asteroids. This is heady stuff, isn't it? The good news is that we have another thirteen million years before the rocks of doom come calling!

Literature Cited

Adams, D. A. 1991. The significance of sternal position and orientation to reconstruction of ceratopsid stance and appearance. *Journal of Vertebrate Paleontology* 11(3, supplement): 14A.

Alexander, R. M. 1989. *Dynamics of Dinosaurs and Other Extinct Giants*. New York: Columbia University Press. 167 pp.

———. 1991. How dinosaurs ran. *Scientific American* 264(4): 130–136.

Alvarez, L. W., W. Alvarez, F. Asaro, and H. V. Michel. 1980. Extraterrestrial cause for Cretaceous/Tertiary extinction. *Science* 208: 1095–1108.

Anderson, J. A., A. Hall-Martin, and D. A. Russell. 1985. Long bone circumference and weight in mammals, birds and dinosaurs. *Journal of Zoology* 207: 53–61.

Andrews, R. C. 1929. *Ends of the Earth*. New York: G. P. Putnam. 355 pp.

———. 1932. *The New Conquest of Central Asia*. New York: American Museum of Natural History. 678 pp.

Archibald, J. D. 1996. *Dinosaur Extinction and the End of an Era*. New York: Columbia University Press. 237 pp.

Bakker, R. T. 1986. *The Dinosaur Heresies*. New York: William Morrow. 481 pp.

———. 1987. The return of the dancing dinosaurs. In: *Dinosaurs Past and Present*, Volume 1, Czerkas, S. J., and E. C. Olson, eds. Los Angeles: Natural History Museum of Los Angeles County, pp. 38–69.

Bakker, R. T., and Galton, P. M. 1974. Dinosaur monophyly and a new class of vertebrates. *Nature* 248: 168–172.

Béland, P., and D. A. Russell. 1978. Paleoecology of Dinosaur Provincial Park (Cretaceous), Alberta, interpreted from the distribution of articulated dinosaur remains. *Canadian Journal of Earth Sciences* 15: 1012–1024.

Benton, M. 1990. Origin and interrelationships of dinosaurs. In: *The Dinosauria*, Weishampel, D. B., P. Dodson, and H. Osmólska, eds. Berkeley: University of California Press, pp. 11–30.

Brown, B. 1914a. *Anchiceratops*, a new genus of horned dinosaurs from the Edmonton Cretaceous of Alberta. With a discussion of the origin of the ceratopsian crest and the brain casts of *Anchiceratops* and *Trachodon. Bulletin of the American Museum of Natural History* 33: 539–548.

———. 1914b. A complete skull of *Monoclonius*, from the Belly River Cretaceous of Alberta. *Bulletin of the American Museum of Natural History* 33: 549–558.

———. 1914c. *Leptoceratops*, a new genus of Ceratopsia from the Edmonton Cretaceous of Alberta. *Bulletin of the American Museum of Natural History* 33: 567–580.

————. 1917. A complete skeleton of the horned dinosaur *Monoclonius*, and description of a second skeleton showing skin impressions. *Bulletin of the American Museum of Natural History* 37: 281–306.

————. 1933a. A gigantic new dinosaur, *Triceratops maximus*, new species. *American Museum Novitates* 649: 1–9.

————. 1933b. A new longhorned Belly River ceratopsian. *American Museum Novitates* 669: 1–3.

Brown, B., and E. M. Schlaikjer. 1937. The skeleton of *Styracosaurus* with the description of a new species. *American Museum Novitates* 955: 1–12.

————. 1940. The structure and relationships of *Protoceratops*. *Annals of the New York Academy of Sciences* 40: 133–266.

————. 1942. The skeleton of *Leptoceratops* with the description of a new species. *American Museum Novitates* 1169: 1–15.

Buffetaut, E., and V. Suteethorn. 1992. A new species of the ornithischian dinosaur *Psittacosaurus* from the Early Cretaceous of Thailand. *Palaeontology* 35: 801–812.

Carpenter, K., K. F. Hirsch, and J. R. Horner, eds. 1994. *Dinosaur Eggs and Babies*. New York: Cambridge University Press. 372 pp.

Chapman, R. E. 1990. Shape analysis in the study of dinosaur morphology. In: *Dinosaur Systematics—Approaches and Perspectives*, Carpenter, K., and P. J. Currie, eds. New York: Cambridge University Press, pp. 21–42.

Chao, S. 1962. Concerning new species of *Psittacosaurus* from Laiyang, Shantung [in Chinese]. *Vertebrata PalAsiatica* 6: 349–360.

Chure, D. J., and J. S. McIntosh. 1989. A bibliography of the Dinosauria (exclusive of Aves) 1677–1986. *Museum of Western Colorado Paleontology Series* 1: 1–226.

Clemens, W. A., and L. G. Nelms. 1993. Paleoecological implications of Alaskan terrestrial vertebrate fauna in latest Cretaceous time at high paleolatitudes. *Geology* 21: 503–506.

Coe, M. J., D. L. Dilcher, J. O. Farlow, D. M. Jarzen, and D. A. Russell. 1987. Dinosaurs and land plants. In: *The Origins of Land Plants and Their Biological Consequences*, Friis, E. M., W. G. Chaloner, and P. R. Crane, eds. Cambridge: Cambridge University Press, pp. 225–258.

Colbert, E. H. 1948. The evolution of horned dinosaurs. *Evolution* 2: 145–163.

————. 1961. *Dinosaurs: Their Discovery and Their World*. New York: Dutton. 300 pp.

————. 1962. The weights of dinosaurs. *American Museum Novitates* 2076: 1–16.

Colbert, E. H., and J. D. Bump. 1947. A skull of *Torosaurus* from South Dakota and a revision of the genus. *Proceedings of the Academy of Natural Sciences of Philadelphia* 99: 93–106.

Coombs, W. P., Jr. 1978. Theoretical aspects of cursorial adaptations in dinosaurs. *Quarterly Review of Biology* 53: 393–418.

————. 1980. Juvenile ceratopsians from Mongolia—the smallest known dinosaur specimens. *Nature* 283: 380–381.

————. 1982. Juvenile specimens of the ornithischian dinosaur *Psittacosaurus*. *Palaeontology* 25: 89–107.

Cope, E. D. 1872. On the existence of the Dinosauria in the transition beds of Wyoming. *Proceedings of the American Philosophical Society* 12: 481–483.

————. 1874. Report on the stratigraphy and Pliocene vertebrate paleontology of northern Colorado. *Bulletin of the U.S. Geological and Geographical Survey of the Territories* 9: 9–28.

————. 1876. Descriptions of some vertebrate remains from the Fort Union beds of Montana. *Proceedings of the Academy of Natural Sciences of Philadelphia* 28: 248–261.

————. 1877. Report on the geology of the region of the Judith River, Montana, and on vertebrate fossils obtained on or near the Missouri River. *Bulletin of the U.S. Geological Survey* 3: 565–597.

————. 1889. The horned Dinosauria of the Laramie. *American Naturalist* 23: 715–717.

Creisler, B. 1992. Why *Monoclonius* Cope was not named for its horn: the etymologies of Cope's dinosaurs. *Journal of Vertebrate Paleontology* 12: 313–317.

————. 1993. Nomenclatural notes. *The Dinosaur Report* 1993 (Spring/Summer): 4–5.

Currie, P. J. 1994. Hunting ancient dragons in China and Canada. In: *Dino Fest*, Rosenberg, G. D., and D.L. Wolberg, eds. Paleontological Society of America Special Publication No. 7. Knoxville, Tenn.: Paleontological Society of America, pp. 387–396.

Currie, P. J., and P. Dodson. 1984. Mass death of a herd of ceratopsian dinosaurs. In: *Third Symposium on Mesozoic Terrestrial Ecosystems, Short Papers*, Reif, W.-E., and F. Westphal, eds. Tübingen: Attempto Verlag, pp. 61–66.

Davitashvili, L. A. 1961. *The Theory of Sexual Selection* [in Russian]. Moscow: Izd-vo, Akademia Nauk, SSSR. 538 pp.

Delair, J. B., and W.A.S. Sarjeant. 1975. The earliest discoveries of dinosaurs. *Isis* 66: 5–25.

Desmond, A. J. 1979. Designing the dinosaur: Richard Owen's response to Robert Edmond Grant. *Isis* 70: 224–234.

Dodson, P. 1975. Taxonomic implications of relative growth in lambeosaurine hadrosaurs. *Systematic Zoology* 24: 37–54.

————. 1976. Quantitative aspects of relative growth and sexual dimorphism in *Protoceratops*. *Journal of Paleontology* 50: 929–940.

————. 1983. A faunal review of the Judith River (Oldman) Formation, Dinosaur Provincial Park, Alberta. *The Mosasaur* 1: 89–118.

————. 1986. *Avaceratops lammersi*: a new ceratopsid from the Judith River Formation of Montana. *Proceedings of the Academy of Natural Sciences of Philadelphia* 138: 305–317.

————. 1990a. On the status of the ceratopsids *Monoclonius* and *Centrosaurus*. In: *Dinosaur Systematics—Approaches and Perspectives*, Carpenter, K., and P. J. Currie, eds. New York: Cambridge University Press, pp. 231–243.

———. 1990b. Counting dinosaurs: how many kinds were there? *Proceedings of the National Academy of Sciences* 87: 7608–7612.

———. 1993. Comparative craniology of the Ceratopsia. *American Journal of Science* 293A: 200–234.

———. 1995. Review of *Dinosaur Eggs and Babies,* edited by Kenneth Carpenter, Karl F. Hirsch, and John R. Horner. *Journal of Vertebrate Paleontology* 15: 863–866.

Dodson, P., and P. J. Currie. 1988. The smallest ceratopsid skull—Judith River Formation of Alberta. *Canadian Journal of Earth Science* 25: 926–930.

———. 1990. Neoceratopsia. In: *The Dinosauria,* Weishampel, D. B., P. Dodson, and H. Osmólska, eds. Berkeley: University of California Press, pp. 593–618.

Dodson, P., and S. D. Dawson. 1991. Making the fossil record of dinosaurs. *Modern Geology* 16: 3–15.

Dodson, P., A. K. Behrensmeyer, R. T. Bakker, and J. S. McIntosh. 1980. Taphonomy and paleoecology of the dinosaur beds of the Upper Jurassic Morrison Formation. *Paleobiology* 6: 208–232.

Dong, Z., and P. J. Currie. 1993. Protoceratopsian embryos from Inner Mongolia, People's Republic of China. *Canadian Journal of Earth Science* 30: 2248–2254.

Eberth, D. A. 1993. Depositional environments and facies transitions of dinosaur-bearing Upper Cretaceous redbeds at Bayan Mandahu (Inner Mongolia, People's Republic of China). *Canadian Journal of Earth Sciences* 30: 2196–2213.

Erickson, B. R. 1966. Mounted skeleton of *Triceratops prorsus* in the Science Museum. *Scientific Publications of the Science Museum* 1: 1–16.

Farlow, J. O., and P. Dodson. 1975. The behavioral significance of frill and horn morphology in ceratopsian dinosaurs. *Evolution* 29: 353–361.

Farlow, J. O., J. G. Pittman, and J. M. Hawthorne. 1989. *Brontopodus birdi,* Lower Cretaceous footprints from the U.S. Gulf Coastal Plain. In: *Dinosaur Tracks and Traces,* Gillette, D. D., and M. G. Lockley, eds. New York: Cambridge University Press, pp. 371–394.

Fiorillo, A. R. 1991. Taphonomy and depositional setting of the Careless Creek Quarry (Judith River Formation), Wheatland County, Montana, U.S.A. *Palaeogeography, Palaeoclimatology, Palaeoecology* 81: 281–311.

Forster, C. A. 1990. The cranial morphology and systematics of *Triceratops* with a preliminary analysis of ceratopsian phylogeny. Philadelphia: University of Pennsylvania Department of Geology, unpublished Ph.D. thesis. 227 pp.

———. 1996. Species resolution in *Triceratops:* cladistic and morphometric approaches. *Journal of Vertebrate Paleontology* 16: 259–270.

Forster, C. A., P. C. Sereno, T. W. Evans, and T. Rowe. 1993. A complete skull of *Chasmosaurus mariscalensis* (Dinosauria: Ceratopsidae) from the Aguja Formation (late Campanian) of West Texas. *Journal of Vertebrate Paleontology* 13: 161–170.

Geist, V. 1966. The evolution of horn-like organs. *Behaviour* 27: 175–214.

314

Gilmore, C. W. 1914. A new ceratopsian dinosaur from the Upper Cretaceous of Montana, with note on *Hypacrosaurus*. *Smithsonian Miscellaneous Collections* 63(3): 1–10.

———. 1917. *Brachyceratops*, a ceratopsian dinosaur from the Two Medicine Formation of Montana with notes on associated reptiles. *U.S. Geological Survey Professional Paper* 103: 1–45.

———. 1920. Reptile reconstructions in the United States National Museum. In: *Smithsonian Report for 1918* (Publication 2561). Washington, D.C.: Smithsonian Institution, pp. 271–280.

———. 1923. A new species of *Corythosaurus*, with notes on other Belly River Dinosauria. *Canadian Field-Naturalist* 37: 1–9.

———. 1930. On dinosaurian reptiles from the Two Medicine Formation of Montana. *Proceedings of the U.S. National Museum* 77(16): 1–39.

———. 1939. Ceratopsian dinosaurs from the Two Medicine Formation, Upper Cretaceous of Montana. *Proceedings of the U.S. National Museum* 87: 1–18.

———. 1946. Reptilian fauna of the North Horn Formation of central Utah. *U.S. Geological Survey Professional Paper* 210C: 29–53.

Godfrey, S. J., and R. Holmes. 1995. Cranial mophology and systematics of *Chasmosaurus* (Dinosauria: Ceratopsidae) from the Upper Cretaceous of Western Canada. *Journal of Vertebrate Paleontology* 15: 726–742.

Gould, S. J. 1987. *Time's Arrow, Time's Cycle*. Cambridge, Mass.: Harvard University Press. 222 pp.

Granger, W. 1936. The story of the dinosaur eggs. *Natural History* 1936(6): 21–25.

Granger, W., and W. K. Gregory. 1923. *Protoceratops andrewsi*, a pre-ceratopsian dinosaur from Mongolia. *American Museum Novitates* 72: 1–9.

Gregory, W. K., and C. C. Mook. 1925. On *Protoceratops*, a primitive ceratopsian dinosaur from the Lower Cretaceous of Mongolia. *American Museum Novitates* 156: 1–9.

Haas, G. 1955. The jaw musculature in *Protoceratops* and in other ceratopsians. *American Museum Novitates* 1729: 1–24.

Hatcher, J. B. 1893. The *Ceratops* beds of Converse County, Wyoming. *American Journal of Science*, ser. 3, 45: 135–144.

———. 1905. Two new Ceratopsia from the Laramie of Converse County, Wyoming. *American Journal of Science*, ser. 4, 20: 413–419.

Hatcher, J. B., O. C. Marsh, and R. S. Lull. 1907. The Ceratopsia. *United States Geological Survey Monograph* 49: 1–300.

Heilmann, G. 1927. *The Origin of Birds*. New York: D. Appleton. 210 pp.

Hennig, E. 1916. *Kentrurosaurus*, non *Doryphorosaurus*. *Zentralblatt für Mineralogie, Geologie, und Paläontologie* 1916: 578.

Hennig, W. 1966. *Phylogenetic Systematics*. Urbana: University of Illinois Press. 263 pp.

Hildebrand, A. R., G. T. Penfield, D. A. Kring, M. Pilkington, Z. A. Camargo, S. B. Jacobsen, and W. V. Boynton. 1991. Chicxulub crater: a possible Cretaceous/Tertiary boundary impact crater on the Yucutan Peninsula. *Geology* 19: 867–871.

Hopson, J. A. 1977. Relative brain size and behavior in archosaurian reptiles. *Annual Review of Systematics and Ecology* 8: 429–448.

Horner, J. R. 1982. Evidence for colonial nesting and "site fidelity" among ornithischian dinosaurs. *Nature* 297: 675–676.

———. 1984. Three ecologically distinct vertebrate faunal communities from the Late Cretaceous Two Medicine Formation of Montana, with discussion of evolutionary pressures induced by interior seaway fluctuations. *Montana Geological Society 1984 Field Conference*, pp. 299–303.

Horner, J. R., and R. Makela. 1979. Nest of juveniles provides evidence of family structure among dinosaurs. *Nature* 282: 296–298.

Horner, J. R., D. J. Varricchio, and M. B. Goodwin. 1992. Marine transgression and the evolution of Cretaceous dinosaurs. *Nature* 358: 59–61.

Hutton, J. 1788. Theory of the Earth. *Transactions of the Royal Society of Edinburgh* 1: 209–305.

Jarzen, D. M. 1982. Palynology of Dinosaur Provincial Park (Campanian), Alberta. *Syllogeus* 38: 1–69.

Jerzykiewicz, T., and D. A. Russell. 1991. Late Mesozoic biostratigraphy and vertebrates of the Gobi Basin, a review. *Cretaceous Research* 12: 345–377.

Johnson, R. E., and J. H. Ostrom. 1995. The forelimb of *Torosaurus* and an analysis of the posture and gait of ceratopsians. In: *Functional Morphology in Vertebrate Paleontology*, Thomason, J., ed. New York: Cambridge University Press, pp. 205–218.

Kielan-Jaworowska, Z., ed. 1968–1984. Results of the Polish-Mongolian Palaeontological Expeditions—Parts I–X. *Palaeontologia Polonica*. Warsaw: Polish Academy of Science.

———. 1969. *Hunting for Dinosaurs*. Cambridge, Mass.: MIT Press. 177 pp.

Kielan-Jaworowska, Z., and R. Barsbold. 1972. Narrative of the Polish-Mongolian paleontological expeditions 1967–1971. *Paleontologia Polonica* 27: 5–13.

Kurzanov, S. M. 1972. Sexual dimorphism in protoceratopsians. *Paleontological Journal* 1972: 91–97.

———. 1990. A new Late Cretaceous protoceratopsid genus [in Russian]. *Paleontological Journal* 24: 85–91.

———. 1992. A giant protoceratopsid from the Upper Cretaceous of Mongolia [in Russian]. *Paleontological Journal* 1992: 81–93.

Lambe, L. M. 1902. On Vertebrata of the Mid-Cretaceous of the North West Territory. 2. New genera and species from the Belly River Series (Mid-Cretaceous). *Geological Survey of Canada Contributions to Palaeontology* 3: 25–81.

———. 1904a. On the squamoso-parietal crest of the horned dinosaurs *Centrosaurus apertus* and *Monoclonius canadensis* from the Cretaceous of Alberta. *Transactions of the Royal Society of Canada* 10: 3–10.

———. 1904b. On the squamoso-parietal crest of two species of horned dinosaurs from the Cretaceous of Alberta. *Ottawa Naturalist* 18: 81–84.

———. 1910. Note on the parietal crest of *Centrosaurus apertus* and a proposed new generic name for *Stereocephalus tutus*. *Ottawa Naturalist* 24: 149–151.

———. 1913. A new genus and species of Ceratopsia from the Belly River Formation of Alberta. *Ottawa Naturalist* 27: 109–116.

———. 1914a. On the fore-limb of a carnivorous dinosaur from the Belly River Formation of Alberta, and a new genus of Ceratopsia from the same horizon, with remarks on the integument of some Cretaceous herbivorous dinosaurs. *Ottawa Naturalist* 27: 129–135.

———. 1914b. On *Gryposaurus notabilis*, a new genus and species of trachodont dinosaur from the Belly River Formation of Alberta, with a description of the skull of *Chasmosaurus belli*. *Ottawa Naturalist* 27: 145–155.

———. 1915. On *Eoceratops canadensis*, gen. nov., with remarks on other genera of Cretaceous horned dinosaurs. *Canada Department of Mines Geological Survey Museum Bulletin* 12: 1–49.

Lambert, D. 1990. *The Dinosaur Data Book*. New York: Avon Books. 320 pp.

Langston, W. J. 1959. *Anchiceratops* from the Oldman Formation of Alberta. *National Museum of Canada Natural History Papers* 3: 1–11.

———. 1967. The thick-headed ceratopsian dinosaur *Pachyrhinososaurus* (Reptilia: Ornithischia), from the Edmonton Formation near Drumheller, Canada. *Canadian Journal of Earth Sciences* 4: 171–186.

———. 1968. A further note on *Pachyrhinososaurus* (Reptilia: Ceratopsia). *Journal of Paleontology* 42: 1303–1304.

———. 1975. The ceratopsian dinosaurs and associated lower vertebrates from the St. Mary River Formation (Maestrichtian) at Scabby Butte, southern Alberta. *Canadian Journal of Earth Sciences* 12: 1576–1608.

Lawson, D. A. 1976. *Tyrannosaurus* and *Torosaurus*, Maestrichtian dinosaurs from trans-Pecos Texas. *Journal of Paleontology* 50: 158–164.

Lawton, R. 1977. Taphonomy of the dinosaur quarry, Dinosaur National Monument. *University of Wyoming Contributions to Geology* 15: 119–126.

Lehman, T. M. 1987. Late Maastrichtian paleoenvironments and dinosaur biogeography in the western interior of North America. *Palaeogeography, Palaeoclimatology, Palaeoecology* 60: 189–217.

Lehman, T. M. 1989. *Chasmosaurus mariscalensis*, sp. nov., a new ceratopsian dinosaur from Texas. *Journal of Vertebrate Paleontology* 9: 137–162.

———. 1990. The ceratopsian subfamily Chasmosaurinae: sexual dimorphism and systematics. In: *Dinosaur Systematics—Approaches and Perspectives*, Carpenter, K., and P. J. Currie, eds. New York: Cambridge University Press, pp. 211–229.

317

———. 1993. New data on the ceratopsian dinosaur *Pentaceratops sternbergii* from New Mexico. *Journal of Paleontology* 67: 229–288.

Leidy, J. 1856. Notice of remains of extinct reptiles and fishes, discovered by Dr. F. V. Hayden in the badlands of the Judith River, Nebraska Territory. *Proceedings of the Academy of Natural Sciences, Philadelphia* 8: 72–73.

Lilligraven, J. A., and M. C. McKenna. 1986. Fossil mammals from the "Mesaverde" Formation (Late Cretaceous, Judithian) of the Bighorn and Wind River basins, Wyoming. With definitions of Late Cretaceous land-mammal "ages." *American Museum Novitates* 2840: 1–68.

Lilligraven, J. A., Z. Kielan-Jaworowska, and W. A. Clemens, eds. 1979. *Mesozoic Mammals. The First Two-Thirds of Mammalian History.* Berkeley: University of California Press. 311 pp.

Lockley, M. G., and A. P. Hunt. 1995. Ceratopsid tracks and associated ichnofauna from the Laramie Formation (Upper Cretaceous: Maastrichtian) of Colorado. *Journal of Vertebrate Paleontology* 15: 592–614.

Lull, R. S. 1905a. *Diceratops hatcheri* Lull, gen. et. sp. nov. *American Journal of Science*, ser. 4, 20: 417–419.

———. 1905b. Two new Ceratopsia from the Laramie of Converse County, Wyoming. *American Journal of Scienc*, ser. 4, 20: 417–419.

———. 1908. The cranial musculature and the origin of the frill in the ceratopsian dinosaurs. *American Journal of Science*, ser. 4, 25: 387–399.

———. 1915. The mammals and horned dinosaurs of the Lance Formation of Niobrara County, Wyoming. *American Journal of Science* 40: 319–348.

———. 1933. A revision of the Ceratopsia or horned dinosaurs. *Peabody Museum of Natural History Memoirs* 3(3): 1–175.

Lull, R. S., and N. E. Wright. 1942. Hadrosaurian dinosaurs of North America. *Geological Society of America Special Paper* 40: 1–242.

Macdonald, D. 1984. *The Encyclopedia of Mammals.* New York: Facts on File. 895 pp.

McIntosh, J. S. 1990. Sauropoda. In: *The Dinosauria*, Weishampel, D. B., P. Dodson, and H. Osmólska, eds. Berkeley: University of California Press, pp. 345–401.

Marsh, O. C. 1887. Notice of new fossil mammals. *American Journal of Science*, ser. 3, 34: 323–324.

———. 1888. A new family of horned dinosaurs from the Cretaceous. *American Journal of Science*, ser. 3, 36: 477–478.

———. 1889a. Notice of new American Dinosauria. *American Journal of Science*, ser. 3, 37: 331–336.

———. 1889b. Notice of gigantic horned Dinosauria from the Cretaceous. *American Journal of Science*, ser. 3, 38: 173–175.

———. 1889c. The skull of the gigantic Ceratopsidae. *American Journal of Science*, ser. 3, 38: 501–506.

———. 1890a. Description of new dinosaurian reptiles. *American Journal of Science*, ser. 3, 39: 81–86.

————. 1890b. Additional characters of the Ceratopsidae, with notice of new Cretaceous dinosaurs. *American Journal of Science,* ser. 3, 39: 418–426.

————. 1891a. The gigantic Ceratopsidae, or horned Dinosaurs, of North America. *American Journal of Science,* ser. 3, 41: 167–178.

————. 1891b. Restoration of *Triceratops. American Journal of Science,* ser. 3, 41: 339–342.

————. 1891c. Notice of new vertebrate fossils. *American Journal of Science,* ser. 3, 42: 265–269.

————. 1892. The skull of *Torosaurus. American Journal of Science,* ser. 3, 43: 81–84.

————. 1895. On the affinities and classification of the dinosaurian reptiles. *American Journal of Science,* ser. 3, 50: 483–498.

————. 1898. New species of Ceratopsia. *American Journal of Science,* ser. 4, 6: 92.

Maryańska, T., and H. Osmólska. 1975. Protoceratopsidae (Dinosauria) of Asia. *Palaeontologia Polonica* 33: 133–181.

Mayor, A. 1991. Griffin bones: ancient folklore and paleontology. *Cryptozoology* 10: 16–41.

————. 1994. Guardians of the gold. *Archaeology* 1994 (Nov./Dec.): 53–58.

Molnar, R. E. 1977. Analogies in the evolution of combat and display structures in ornithopods and ungulates. *Evolutionary Theory* 3: 165–190.

Nessov, L. A., L. F. Kaznyshkina, and G. O. Cherepanov. 1989. Mesozoic ceratopsian dinosaurs and crocodiles of Central Asia [in Russian]. In: *Theoretical and Applied Aspects of Modern Paleontology,* Bogdanova, T. N., and L. I. Khozatsky, eds. Leningrad: Nauka, pp. 144–154.

Norell, M. A., J. M. Clark, L. M. Chiappe, and D. Dashzeveg. 1995. A nesting dinosaur. *Nature* 378: 774–776.

Norell, M. A., J. M. Clark, D. Dashzeveg, R. Barsbold, L. M. Chiappe, A. R. Davidson, M. C. McKenna, A. Perle, and M. J. Novacek. 1994. A theropod dinosaur embryo and the affinities of the Flaming Cliffs dinosaur eggs. *Science* 266: 779–782.

Norman, D. B., and D. B. Weishampel. 1985. Ornithopod feeding mechanisms: their bearing on the evolution of herbivory. *American Naturalist* 126: 151–164.

Osborn, H. F. 1902. On Vertebrata of the Mid-Cretaceous of the North West Territory. 1. Distinctive characters of the Mid-Cretaceous fauna. *Geological Survey of Canada Contributions to Palaeontology* 3:9–21.

————. 1923a. A new genus and species of Ceratopsia from New Mexico, *Pentaceratops sternbergii. American Museum Novitates* 93: 1–3.

————. 1923b. Two Lower Cretaceous dinosaurs of Mongolia. *American Museum Novitates* 95: 1–10.

————. 1924a. Three new Theropoda, *Protoceratops* zone, central Mongolia. *American Museum Novitates* 144: 1–12.

————. 1924b. *Psittacosaurus* and *Protiguanodon:* two Lower Cretaceous iguanodonts from Mongolia. *American Museum Novitates* 127: 1–16.

————. 1931. *Cope: Master Naturalist.* Princeton, N.J.: Princeton University Press. 741 pp.

————. 1933. Mounted skeleton of *Triceratops elatus*. *American Museum Novitates* 654: 1–14.

Osmólska, H. 1980. The Late Cretaceous vertebrate assemblages of the Gobi Desert, Mongolia. *Mémoires de la Société Géologique de France* 139: 145–150.

Ostrom, J. H. 1964. A functional analysis of jaw mechanics in the dinosaur *Triceratops*. *Postilla* 88: 1–35.

————. 1966. Functional morphology and evolution of the ceratopsian dinosaurs. *Evolution* 20: 290–308.

————. 1978. *Leptoceratops gracilis* from the "Lance" Formation of Wyoming. *Journal of Paleontology* 52: 697–704.

Ostrom, J. H., and P. Wellnhofer. 1986. The Munich specimen of *Triceratops* with a revision of the genus. *Zitteliana* 14: 111–158.

Parks, W. A. 1921. The head and fore limb of a specimen of *Centrosaurus apertus*. *Transactions of the Royal Society of Canada* 15: 53–62.

————. 1925. *Arrhinoceratops brachyops*, a new genus and species of Ceratopsia from the Edmonton Formation of Alberta. *University of Toronto Studies, Geological Series* 19: 5–15.

Paul, G. S. 1987. The science and art of restoring the life appearance of dinosaurs and their relatives. In: *Dinosaurs Past and Present*, Volume 2, Czerkas, S. J., and E. C. Olson, eds. Los Angeles: Natural History Museum of Los Angeles County, pp. 4–49.

————. 1991. Giant horned dinosaurs really did have fully erect forelimbs. *Journal of Vertebrate Paleontology* 11(3, supplement): 50A.

Penkalski, P. G. 1994. The morphology of *Avaceratops lammersi*, a primitive ceratopsid from the Judith River Formation (Late Campanian) of Montana. Philadelphia: University of Pennsylvania Department of Geology, unpublished M.S. thesis. 55 pp.

Rainger, R. 1991. *An Agenda for Antiquity: Henry Fairfield Osborn and Vertebrate Paleontology at the American Museum of Natural History, 1890–1935*. Tuscaloosa: University of Alabama Press. 360 pp.

Raup, D. M. 1986. *The Nemesis Affair—A Story of the Death of Dinosaurs and the Ways of Science*. New York: Norton. 220 pp.

Rogers, R. R. 1990. Taphonomy of three dinosaur bone beds in the Upper Cretaceous Two Medicine Formation of northwestern Montana: evidence for drought-related mortality. *Palaios* 5: 394–413.

Rogers, R. R., and S. D. Sampson. 1989. A drought-related mass death of ceratopsian dinosaurs (Reptilia: Ornithischia) from the Two Medicine Formation (Campanian) of Montana: behavioral implications. *Journal of Vertebrate Paleontology* 9(3, supplement): 36A.

Romer, A. S. 1956. *Osteology of the Reptiles*. Chicago: University of Chicago Press. 772 pp.

————. 1966. *Vertebrate Paleontology*. Chicago: University of Chicago Press. 468 pp.

———. 1968. *Notes and Comments on Vertebrate Paleontology.* Chicago: University of Chicago Press. 304 pp.

Rowe, T., E. H. Colbert, and J. D. Nations. 1981. The occurrence of *Pentaceratops* with a description of its frill. In: *Advances in San Juan Basin Paleontology,* Lucas, S. G., J. K. Rigby, and B. S. Kues, eds. Albuquerque: University of New Mexico Press, pp. 29–48.

Rudwick, M.J.S. 1985. *The Meaning of Fossils,* 2nd ed. Chicago: University of Chicago Press. 287 pp.

Russell, D. A. 1970. A skeletal reconstruction of *Leptoceratops gracilis* from the Upper Edmonton Formation (Cretaceous) of Alberta. *Canadian Journal of Earth Sciences* 7: 181–184.

———. 1977. *A Vanished World. The Dinosaurs of Western Canada.* Ottawa: National Museum of Canada. 142 pp.

———. 1989. *An Odyssey in Time: The Dinosaurs of North America.* Toronto: University of Toronto Press. 240 pp.

Russell, L. S. 1935. Musculature and functions in the Ceratopsia. *National Museum of Canada Bulletin* 77: 39–48.

Sabath, K. 1991. Upper Cretaceous amniotic eggs from the Gobi Desert. *Acta Palaeontologia Polonica* 36: 151–192.

Sampson, S. D. 1993. Cranial ornamentations in ceratopsid dinosaurs: systematic, behavioural, and evolutionary implications. Toronto: University of Toronto Department of Zoology, unpublished Ph.D. dissertation. 299 pp.

———. 1994. Two new horned dinosaurs (Ornithischia: Ceratopsidae), from the Upper Cretaceous Two Medicine Formation of Montana, USA. *Journal of Vertebrate Paleontology* 14(3, supplement): 44A.

———. 1995a. Horns, herds, and hierarchies. *Natural History* 104(6): 36–40.

———. 1995b. Two new horned dinosaurs from the Upper Cretaceous Two Medicine Formation of Montana; with a phylogenetic analysis of the Centrosaurinae (Ornithischia: Ceratopsidae). *Journal of Vertebrate Paleontology* 15: 743–760.

Sattler, H. R. 1990. *The New Illustrated Dinosaur Dictionary.* New York: Lothrop, Lee and Shepard. 363 pp.

Schlaikjer, E. M. 1935. The Torrington member of the Lance Formation and a study of a new *Triceratops. Bulletin of the Museum of Comparative Zoology* 76: 31–68.

Schuchert, C. 1905. The mounted skeleton of *Triceratops prorsus* in the U.S. National Museum. *American Journal of Science,* ser. 4, 20: 458–459.

Schuchert, C., and C. M. LeVene. 1940. *O. C. Marsh, Pioneer in Paleontology.* New Haven, Conn.: Yale University Press. 541 pp.

Sereno, P. C. 1986. Phylogeny of the bird-hipped dinosaurs. *National Geographic Research* 2: 234–236.

———. 1990a. Psittacosauridae. In: *The Dinosauria,* Weishampel, D. B., P. Dodson, and H. Osmólska, eds. Berkeley: University of California Press, pp. 579–592.

———. 1990b. New data on parrot-beaked dinosaurs (*Psittacosaurus*). In: *Dinosaur Systematics—Approaches and Perspectives,* Carpenter, K., and P. J. Currie, eds. New York: Cambridge University Press, pp. 203–210.

———. 1990c. Clades and grades in dinosaur systematics. In: *Dinosaur Systematics—Approaches and Perspectives,* Carpenter, K., and P. J. Currie, eds. New York: Cambridge University Press, pp. 9–20.

Sereno, P. C., and S. Chao. 1988. *Psittacosaurus xinjiangensis* (Ornithischia: Ceratopsia), a new psittacosaur from the Lower Cretaceous of Northwestern China. *Journal of Vertebrate Paleontology* 8: 353–365.

Sereno, P. C., S. Chao, Z. Cheng, and C. Rao. 1988. *Psittacosaurus meileyingensis* (Ornithischia: Ceratopsia), a new psittacosaur from the Lower Cretaceous of Northeastern China. *Journal of Vertebrate Paleontology* 8: 366–377.

Seymour, R. S. 1979. Dinosaur eggs: gas conductance through the shell, water loss during incubation and clutch size. *Paleobiology* 5: 1–11.

Simpson, G. G. 1980. *Splendid Isolation.* New Haven, Conn.: Yale University Press. 266 pp.

Spotila, J. R., M. P. O'Connor, P. Dodson, and F. V. Paladino. 1991. Hot and cold running dinosaurs: body size, metabolism and migration. *Modern Geology* 16: 203–227.

Sternberg, C. H. 1909. *Life of a Fossil Hunter.* New York: Holt (University of Indiana Press reprint, 1990). 286 pp.

Sternberg, C. M. 1925. Integument of *Chasmosaurus belli. Canadian Field-Naturalist* 34: 108–110.

———. 1927. Horned dinosaur group in the National Museum of Canada. *Canadian Field-Naturalist* 41: 67–73.

———. 1929. A new species of horned dinosaur from the Upper Cretaceous of Alberta. *National Museum of Canada Bulletin* 54: 34–37.

———. 1938. *Monoclonius* from southeastern Alberta compared with *Centrosaurus. Journal of Paleontology* 12: 284–286.

———. 1940. Ceratopsidae from Alberta. *Journal of Paleontology* 14: 468–480.

———. 1949. The Edmonton fauna and description of a new *Triceratops* from the Upper Edmonton member; phylogeny of the Ceratopsidae. *National Museum of Canada Bulletin* 113: 33–46.

———. 1950. *Pachyrhinosaurus canadensis,* representing a new family of the Ceratopsia, from southern Alberta. *National Museum of Canada Bulletin* 118: 109–120.

———. 1951. Complete skeleton of *Leptoceratops gracilis* Brown from the Upper Edmonton member on the Red Deer River, Alberta. *National Museum of Canada Bulletin* 123: 225–255.

Tait, J., and B. Brown. 1928. How the Ceratopsia carried and used their head. *Transactions of the Royal Society of Canada* 22: 13–23.

Tanke, D. 1988. Ontogeny and dimorphism in *Pachyrhinosaurus* (Reptilia, Ceratopsidae), Pipestone Creek, N.W. Alberta, Canada. *Journal of Vertebrate Paleontology* 8(3, supplement): 27A.

—————. 1989. K/U centrosaurine (Ornithischia: Ceratopsidae) paleopathologies and behavioral implications. *Journal of Vertebrate Paleontology* 9(3, supplement): 41A.

Thulborn, R. A. 1982. Speeds and gaits of dinosaurs. *Palaeogeography, Palaeoclimatology, Palaeoecology* 38: 227–256.

Tokaryk, T. 1986. Ceratopsian dinosaurs from the Frenchman Formation (Upper Cretaceous) of Saskatchewan. *Canadian Field-Naturalist* 100: 192–196.

Torrens, H. 1992. When did the dinosaur get its name? *New Scientist,* April 4, 1992: 40–44.

Tyson, H. 1981. The structure and relationships of the horned dinosaur *Arrhinoceratops* Parks (Ornithischia: Ceratopsidae). *Canadian Journal of Earth Sciences* 18: 1241–1247.

Van Straelen, V. 1925. The microstructure of dinosaurian egg-shells from the Cretaceous beds of Mongolia. *American Museum Novitates* 173: 1–4.

Weishampel, D. B. 1984. Evolution of jaw mechanisms in ornithopod dinosaurs. *Advances in Anatomy, Embryology and Cell Biology* 87: 1–110.

Weishampel, D. B., P. Dodson, and H. Osmólska, eds. 1990. *The Dinosauria.* Berkeley: University of California Press. 733 pp.

Weishampel, D. B., and J. R. Horner. 1987. Dinosaurs, habitat bottlenecks, and the St. Mary River Formation. In: *Fourth Symposium on Mesozoic Terrestrial Ecosystems: Short Papers,* Currie, P. J., and E. H. Koster, eds. Drumheller, Alberta: Tyrrell Museum of Palaeontology, pp. 222–227.

Weishampel, D. B., and D. B. Norman. 1989. Vertebrate herbivory in the Mesozoic: jaws, plants, and evolutionary metrics. *Geological Society of America Special Papers* 238: 87–100.

Wilford, J. N. 1985. *The Riddle of the Dinosaur.* New York: Alfred A. Knopf. 304 pp.

Wiman, C. 1929. Über Ceratopsia aus der Oberen Kreide in New Mexico. *Nova Acta Regiae Societas Scientarum Upsaliensis* 7: 1–19.

Wing, S. L., and B. H. Tiffney. 1987. The reciprocal interaction of angiosperm evolution and tetrapod herbivory. *Review of Palaeobotany and Palynology* 50: 179–210.

Witmer, L. M. 1995. The extant phylogenetic bracket and the importance of reconstructing soft tissues in fossils. In: *Functional Morphology in Vertebrate Paleontology,* Thomason, J., ed. New York: Cambridge University Press, pp. 19–33.

Young, C. C. 1931. On some new dinosaurs from western Suiyan, Inner Mongolia. *Bulletin of the Geological Survey of China* 2: 159–166.

—————. 1958. The dinosaurian remains of Laiyang, Shantung [in Chinese]. *Palaeontologica Sinica* 16: 1–138.

Index

Entries suffixed by an f denote citations within figure captions; those suffixed by an m, within map captions; those suffixed by an n, within the note whose number immediately follows the n; and those suffixed by a t, within tables.

325

Ostrom, John, 274; on jaw muscles, 266, 269–270; *Leptoceratops gracilis* and, 205, 206; on teeth, 263, 265; *Triceratops* and, 82, 83–84, 85, 248

Ottawa. *See* Geological Survey of Canada; National Museum of Canada

Outgroups, 253

Oviraptor philoceratops, 209, 217, 219

Owen, Richard, 6, 33, 74, 123, 284n6

Pachycephalosauria, 254

Pachycephalosaurus, 279

Pachyrhinosaurus, 123, 180–181; *Avaceratops lammersi* compared with, 190; bonebed data on, 261; classification of, 4t; fossil record of, 14; geographic scope of finds, 181–182; size of, 16, 180

Pachyrhinosaurus canadensis, 9, 20f, 21f, 170–175, 176f, 183, 186, 194, 195, 248, 309n2; *Monoclonius* overlap with, 173, 175, 197; phylogenetic reconstruction of, 249, 256–257; size of, 172

Pachyrhinosauridae, 171

Palatine, 53, 255

Palate, 52–53, 213

Paleontology, 6

Paleoscincus, 124

Panoplosaurus, 99

Paraphyletic assemblages, 253

Parasaurolophus walkeri, 120

Parietal bone, 301n41; of *Anchiceratops*, 113, 114; of *Avaceratops lammersi*, 185, 186, 189–190; of *Brachyceratops mon-tanensis*, 155, 156, 157; of *Centrosaurus apertus*, 198; of *Chasmosaurus*, 95; of *Chasmosaurus belli*, 49; of *Chasmosaurus mariscalensis*, 46f, 106; of *Chasmosaurus russelli*, 104; of *Diceratops hatcheri*, 76, 77, 85; jaw muscles and, 269, 270; of Marginocephalia, 254; of *Monoclonius*, 132, 134–135; of *Monoclonius crassus*, 145; of *Monoclonius dawsoni*, 138; of *Monoclonius flexus*, 142, 151; of *Monoclonius lowei*, 161; of newly discov-ered styracosaur, 195; of Pachycephalo-sauria, 254; of *Pachyrhinosaurus*, 181; of *Pachyrhinosaurus canadensis*, 171, 175; of

Pentaceratops, 117; of *Protoceratops*, 212; of *Styracosaurus albertensis*, 165, 166; of *Styracosaurus parksi*, 168; of *Triceratops*, 66; of *Triceratops eurycephalus*, 78

Parietal crest, 299n29; of *Sterrholophus*, 64; of *Torosaurus*, 90

Parietal fenestra: of *Anchiceratops*, 257; of *Arrhinoceratops brachyops*, 121; of *Brachyceratops montanensis*, 155, 158; of centrosaurines versus chasmosaurines, 251; of *Centrosaurus apertus*, 138–139; of *Chasmosaurus*, 95–96, 267; of *Chasmo-saurus belli*, 49; of *Diceratops hatcheri*, 76, 85; jaw muscles and, 270; of *Microceratops gobiensis*, 239; of *Monoclonius*, 134, 160; of *Monoclonius nasicornus*, 199; of *Pachy-rhinosaurus canadensis*, 173, 175; of *Penta-ceratops*, 116, 117, 267; of *Styracosaurus albertensis*, 166; of *Torosaurus*, 89–90; of *Torosaurus latus*, 91; of *Triceratops*, 248

Parietal fontanelle, 111

Parietal frill: of *Arrhinoceratops brachyops*, 122; of *Avaceratops lammersi*, 185; of *Cen-trosaurus apertus*, 139f, 198, 199; of *Leptoceratops gracilis*, 201; of *Mono-clonius*, 193; of *Monoclonius crassus*, 135f; of Neoceratopsia, 255; of newly discov-ered styracosaur, 194; of *Pachyrhino-saurus*, 181; of *Pachyrhinosaurus canadensis*, 175; of *Protoceratops*, 213; of *Triceratops serratus*, 67

Parks, William A., 9, 119–122, 149–150, 167

Parsimony, 309n14

Patagonia, 59, 284n15

Paul, Gregory, 33, 101f, 118f, 146f, 214f, 271f, 273

Peabody, George, 7, 57

Peabody Museum. *See* Yale Peabody Museum

Peace River Canyon, British Columbia, 178

Pectinodon, 179

Pectoral crest, 278

Pelvis, 38–39; of *Brachyceratops montanen-sis*, 156; of *Breviceratops kozlowskii*, 238; of *Microceratops gobiensis*, 239; of *Monoclonius*, 132; of *Monoclonius cutleri*, 148; of *Montanoceratops cerorhynchus*,